T0250595

Lecture Notes in Computer Science 1215

Edited by G. Goos, J. Hartmanis and J. van Leeuwen

Advisory Board: W. Brauer D. Gries J. Stoer

Springer

Berlin
Heidelberg
New York
Barcelona
Budapest
Hong Kong
London
Milan
Paris
Santa Clara
Singapore
Tokyo

José M.L.M. Palma Jack Dongarra (Eds.)

Vector and Parallel Processing – VECPAR'96

Second International Conference
on Vector and Parallel Processing –
Systems and Applications
Porto, Portugal, September 25-27, 1996
Selected Papers

 Springer

Series Editors

Gerhard Goos, Karlsruhe University, Germany

Juris Hartmanis, Cornell University, NY, USA

Jan van Leeuwen, Utrecht University, The Netherlands

Volume Editors

José M.L.M. Palma
Faculdade de Engenharia da Universidade do Porto
Rua dos Bragas, 4099 Porto Codex, Portugal
E-mail: jpalma@fe.up.pt

Jack Dongarra
University of Tennessee, Department of Computer Science
Knoxville, TN 37996-1301, USA
and
Oak Ridge National Laboratory
Oak Ridge, TN 37821, USA
E-mail: dongarra@cs.utk.edu

Cataloging-in-Publication data applied for

Die Deutsche Bibliothek - CIP-Einheitsaufnahme

Vector and parallel processing : selected papers / VECPAR '96,
Second International Conference on Vector and Parallel Processing -
Systems and Applications, Porto, Portugal, September 25 - 27, 1996.
Jose M. L. M. Palma ; Jack Dongarra (ed.). - Berlin ; Heidelberg ;
New York ; Barcelona ; Budapest ; Hong Kong ; London ; Milan ;
Paris ; Santa Clara ; Singapore ; Tokyo : Springer, 1997
   (Lecture notes in computer science ; Vol. 1215)
   ISBN 3-540-62828-2 kart.

CR Subject Classification (1991): G.1-2, D.1.3, C.2,F.1.2, F.2

ISSN 0302-9743
ISBN 3-540-62828-2 Springer-Verlag Berlin Heidelberg New York

Typesetting: Camera-ready by author
SPIN 10549420   06/3142 – 5 4 3 2 1 0   Printed on acid-free paper

Preface

This book contains the invited talks and a selection of papers which were presented at *VECPAR'96 - 2nd International Meeting on Vector and Parallel Processing (Systems and Applications)*. This meeting which took place in Porto, September 25–27, 1996, was the second organized by the Faculty of Engineering of the University of Porto, and its main objective was the dissemination of present knowledge on vector and parallel processing.

The field of parallel computing has undergone an exceptional growth during the last few years; nearly all computer manufacturers are developing computer hardware with some degree of parallelism. HPCN programs are part of the research strategy of major research financing institutions all over the world. The various talks and articles in this volume reflect a broad spectrum of applications and areas in which parallelism is making an impact, such as parallel linear algebra and computational fluid dynamics.

The meeting was organized around scientific sessions initiated by thematic key invited lectures, followed by contributed papers. All papers selected for the conference were submitted to prior review based on an extended abstract. Papers included in this volume were submitted to further review based on the full text.

We hope that those interested in the topic find in this book a very valuable reference.

February 1997   *José M.L.M. Palma* and *Jack Dongarra*

Committees

Organizing Committee

José M. Laginha M. Palma
Lígia Maria Ribeiro

Advisory Committee

P. Amestoy	France
P. Berger	France
A. Campilho	Portugal
J. Cunha	Portugal
J. Damas	Portugal
L. Décamps	France
J. Diaz	Spain
J. Duarte	Portugal
F. Ferreira	Portugal
G. Golub	USA
S. Hammarling	UK
A. Hansen	Norway
D. Heermann	Germany
W. Janke	Germany
J. C. Lopes	Portugal
J. P. Lopes	Portugal
J. McGuirk	UK
J. Martins	Portugal
F. Moura	Portugal
M. Novotny	USA
A. Padilha	Portugal
J. Palma	Portugal
R. Pandey	USA
A. Proença	Portugal
R. Ralha	Portugal
A. Ruano	Portugal
M. Ruano	Portugal
D. Ruiz	France
H. Ruskin	Ireland
M. Ryan	Ireland
M. Valero	Spain
J.-S. Wang	Singapore

Sponsoring Organizations

FEUP–Faculdade de Engenharia da Universidade do Porto
European Commission (DG III)
JNICT–Junta Nacional de Investigação Científica
Reitoria da Universidade do Porto
FCCN–Fundação para o Desenvolvimento
    dos Meios Nacionais de Cálculo Científico
Câmara Municipal do Porto
Fundação Eng. António de Almeida
IEEE (Portuguese Section)
Digital Equipment Portugal
Porto Convention Bureau

Invited Lecturers

- Michel J. Daydé
  ENSEEIHT-IRIT, Toulouse (France)
- Jean-Luc Dekeyser
  LIFL, Lille (France)
- Filomena Dias d'Almeida
  FEUP, Porto (Portugal)
- Jack Dongarra
  University of Tennessee, Knoxville (USA)
  and Oak Ridge National Laboratory, Oak Ridge (USA)
- Iain S. Duff
  Rutherford Appleton Laboratory, Oxfordshire (England)
  and CERFACS, Toulouse (France)
- Djalma M. Falcão
  Universidade Federal do Rio de Janeiro (Brasil)
- Vicente Hernandez
  Universidad Politécnica de Valencia, Valencia (Spain)
- Doyle D. Knight
  Rutgers University, Piscataway, New Jersey (USA)
- Rafael Dueire Lins
  Universidade Federal de Pernambuco em Recife (Brasil)
- Kai Nagel
  Los Alamos National Laboratory, Los Alamos (USA)
- Eugenio Oñate
  ICNME, Barcelona (Spain)
- Heitor Pina
  FCCN, Lisboa (Portugal)
- Thierry van Der Pyl
  European Commission (DG III), Brussels (Belgium)

Table of Contents
List of Invited Talks (bold face) and Articles

High Performance Computing in Power System Applications

Djalma M. Falcão

COPPE, Federal University of Rio de Janeiro
C.P. 68504, 21945-970, Rio de Janeiro RJ
Brazil

Abstract. This paper presents a review of the research activities developed in recent years in the field of High Performance Computing (HPC) application to power system problems and a perspective view of the utilization of this technology by the power industry. The paper starts with a brief introduction to the different types of HPC platforms adequate to power system applications. Then, the most computer intensive power system computation models are described. Next, the promising areas of HPC application in power system are commented. Finally, a critical review of the recent developed research work in the field, along with prospective developments, is presented.

1 Introduction

Power system simulation, optimization, and control can be included in the category of highly computer intensive problems found in practical engineering applications. Modern power system studies require more complex mathematical models owing to the use of power electronic based control devices and the implementation of deregulation policies which leads to operation close to the system limits. New computing techniques, such as those based on artificial intelligence and evolutionary principles, are also being introduced in these studies. All these facts are increasing even further the computer requirements of power system applications. High Performance Computing (HPC), encompassing parallel, vector, and other processing techniques, have achieved a stage of industrial development which allows economical use in this type of application. This paper presents a review of the research activities developed in recent years in the field of HPC application to power system problems and a perspective view of the utilization of this technology by the power industry. The paper starts with a brief introduction to the different types of high performance computing platforms adequate to power system applications. Then, the most computer intensive power system computation models are described. Next, the promising areas of HPC application in power system are commented. Finally, a critical review of the recent developed research work in the field, along with prospective developments, is presented.

2 High Performance Computing

The processing capabilities of single processors, despite the substantial increase achieved in the last years, is not high enough to cope with the rising demand observed in in several fields of science and engineering. For that reason, HPC has been relying more and more on the exploitation of concurrent tasks in the programs which can be executed in parallel on computer systems with multiplicity of hardware components. Several types of computer architectures are available for this purpose [1, 2]:

1) *Superscalar Processors* are single processors able to execute concurrently more than one instruction per clock cycle. Its efficiency depends on the ability of compilers to detect instructions that can be executed in parallel. They are used in high performance workstations and in some multiprocessor systems. Examples of superscalar processors are the IBM Power2, DEC Alpha, MIPS R10000, etc.

2) *Vector Processors* are processors designed to optimize the execution of arithmetic operations in long vectors. These processors are mostly based on the *pipeline architecture*. Almost all of the so called supercomputers, like the ones manufactured by Cray, Fujitsu, NEC, etc., are based on powerful vector processors.

3) *Shared Memory Multiprocessors* are machines composed of several processors which communicate among themselves through a global memory shared by all processors. Some of these machines have a few (2-16) powerful vector processors accessing high speed memory. Examples of such architectures are the Cray T90 and J90 families. Others, like the SGI Power Challenge, may have a larger (up to 32) number of less powerful superscalar processors.

4) *SIMD*[1] *Massively Parallel Machines* are composed of hundreds or thousands of relatively simple processors which execute, synchronously, the same instructions on different sets of data (*data parallelism*) under the command of central control unity.

5) *Distributed Memory Multicomputers* are machines composed of several pairs of memory-processor sets, connected by a high speed data communication network, which exchange information by message passing. The processors have a relatively high processing capacity and the number of processors may be large (2-1024). Owing to the possible high number of processors, this type of architecture may also be referred to as massively parallel. Examples of multicomputers are the IBM SP-2, Cray T3D/3E, Intel Paragon, etc.

6) *Heterogeneous Network of Workstations* may be used as a virtual parallel machine to solve a problem concurrently by the use of specially developed communication and coordination software like PVM and MPI [3]. From the point of view of applications development, this computer system is similar to the distributed memory multicomputers but its efficiency and reliability is usually inferior. On the other hand, the possibility of using idle workstations, already available in a company for other purposes, as a virtual parallel machine is attractive from a an economical point of view.

---

[1] Single Instruction Stream Multiple Data Stream

The development of applications on the HPC architectures described above may follow different programming paradigms and procedures. Parallelism can be exploited at one ore more granularity levels ranging from instruction-level (fine grain parallelism) to subprogram-level (coarse grain parallelism). Superscalar and vector processors, as well as SIMD machines, are more adequate to instruction level parallelism while multiprocessor and multicomputer architectures adapt better to subprogram parallelism. Coarse grain parallelism can be implemented using the shared memory architecture of multicomputers or the message passing paradigm used in multicomputers and network of workstations. The first model is conducive to programs easy to develop and maintain but shared memory multiprocessors are usually more expensive and less scalable than multicomputers. For that reason, HPC manufacturers have been trying to develop smart operating systems that could mimic a shared memory environment on a physically distributed memory system. The detection of parallelism in the code is mostly performed manually by the applications developers. Automatic detection of parallelism is still a challenge for HPC except in the case of superscalar and vector processors.

The gain obtained in moving an application to a parallel computer is measured in terms of the speedup and efficiency of the parallel implementation compared with the *best* available sequential code. *Speedup* is defined as the ratio between the execution time of the best available sequential code in one processor of the parallel machine to the time to run the parallel code in p processors. *Efficiency* of the parallelization process is defined as the ratio between the speedup achieved on p processors to p. In the early stages of applications development for parallel computers these two indexes were almost exclusively the determinants of the quality of parallel algorithms. As more parallel machines became commercially available, and practical applications begin to be actually implemented, other aspects of the problem started to become important. For instance, the cost/performance ratio (Mflops/\$; Mflops=10^6 floating point operations per second) attainable in real-life applications. In other cases, although speedup and efficiencies are not so high, the implementation in a parallel machine is the only way to achieve the required speed in the computations.

3 Potential Areas for HPC Applications

The major impact of HPC application in power systems may occur in problems for which conventional computers have failed so far to deliver satisfactory performance or in areas in which the requirement of more complex models will demand extra computational performance in the future. Another possibility is in the development of a new generation of analysis and synthesis tools exploiting the potential offered by modern computer technology: intelligent systems, visualization, distributed data basis, etc. Some candidate areas are commented in the following.

3.1 Real-time Control

The complexity and fast response requirement of modern Energy Management System software, particularly the components associated with security assessment, make this area a potential candidate for HPC use [4, 5]. In most of the present implementations of security control functions only static models are considered. This deficiency imposes severe limitations to their ability of detecting potentially dangerous situations in system operation. The consideration of dynamic models, associated with the angle and voltage stability phenomena, require a computational power not yet available in power system computerized control centers. Even considering only static models, problems like security constrained optimal power flow are too demanding for the present control center hardware. Taking into consideration the present trend towards a distributed architecture in control center design, the inclusion of parallel computers as number crunching servers in this architecture may offer a possibility to attend this high computing requirement. Another possibility would be the utilization of the control center network of workstations as a virtual parallel machine to solve problems requiring computational power above the one available in each of the individual workstations.

3.2 Real-time Simulation

The capacity to simulate the dynamic behavior of the power system, taking into consideration electromechanical and electromagnetic transients, in the same time scale of the physical phenomena, is of great importance in the design and testing of new apparatus, control and protection schemes, disturbance analysis, training and education, etc. [6]. Real-time simulation can be performed using analog devices (reduced model or electronic devices) or digital simulators. Hybrid simulators combine these two type of simulation technique. Digital simulators are more flexible and smaller in size due to the processors very large scale integration technology. Another advantage of digital simulators is the facility to manipulate and display results using sophisticated graphical interfaces. For a long period, analog simulation was the only way to obtain real-time performance of fast phenomena in practical size systems. The progress in digital hardware technology, however, is changing this scenario: massively parallel computer systems are nowadays able to deliver the computer power necessary to drive fully digital or hybrid real-time power system simulators. In this type of application, the high performance computer must be dedicated to the process owing to the need to interface it physically to the equipment being tested.

3.3 Optimization

Power system is a rich field for the application of optimization techniques. Problems range from the classical economic dispatch, which can be modeled straightforwardly as a non-linear programming problem and solved by gradient techniques, to the stochastic dynamic programming formulation of the multireservoir optimization problem. Other interesting and complex problems are the transmission and distribution network expansion planning and contingency constrained optimal power flow, to cite only the most reported studies in the literature. In most of these problems, a realistic formulation leads to highly nonlinear relationships, nonconvex functions, discrete and integer variables, and many other ill-behaved characteristics of the mathematical models. Some of these problems are combinatorial optimization problems with exponential increase in computer requirement. Another difficulty always present is high dimension: problems involving thousands of constraints are quite common. Most of the problems formulations are adequate for the application of decomposition techniques what allows efficient utilization of parallel computers in their solution. Recently, heuristic search optimization techniques, like Simulated Annealing, Genetic Algorithms, Evolutionary Computation, etc., have been proposed to solve some of these problems. These techniques, also, present great potential for efficient implementation on HPC platforms.

3.4 Probabilistic Assessment

This type of power system performance assessment is becoming more and more accepted as practical tools for expansion and operational planning. Some studies involving probabilistic models, like the composite reliability assessment of generation-transmission systems, require great computational effort to analyze realistic size power system even if only simplified models are used such that static representation, linearization, etc. [7, 8]. The inclusion of more realistic models make the problem almost intractable in present conventional computers. On the other hand, most of the methods used in such calculations (Monte Carlo simulation, enumeration techniques, etc) are adequate for massively parallel processing. This is one of the most promising areas of HPC application to power systems.

3.5 Intelligent Tools for Analysis and Synthesis

Power system operational and expansion planning require a time consuming and tedious cycle of scenario data preparation, simulation execution, analysis of results, and decision to choose other scenarios. A complete study involves several different types of simulation software (power flow, transient stability, electromagnetic transients, etc.) not well integrated and compatible. Present day software tools, although taking advantage of some of the modern computer facilities like graphic interfaces and integrated data bases, does not fully exploit all the hardware and software resources made available by the computer industry.

Taking into consideration the scenario depicted in the introduction of this paper, it is believed that the power system computational tools of the future will need to fulfill the following requirements:

- *Robustness* to cope with analysis of stressed systems;
- *Friendliness* to relieve engineers from routine work;

- *Integration* to broad the engineers ability of analysis;
- *Learning capability* to automatically accumulate experience;
- *Fast response* to speed up analysis and decision making.

Robustness can be achieved by better modeling and *robust algorithms* specially developed to cope with analysis of extreme operating conditions. Friendliness can be greatly improved by the use of the highly sophisticated *visualization tools* presently available provided that the power system engineers could find efficient ways to synthesize graphically the results of power system studies. Integration and learning capabilities can be achieved by the use of *intelligent systems* techniques like Expert Systems, Artificial Neural Networks, Fuzzy Logics, etc. The integration of all these computational tools to perform studies in large scale power system models would certainly need a *HPC environment* to achieve the required fast response. A visual summary of the structure of such computational tool is shown in Figure 1.

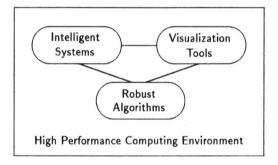

Fig. 1. Power system computational tool of the future

4   Literature Review

This section reviews the main areas of HPC application to power systems problems. The review is not meant to be exhaustive. Only areas in which a substantial amount of work has been published or areas that, in the author's opinion, has a great chance of becoming relevant in the near future, are covered.

4.1   Simulation of Electromechanical Transients

The simulation of electromechanical transients has been one of the most studied areas of application of HPC in power systems. This interest comes from the possibility it opens to real-time dynamic security assessment and the development of real-time simulators. The mathematical model usually adopted in this kind of simulation consists of a set of ordinary non-linear differential equations, associated to the synchronous machine rotors and their controllers, constrained by a set of non-linear algebraic equations associated to the transmission network, synchronous machine stators, and loads [9, 10]. These equations can be expressed as:

$$\dot{x} = f(x, z) \tag{1}$$
$$0 = g(x, z) \tag{2}$$

where f and g are non-linear vector functions; x is the vector of state variables; and z is the vector the algebraic equations variables.

In the model defined in (1) and (2), the differential equations representing one machine present interaction with the equations representing other machines only via the network equations variables. From a structural point of view, this model can be visualized as shown in Figure 2: clusters of generators are connected by local transmission and subtransmission networks and interconnected among themselves and to load centers by tie-lines. In the sequential computer context, several solution schemes have been used to solve the dynamic simulation problem. The main differences between these schemes are in the numerical integration approach (implicit or explicit) and in the strategy to solve the differential and algebraic set of equations (simultaneous or alternating). Implicit integration methods, particularly the trapezoidal rule, have been mostly adopted for this application. The most used schemes are the Alternating Implicit Scheme (AIS) and the Simultaneous Implicit Scheme (SIS)[9].

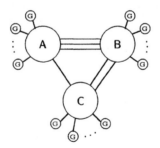

Fig. 2. Dynamic Simulation Model Decomposition

The difficulties for the parallelization of the dynamic simulation problem in the AIS are concentrated on the network solution. The differential equations associated with the synchronous machines and their controllers are naturally decoupled and easy to parallelize. On the other hand, the network equations constitute a *tightly coupled* problem requiring ingenious decomposition schemes and solution methods suitable for parallel applications. The SIS also requires the parallel solution of linear algebraic equations sets in every integration step, with difficulties similar to the ones described for the AIS.

A basic numerical problem in both simulation schemes, as well as in several other power system problems, is the parallel solution of sets of linear algebraic equations. Direct methods, like LU factorization, have been dominating this application on conventional computers. Several schemes for the efficient solution of linear algebraic equations on vector computer have also been proposed [11, 12, 13] in the context of power system simulation. In most of these schemes, only the substitution phase of the direct methods are vectorized. If parallel computers are considered, however, the hegemony of direct methods is no more guaranteed. In several other engineering and scientific fields, parallel implementations of iterative methods have shown superior performance. Among the most successful iterative methods are the ones belonging to the Conjugate Gradient (CG) category [14, 15, 16, 17, 18]. The parallelization of the network equations solution requires the decomposition of the set of equations in a number of subsets equal to the number of processors used in the simulation. An adequate decomposition is fundamental to the success of the parallel solution and need to take into consideration factors like computation load balancing, convergence rate of the iterative algorithms, etc. [19].

In the last decade or so, several parallel methods were proposed for the solution of the dynamic simulation problem. In the following sections, some of these methods are reviewed.

Spatial Parallelization Methods in this category exploit the structural properties of the the equations to be solved in each integration step of the conventional simulation schemes (AIS or SIS). Four methods are briefly described below:

1) *The Parallel VDHN* [20] consists in a straightforward parallelization of the *Very Dishonest Newton Method (VDHN)*, applied to the SIS, simply identifying tasks that can be performed concurrently and allocating them among the processors. This method was implemented on the parallel computers Intel iPSC/2 (distributed memory) and Alliant FX/8 (shared memory) and tests performed with the IEEE 118 bus and US Midwestern system with 662 buses. The results show speedups slightly superior for the iPSC/2 with a strong saturation with the increase in the number of processors. The maximum obtained speedup was 5.61 for 32 processors (efficiency = 17.5%).

2) *The Parallel Newton-W matrix Method* [21] uses a parallel version of the Sparse Matrix Inverse Factors [22] in the SIS. The method was tested on the shared memory Symmetry parallel computer and the same test systems used in the work cited in the previous item. The results show a worse performance of this method when compared to the parallel VHDN with an slowdown of 10% to 30% depending on the chosen partitions.

3) *The Parallel Real-Time Digital Simulator* [23] is based on the AIS using the trapezoidal integration method. One processor is associated to each network bus. The differential equations corresponding to each generator and its controllers are solved on the processor assigned to the bus in which the generator is connected. The network equations are solved by a Gauss-Seidel like method also allocating one equation to each processor. Therefore, the number of processors required

to perform the simulation is equal to the number of network buses. Reported results with a 261 buses network, on a 512 node nCube parallel computer, show that the required cpu time is not affected by the system dimensions. However, it is doubtful weather this property can be kept valid for larger system taking into consideration that the number of iterations required by the Gauss-Seidel algorithm increases considerably with system size. This approach exhibits low speedup and efficiency measured by the traditional indexes. However, its impact in the power system research community was considerable as it has demonstrated the usefulness of parallel processing in solving a real world problem.

4) *The Hybrid CG-LU Approach* [14, 16] is based on the AIS, the decomposition of the network equations in a Block Bordered Diagonal Form (BBDF), and a hybrid solution scheme using LU decomposition and the CG method. The equations are solved by Block-Gaussian Elimination in a two phase scheme: firstly, the interconnection equations are solved by the CG method; secondly, a number of independent sets of equations, corresponding to the diagonal blocks of the BBDF, are solved by LU factorization one in each processor. This method has the disadvantage of applying the CG method to a relatively small system of equations (interconnection block). Owing to the BBDF characteristics, the interconnection matrix is usually well-conditioned. However, the use of a Truncated Series preconditioner improves the performance of the method. Results of experiments performed with this method, as well as with other simulation methods based on the CG's methods, are presented in a later section of this paper.

5) *The Full CG Approach* [16] solves the network equations as a whole by a block-parallel version of the Preconditioned CG method. The network matrix is decomposed in such a way that the blocks in the diagonal are weakly coupled to each other, i.e., in a Near Block Diagonal Form (NBDF). The NBDF is equivalent to the decomposition of the network in subnetworks weakly coupled. A block-diagonal matrix, obtained from the NBDF neglecting the off-diagonal blocks, is used as a preconditioner.

Waveform Relaxation This method [24, 25, 26] consists in the decomposition of the set of equations describing the power system dynamics into subsystems weakly coupled and to solve each subsystem independently for several integration steps to get a first approximation of the time response. The results are, then, exchanged and the process repeated. The advantages of this method are the possibility of using different integration steps for each subsystem (multi-rate integration) and to avoid of the need to solve large sets of linear algebraic equations. However, the difficulty to obtain an efficient decomposition of the differential equations set is a major drawback in the practical application of this method.

Space and Time Parallelization This class of methods follows the idea introduced in [27] in which the differential equations are algebrized for several integration steps, called integration windows, and solved together with the algebraic equations of this window by the Newton method. Two methods are briefly described below:

1) *The Space and Time CG Approach* [16] uses two versions of the CG method (Bi-CG and Bi-CGSTAB), suitable for asymmetric sets of linear equations, to solve the resulting set of equations which presents a *stair-like* coefficient matrix. The parallelization process follows a natural choice: the equations corresponding to each integration step are assigned to different processors. Therefore, a number of integration steps equal to the number of processors available in the parallel machine can be processed concurrently. A block-diagonal preconditioning matrix, derived from the coefficient matrix, was found to be effective for both CG methods.

2) *The Space and Time Gauss-Jacobi-Block-Newton Approach* [28, 29] uses a slightly different formulation of the waveform relaxation concept. The discretization of the differential equations is performed for all integration steps simultaneously resulting in an extremely large set of algebraic equations. In a first work [28], this set of equations was solved by a parallel version of the Gauss-Jacobi method with a poor performance. In a second work [29], a method called Gauss-Jacobi-Block-Newton Approach was used with better results. This method consists, essentially, in the application of the VDHN method to the equations associated to each integration step and, then, to apply the Gauss-Jacobi globally to all integration steps. Both works present results only for simulations of parallel implementation.

Conjugate Gradient Approach Results The Hybrid CG-LU, Full CG, and Space and Time CG methods described above [14, 16] were tested using different test systems, including a representation of the South-Southern Brazilian interconnected system with 80 machines and 616 buses. The tests were performed on the iPSC/860 computer and in a prototype parallel computer using the Transputer T800 processor. Despite the difficulties in parallelizing this application, the results obtained in these tests showed a considerable reduction in computation time. The CG methods presented adequate robustness, accuracy, and computation speed establishing themselves firmly as an alternative to direct methods in parallel dynamic simulation. Moderate efficiencies and speedups were achieved, particularly in the tests performed on the iPSC/860, which are partially explained by the relatively low communication/computation speed ratio of the machines used in the tests. It is believed that in other commercially available parallel machines, the studied algorithms will be able to achieve higher levels of speedup and efficiency.

4.2 Simulation of Electromagnetic Transients

In the usual model of the power network for electromagnetic transient simulation, all network components, except transmission lines, are modeled by lumped parameter equivalent circuits composed of voltage and current sources, linear and non-linear resistors, inductors, capacitors, ideal switches, etc. These elements are described in the mathematical model by ordinary differential equations which are solved by step-by-step numerical integration, often using the trapezoidal rule, leading to equivalent circuits consisting of resistors and current sources [30].

Transmission lines often have dimensions comparable to the wave-length of the high frequency transients and, therefore, have to be modeled as distributed parameter elements described mathematically by partial differential equations (wave equation). For instance, in a transmission line of length ℓ, the voltage and current in a point at a distance x from the sending end, at a time t, are related through the following equation:

$$-\frac{\partial E(x,t)}{\partial x} = L\,\frac{\partial I(x,t)}{\partial t} + R\,I(x,t) \tag{3}$$

$$-\frac{\partial I(x,t)}{\partial x} = C\,\frac{\partial E(x,t)}{\partial t} + G\,E(x,t) \tag{4}$$

where $E(x,t)$ and $I(x,t)$ are $p \times 1$ vectors of phase voltage and currents (p is the number of phases); R, G, L and C are $p \times p$ matrices of the transmission line parameters.

The wave equation does not have an analytic solution in the time domain, in the case of a lossy line, but it has been shown that it can be adequately represented by a traveling wave model consisting of two disjoint equivalent circuits containing a current source in parallel with an impedance in both ends of the line as shown in Figure 3. The value of the current sources are determined by circuit variables computed in past integration steps (*history* terms).

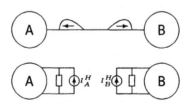

Fig. 3. Transmission Line Model

This model is nicely structured for parallel processing: subnetworks of lumped parameter circuit elements connected by transmission lines, representing a group of devices in a substation for instance, can be represented by sets of nodal equations that interface with other groups of equations by the variables required to calculate the current sources in the transmission line equivalent circuits. The exploitation of this characteristic of the network model, in the partitioning of the set of equations for parallel processing, often correspond to a geographical mapping of the power system onto the multiprocessor topology as shown below for a two subnetwork example:

$$\begin{bmatrix} G_A & 0 \\ 0 & G_B \end{bmatrix}\begin{bmatrix} E_A \\ E_B \end{bmatrix} + \begin{bmatrix} \mathcal{F}_A(E_A) \\ \mathcal{F}_B(E_B) \end{bmatrix} = \begin{bmatrix} I_A^S \\ I_B^S \end{bmatrix} + \begin{bmatrix} I_A^L \\ I_B^L \end{bmatrix} + \begin{bmatrix} I_A^C \\ I_B^C \end{bmatrix} + \begin{bmatrix} I_A^H \\ I_B^H \end{bmatrix} \tag{5}$$

where G_A and G_B are conductance matrices related to linear branch elements; $\mathcal{F}_A$ and $\mathcal{F}_B$ are non-linear functions related to non-linear branch elements; E_A and E_B are vectors of the unknown node voltages; I_A^S, I_B^S are nodal current injections corresponding to independent sources, I_A^L, I_B^L, I_A^C, I_B^C are the nodal injection currents related to the equivalent circuits of inductors and capacitors, and I_A^H, I_B^H are the nodal current injections present in the transmission line models.

Since $I^S(t)$ is known and $I^H(t)$, $I^L(t)$, and $I^C(t)$ depend only on terms computed in previous integration steps, $E_A(t)$ and $E_B(t)$ can be computed independently in different processors. The computation of the terms $I_A^S(t)$, $I_B^S(t)$, $I_A^L(t)$, $I_B^L(t)$, $I_A^C(t)$, and $I_B^C(t)$ can also be executed in parallel, since the equations related to branches in a particular subnetwork depend only on nodal voltages belonging to the same subnetwork. However, the term $I_A^H(t)$ depend on the past terms $I_B^H(t - \tau)$ and $E_B(t - \tau)$, as well as $I_B^H(t)$ depend on the past terms $I_A^H(t - \tau)$ and $E_a(t - \tau)$. Since such terms have already been evaluated in previous integration steps, the processors must exchange data in order to each one be able to compute its part of the vector $I^H(t)$.

Several parallel implementations of the electromagnetic transients simulation methodology described above are reported in the literature. In [31], [32], and [33], prototypes of parallel machines based on different networks of Transputer processors were used for these implementations, with excellent results in terms of speedup for some realistic size test systems. In [34], the implementation is performed on a workstation based on a superscalar computer architecture (IBM RISC System/6000 Model 560). The results obtained in this implementation, for medium size systems, indicate the possibility of achieving real-time simulation.

4.3 Small-Signal Stability

Power system oscillations are the result of insufficient damping torque between generators and groups of generators. This situation may arise as a consequence of heavily loaded lines, weak interconnections, high gain excitation systems, etc. Oscillations caused by small disturbances, like the normal load variation, may reach amplitudes high enough to cause protective relays to trip lines and generators which in turn causes partial or total system collapse. This type of problem can be studied using linearized versions of the power system dynamic model given by (1) and (2). The great advantage of this approach is the possibility of the performance assessment of control schemes without time simulation. This assessment is conducted through linear control systems analysis methods. A large scale numerical problem resulting from the application of these techniques is the computation of eigenvalues and eigenvectors associated with the state matrix of the linearized system model [10].

A linearized version of (1) and (2) at an operating point (x_0, z_0) is given by

$$\begin{bmatrix} \Delta \dot{x} \\ 0 \end{bmatrix} = \begin{bmatrix} J_1 & J_2 \\ J_3 & J_4 \end{bmatrix} \begin{bmatrix} \Delta x \\ \Delta z \end{bmatrix} \tag{6}$$

where $J_1, \ldots, J_4$ are Jacobian matrices evaluated at the linearization point. The power system state transition equation can be obtained eliminating Δz from (6):

$$\Delta \dot{x} = (J_1 - J_2 J_4^{-1} J_3)\, \Delta x = A\, \Delta x \tag{7}$$

where A is the system state matrix whose eigenvalues provide information on the local stability of the nonlinear system. Efficient algorithms to obtain the dominant eigenvalues and eigenvectors of A for large scale systems do not require the explicit calculation of this matrix [35]. These algorithms can be directly applied to (6), named the *augmented* system, whose sparse structure can be fully exploited to reduce both cpu time and memory requirements. These methods require repeated solutions of linear equation sets of the form [36]:

$$\begin{bmatrix} J_1 - qI & J_2 \\ J_3 & J_4 \end{bmatrix} \begin{bmatrix} w \\ v \end{bmatrix}^{(k)} = \begin{bmatrix} r \\ 0 \end{bmatrix}^{(k)} \tag{8}$$

where w, v are unknown vectors; q is a complex shift used to make dominant the eigenvalues close to q; I is the identity matrix; r is a complex vector; and k is the iteration counter. These sets of equation are independent and their solution can be obtained concurrently on different processors. This property make the eigenvalue problem well suited for parallel processing.

In the work reported in [36] and [37], algorithms for the parallel solution of the eigenvalue problem for small-signal stability assessment, using the above formulation, are described and the results of tests with models of a large practical power systems are presented. A first investigatory line of research was based on the parallelization of the Lop-sided Simultaneous Iterations method [36]. The obvious parallel stratagem used was to carry out each trial vector solution on a different processor. Results obtained in tests performed on the iPSC/860 parallel computer, using two large scale representations of the Brazilian South-Southern interconnected power system, presented computation efficiencies around 50%. A second approach to the problem uses a Hybrid Method [37] resulting from the combination of the Bi-Iteration version of the Simultaneous Iteration algorithm and the Inverse Iteration method. The Hybrid algorithm exploits the fast eigenvalue estimation of the Bi-Iteration algorithm and the fast eigenvector convergence of the Inverse Iteration algorithm whenever the initial shift is close to an eigenvalue. In the Inverse Iteration stage, the Hybrid algorithm allows perfect parallelization. The results obtained indicate a superior performance of this method both in terms of computation speedup and robustness. In [38], it is described a new method for partial eigensolution of large sparse systems named the Refactored Bi-Iteration Method (RBI). A parallel version of this method was tested using the same test system and parallel computers cited above and the results indicate a possible advantage of using the RBI method in the parallel computation of eigenvalues.

4.4 Security Constrained Optimal Power Flow

The Security Constrained Optimal Power Flow (SCOPF) is usually formulated as a nonlinear programming problem of the form [4, 39]:

$$\min_{z_0, z_i} f(z_0) \tag{9}$$

subject to

$$g_i(z_i) = 0, \qquad i = 0, ..., n \tag{10}$$

$$h_i(z_i) \le 0, \qquad i = 0, ..., n \tag{11}$$

$$\phi(u_i - u_0) \le \theta_i, \qquad i = 1, ..., n \tag{12}$$

where $z_i = [u_i \ x_i]^T$ is a vector of decision variables, the components of which are the vectors of state or dependent variables (x_i) and the vector of control or independent variables (u_i); $z_0, ..., z_n$ corresponds to the base case (z_0) and post-contingency configurations $(z_i, \ i = 1, ..., n)$, respectively; f is the objective function which depends on the particular application; g_i is a nonlinear vector function representing the power flow constraints for the ith configuration; h_i is a nonlinear vector function representing operating constraints such as limits on line flows or bus voltages for the ith configuration; $\phi(.)$ is a distance metric; and θ_i is a vector of upper bounds reflecting ramp-rate limits. Typical problems involves, for each configuration, around 2000 equality constraints and 4000 inequality constraints. The number of different post-contingency configurations considered (n) may reach several hundreds. An efficient way of deal with the high dimensionality of the problem defined in (9) to (12) is by the use of decomposition techniques [39].

One of first proposed decomposition techniques for the SCOPF is based on the Benders approach [40]. In this method, the problem is divided into a master problem (base case) and subproblems (post-contingency configurations). The solution approach starts solving the base case optimization problem $(i = 0)$ and testing weather this solution satisfies the subproblems constraints $(i = 1, ..., n)$. If necessary, corrective rescheduling is performed in the subproblems. If all subproblems are feasible, then the overall problem is solved. In the case that rescheduling alone is not able to relieve constraint violations in the subproblems, then linear inequality constraints, known as Benders cuts, are incorporated to the base case and the process starts again.

In the Benders decomposition approach to SCOPF, the $n + 1$ subproblems associated with base case and the post-contingency states are independent of each other and can, therefore, be solved in parallel. These subproblems are *loosely coupled* since the amount of information exchanged between the base case and each subproblem is small compared with the local processing effort. This fact has been exploited in the work reported in [41] in synchronous and asynchronous implementations of an algorithm for the solution of a linearized version of (9) to (12). In these implementations, one of the available processors solves the base case while the others solve the subproblems. In the synchronous case, the master problem is idle when the subproblems are being solved, and vice-versa, which leads to a low efficiency use of the multiprocessor system. In the asynchronous case, the latest information available in the subproblems is communicated to the master problem enhancing the use of the processors and, therefore, the overall efficiency of the process. Efficiency up to 82 % has been reported in a test system with 504 buses, 880 circuits, and 72 controllable generators. The parallel machine used was a common-bus 16 cpu system (iAPX-286/287 processor).

In [42], an asynchronous version of a parallel solution of the SCOPF, fairly similar to the one described above, is proposed. The solution method is embedded in a general programming model for exchange of messages and data among processors which allows different problems formulation and facilitates the mapping of the application onto different computer architectures. The method was tested using two test systems: the first one with 725 buses, 1212 branches, 76 adjustable generators, and 900 post-contingency states; and the second one with 1663 buses, 2349 branches, 99 adjustable generators, and 1555 post-contingency state. Tests with the smaller system, on a shared-memory common-bus machine with 9 nodes, achieved efficiency values similar to the ones reported in [41]. In the tests with the larger system, in a 64 node distributed memory nCube machine, the achieved efficiency was around 65 %.

In [43], an asynchronous decomposed version of the SCOPF, based on the technique proposed in [44], was implemented in a network of DEC5000 workstations using PVM. The method allows the representation of soft constraints to model operating limits which need not to be enforced sharply. Reported results indicate that the accuracy of the results is not affected by the lag in communication.

4.5   State Estimation

State estimation is a basic module in the Energy Management System (EMS) advanced application software. Its main function is to provide reliable estimates of the quantities required for monitoring and control of the electric power system. In almost all state estimation implementations, a set of measurements obtained by the data acquisition system throughout the whole supervised network, at approximately the same time instant, is centrally processed by a static state estimator at regular intervals or by operator's request. Modern high speed data acquisition equipment is able to obtain new sets of measurements every 1-10 seconds but the present EMS hardware and software allow state estimation processing only every few minutes. It has been argued that a more useful state estimation operational scheme would be achieved by shortening the time interval between consecutive state estimations to allow a closer monitoring of the system evolution particularly in emergency situations in which the system state changes rapidly. Another industry trend is to enlarge the supervised network by extending state estimation to low voltage subnetworks. These trends pose the challenge of performing state estimation in a few seconds for networks with thousands of nodes.

The higher frequency in state estimation execution requires the development of faster state estimation algorithms. The larger size of the supervised networks will increase the demand on the numerical stability of the algorithms. Conventional centralized state estimation methods have reached a development stage in which substantial improvements in either speed or numerical robustness are not likely to occur. These facts, together with the technical developments on distributed EMS, based on fast data communication network technology, opens up the possibility of parallel and distributed implementations of the state estimation function.

The information model used in power system state estimation is represented by the equation

$$z = h(x) + \omega \tag{13}$$

where z is a $(m \times 1)$ measurement vector, x is a $(n \times 1)$ true state vector, $h(.)$ is a $(m \times 1)$ vector of nonlinear functions, ω is a $(m \times 1)$ measurement error vector, m is the number of measurements, and n is the number of state variables. The usual choice for state variables are the voltage phase angles and magnitudes while the measurements are active and reactive power flows and node injections and voltage magnitudes.

A distributed state estimation algorithm, based on dual recursive quadratic programming, is reported in [45]. The algorithm is aimed to perform distributed estimation at the bus level. Reported results indicate a limited computational performance. An improved version of this distributed estimator, including a distributed bad data processing scheme, is proposed in [46]. In the work reported in [47], the possibility of parallel and distributed state estimation implementation was exploited leading to a solution methodology based on conventional state estimation algorithms and a coupling constraint optimization technique. The proposed methodology performs conventional state estimation at the area level and combines these distributed estimations in a way to eliminate discrepancies in the boundary buses. The proposed method was tested on a simulated distributed environment with considerable speed up of the estimation process.

4.6 Composite Generation-Transmission Reliability Evaluation

The reliability assessment of a composite generation-transmission system consists in the evaluation of several probabilistic indices such as the loss of load probability, expected power not supplied, frequency and duration, etc., using stochastic simulation models of the power system operation. A conceptual algorithm for reliability evaluation can be stated as follows [7, 8]:

1. Select a system state x, or a system scenario, corresponding to a particular load level, equipment availability, operating conditions, etc.

2. Calculate the value of a *test function* $F(x)$ which verifies whether there are system limits violations in this specific scenario. The effect of remedial actions, such as generation rescheduling, load curtailment, etc., may be included in this assessment.

3. Update the expected value of the reliability indices based on the result obtained in 2.

4. If the accuracy of the estimates is acceptable, stop. Otherwise, go back to 1.

Step 1 in the algorithm above is usually performed by one of the following methods: enumeration or Monte Carlo sampling. In both approaches, the number of selected scenarios may reach several thousands for practical size systems. Step 2 requires the evaluation of the effect of forced outages in the system behavior for each of the selected scenarios. Static models (power flow) have been mostly used in these evaluations although some dynamic models have also been proposed. Remedial actions may be simulated by special versions of an optimal power flow program.

Step 2 of the conceptual algorithm above is by far the most computer demanding part of the composite reliability evaluation function. It requires the computation of thousands of power flow solutions. Fortunately, these computations are independent and can be carried out easily in parallel. Step 1, also, can be parallelized.

One of the first attempts to parallelize the composite reliability evaluation is the work described in [48]. In this work, a computer package developed for the Electric Power Research Institute (Palo Alto, USA), named Syrel, was adapted to run on multicomputers with hypercube topology (Intel iPSC/1 and iPSC/2). Syrel uses the enumeration approach to perform step 1 of the conceptual composite reliability algorithm. Reported tests with medium size systems (101 and 140 buses) show efficiencies around 70% on the iPSC/1 and 46% on the iPSC/2 (both machines with 16 processors). It should be pointed out that these relatively low efficiencies may be explained by the difficulty in parallelizing a large code (20,000 lines, 148 subroutines) originally developed for sequential computers without substantial changes in the code.

In [49], a parallel version of the Monte Carlo reliability evaluation algorithm was implemented in a 16 node multiprocessor system based on the iAPX 286/287 processor and a common bus shared memory architecture. Tests performed with a large scale model of an actual power system achieved an efficiency close to theoretical maximum efficiency.

In [50], an extensive investigation of topologies for scheduling processes in a parallel implementation of a composite reliability evaluation method based on Monte Carlo simulation is reported. Also, the important issue of generating independent random sequences in each processor is discussed. The schemes studied were implemented in two computer architectures: a distributed memory 64 nodes nCube 2 and a shared memory 10 nodes Sequence Balance. The power system model used in the tests is a synthetic network made up of three areas each of which is the IEEE Reliability Test System. Efficiencies around 50% was achieved in the nCube2 and closer to 100% on the Sequence Balance.

4.7 Power Flow and Contingency Analysis

Power flow is a fundamental tool in power system studies. It is by far the most often used program in evaluating system security, configuration adequacy, etc., and as a starting point for other computations such as short circuit, dynamic simulation, etc. Its efficient solution is certainly a fundamental requirement for the overall efficiency of several integrated power system analysis and synthesis programs. Therefore, it should be expected a great research effort in the parallelization of the power flow algorithms. That has not been the case, however, for two main reasons:

- The practical power flow problem is much more difficult to parallelize than other similar problems owing to the constraints added to the basic system of non-linear algebraic equations.
- Very efficient algorithms are already available which can solve large power flow problems (more than 1000 nodes) in a few seconds on relatively inexpensive computers.

More interesting investigatory lines are the parallelization of multiple power flow solutions (contingency analysis, for instance) and the speed up of power flow programs on vector and superscalar processors. In [51], it is proposed a version of the Newton-Raphson power flow method in which the linearized system of equations is solved by a variant of the Conjugate Gradient Method (Bi-CGSTAB method) with variable convergence tolerance. Results of tests performed in a Cray EL96 computer and a 616 buses model of the Brazilian power system indicates a substantial speedup when compared with the conventional approach.

4.8 Heuristic Search Techniques

The use of heuristic search techniques, such as Simulated Annealing, Genetic Algorithms, Evolutionary Computing, etc., in power system optimization problems has been growing steadily in the last few years. The motivation for the use of such techniques originates in the combinatorial nature of some problems combined with difficult mathematical models (multimodal search space, discontinuities , etc.). These technique have been applied to a variety of power system problems: generation, transmission, and distribution expansion planning, reactive power optimization, unit commitment, economic dispatch, etc. The results reported in the literature indicate that these heuristic search procedure have a great potential for finding global optimal solution to power system problems. However, the computational requirements are usually high in the case of large scale systems.

Parallel implementations of these heuristic search methods have been proposed to overcome this difficulty. In [52], it is reported an implementation of a parallel genetic algorithm for the optimal long-range generation expansion planning problem. The proposed method was tested on a network of Transputers and presented a considerable reduction in computation time in comparison with a conventional approach using dynamic programming. In [53], a parallel simulated annealing method is proposed for the solution of the transmission expansion planning problem. The results obtained show a considerable improvement in terms of reduction of the computing time and quality of the obtained solution.

5   Industrial Implementations

Most of the applications of HPC in power systems effectively used in practice are in the development of real-time simulators. In the following, some of these implementations are described.

5.1   Real-Time Digital Simulator at TEPCO

This simulator, already referred to in section 4.1 of this paper, was developed by Mitsubishi for Tokyo Electric Power Company [23]. The simulator is based on a multicomputer with 512 nodes developed by nCube with a hypercube topology. The multicomputer is interfaced with electronic apparatus through high speed A/D converters. This simulator was able to simulate in real-time the electromechanical transients of a system with 491 busses . The parallel algorithm used in this simulator allocates one processor for each network bus. In this way, the differential equations representing the dynamic behavior of system components connected to a bus are solved in the corresponding processor. The algebraic equations representing the network model are allocated one for each processor and solved by a Gauss-Seidel like procedure. The efficiency achieved in the process is very low as most of the processors time is spent in data communication. However, owing to the large number of processors available, it was possible to achieve the real-time simulation of a practical size power system.

5.2   RTDS of Manitoba HVDC Research Center

This simulator was developed with the objective of real-time simulation of electromagnetic transients in HVDC transmission systems [54, 55]. The simulator uses a parallel architecture based on state-of-the-art DSPs (Digital Signal Processors). A DSP is a processor specially designed for signal processing which is able to simulate power system transients with time steps in the order of 50ms to 100ms. This allows the simulation of high frequency transients which are almost impossible to simulate with the standard processors available in general purpose parallel machines owing to the clock speed of these processors. The software used in this simulator is based on the same mathematical formulation described in section 4.1 and used in most modern digital electromagnetic transient programs [30].

5.3   Supercomputing at Hydro-Quebec

Hydro-Quebec commissioned a Cray X-MP/216 supercomputer in its Centre d'Analyse Numerique de Reseaux in 1991 to be used as a number crunching server for a network of Sun workstations [56]. This supercomputer has been used for transient stability studies using the PSS/E package and electromagnetic transients computations using the EMTP program. In the case of transient stability, models of the Hydro-Quebec system with 12000 buses, which required up to 45 hours of cpu time for a complete study in a workstation, run in the supercomputer in less than 2 hours.

6 Conclusions

High performance computing may be the only way to make viable some power system applications requiring computing capabilities not available in conventional machines, like real-time dynamic security assessment, security constrained optimal power flow, real-time simulation of electromagnetic and electromechanical transients, composite reliability assessment using realistic models, etc. Parallel computers are presently available in a price range compatible with power system applications and presenting the required computation power.

Two main factors are still impairments to the wide acceptance of these machines in power system applications: the requirements for reprogramming or redevelopment of applications and the uncertainty about the prevailing parallel architecture. The first problem is inevitable, as automatic parallelization tools are not likely to become practical in the near future, but has been minimized by the research effort in parallel algorithms and the availability of more efficient programming tools. The second difficulty is becoming less important with the maturity of the high performance computing industry.

Likewise in the history of sequential computer evolution, a unique and overwhelming solution to the parallel computer architecture problem is not to be expected. It is more likely that a few different architectures will be successful in the next few years and the users will have to decide which one is the most adequate for their application. Moreover, it is likely that commercial processing applications, which are now turning towards parallel processing, are the ones that will shape the future parallel computer market. However, to make this scenario a little bit less uncertain, it should be pointed out the tendency in the parallel computer industry to make their products follow open system standards and the possibility of developing applications less dependent on a particular architecture.

References

1. T.G. Lewis and H. El-Rewini. *Introduction to Parallel Computing*. Prentice Hall, New York, 1992.
2. M.J. Quinn. *Parallel Computing: Theory and Practice*. McGraw-Hill, New York, 1994.
3. G.A. Geist and V.S. Sunderam. Network-based concurrent computing on the PVM system. *Concurrency: Practice and Experience*, 4(4):293–311, June 1992.
4. B. Stott, O. Alsac, and A. Monticelli. Security analysis and optimization. *Proceedings of the IEEE*, 75(12):1623–1644, December 1987.
5. N.J. Balu and et al. On-line power system security analysis. *Proceedings of the IEEE*, 80(2):262–280, February 1992.
6. Y. Sekine, K. Takahashi, and T. Sakaguchi. Real-time simulation of power system dynamics. *Int. J. of Electrical Power and Energy Systems*, 16(3):145–156, 1994.
7. M.V.F. Pereira and N.J. Balu. Composite generation/transmission reliability evaluation. *Proceedings of the IEEE*, 80(4):470–491, April 1992.
8. R. Billinton and W. Li. *Reliability Assessment of Electric Power Systems Using Monte Carlo Method*. Plenum Press, New York, 1994.

9. B. Stott. Power system dynamic response calculations. *Proceedings of the IEEE*, 67(2):219–241, February 1979.
10. P. Kundur. *Power System Stability and Control*. McGraw-Hill, New York, 1994.
11. J.Q. Wu and A. Bose. Parallel solution of large sparse matrix equations and parallel power flow. *IEEE Transactions on Power Systems*, 10(3):1343–1349, August 1995.
12. J.Q. Wu and A. Bose. A new successive relaxation scheme for the W-matrix solution method on a shared-memory parallel computer. *IEEE Transactions on Power Systems*, 11(1):233–238, February 1996.
13. G.T. Vuong, R. Chahine, G.P. Granelli, and R. Montagna. Dependency-based algorithm for vector processing of sparse matrix forward/backward substitutions. *IEEE Transactions on Power Systems*, 11(1):198–205, February 1996.
14. I.C. Decker, D.M. Falcão, and E. Kaszkurewicz. Parallel implementation of a power system dynamic simulation methodology using the conjugate gradient method. *IEEE Transactions on Power Systems*, 7(1):458–465, February 1992.
15. F.D. Galiana, H. Javidi, and S. McFee. On the application of pre-conditioned conjugate gradient algorithm to power network analysis. *IEEE Transactions on Power Systems*, 9(2):629–636, May 1994.
16. I.C. Decker, D.M. Falcão, and E. Kaszkurewicz. Conjugate gradient methods for power system dynamic simulation on parallel computers. *IEEE Transactions on Power Systems*, 11(3):1218–1227, August 1996.
17. H. Dağ and F.L. Alvarado. Computation-free preconditioners for the parallel solution of power system problems. *IEEE Transactions on Power Systems*, to appear.
18. H. Dağ and F.L. Alvarado. Toward improved use of the conjugate gradient method for power system applications. *IEEE Transactions on Power Systems*, to appear.
19. M.H.M Vale, D.M. Falcão, and E. Kaszkurewicz. Electrical power network decomposition for parallel computations. In *Proceedings of the IEEE Symposium on Circuits and Systems*, pages 2761–2764, San Diego, USA, May 1992.
20. J.S. Chai, N. Zhu, A. Bose, and D.J. Tylavsky. Parallel newton type methods for power system stability analysis using local and shared memory multiprocessors. *IEEE Transactions on Power Systems*, 6(4):1539–1545, November 1991.
21. J.S. Chai and A. Bose. Bottlenecks in parallel algorithms for power system stability analysis. *IEEE Transactions on Power Systems*, 8(1):9–15, February 1993.
22. A. Padilha and A. Morelato. A W-matrix methodology for solving sparse network equations on multiprocessor computers. *IEEE Transactions on Power Systems*, 7(3):1023–1030, 1992.
23. H. Taoka, I. Iyoda, H. Noguchi, N. Sato, and T. Nakazawa. Real-time digital simulator for power system analysis on a hypercube computer. *IEEE Transactions on Power Systems*, 7(1):1–10, February 1992.
24. M.Ilić-Spong, M.L. Crow, and M.A. Pai. Transient stability simulation by waveform relaxation methods. *IEEE Transactions on Power Systems*, 2(4):943–952, November 1987.
25. M.L. Crow and M. Ilić. The parallel implementation of the waveform relaxation method for transient stability simulations. *IEEE Transactions on Power Systems*, 5(3):922–932, August 1990.
26. L. Hou and A. Bose. Implementation of the waveform relaxation algorithm on a shared memory computer for the transient stability problem. *IEEE Transactions on Power Systems*, to appear.
27. F. Alvarado. Parallel solution of transient problems by trapezoidal integration. *IEEE Transactions on Power Apparatus and Systems*, 98(3):1080–1090, May/June 1979.

28. M. LaScala, A. Bose, D.J. Tylavsky, and J.S. Chai. A highly parallel method for transient stability analysis. *IEEE Transactions on Power Systems*, 5(4):1439–1446, November 1990.

29. M. LaScala, M. Brucoli, F. Torelli, and M. Trovato. A Gauss-Jacobi-block-Newton method for parallel transient stability analysis. *IEEE Transactions on Power Systems*, 5(4):1168–1177, November 1990.

30. H.W. Dommel and W.S. Meyer. Computation of electromagnetic transients. *Proceedings of the IEEE*, 62:983–993, July 1974.

31. D.M. Falcão, E. Kaszkurewicz, and H.L.S. Almeida. Application of parallel processing techniques to the simulation of power system electromagnetic transients. *IEEE Transactions on Power Systems*, 8(1):90–96, February 1993.

32. R.C. Durie and C. Pottle. An extensible real-time digital transient network analyzer. *IEEE Transactions on Power Systems*, 8(1):84–89, February 1993.

33. K. Werlen and H. Glavitsch. Computation of transients by parallel processing. *IEEE Transactions on Power Delivery*, 8(3):1579–1585, July 1993.

34. J.R. Marti and L.R. Linares. A real-time EMTP simulator. *IEEE Transactions on Power Systems*, 9(3):1309–1317, August 1994.

35. N. Martins. Efficient eigenvalue and frequency response methods applied to power system small-signal stability studies. *IEEE Transactions on Power Systems*, 1(1):217–226, February 1986.

36. J.M. Campagnolo, N. Martins, J.L.R. Pereira, L.T. G. Lima, H.J.C.P. Pinto, and D.M.Falcão. Fast small-signal stability assessment using parallel processing. *IEEE Transactions on Power Systems*, 9(2):949–956, May 1994.

37. J.M. Campagnolo, N. Martins, and D.M. Falcão. An efficient and robust eigenvalue method for small-signal stability assessment in parallel computers. *IEEE Transactions on Power Systems*, 10(1):506–511, February 1995.

38. J.M. Campagnolo, N. Martins, and D.M. Falcão. Refactored bi-iteration: A high performance eigensolution method for large power system matrices. *IEEE Transactions on Power Systems*, 11(3):1228–1235, August 1996.

39. S.M. Shahidehpour and V.C. Ramesh. Nonlinear programming algorithms and decomposition strategies for OPF. In *Optimal Power Flow: Solution Techniques, Requirements, and Challenges*, pages 10–25. IEEE Tutorial Course 96 TP 111-0, 1996.

40. A. Monticelli, M.V.F. Pereira, and S. Granville. Security constrained optimal power flow with post-contingency corrective rescheduling. *IEEE Transactions on Power Systems*, 2(1):175–182, February 1987.

41. H.J.C. Pinto, M.V.F. Pereira, and M.J. Teixeira. New parallel algorithms for the security-constrained dispatch with post-contingency corrective actions. In *Proceedings of the 10th Power Systems Computation Conference*, pages 848–853, Graz, Austria, August 1990.

42. M. Rodrigues, O.R. Saavedra, and A. Monticelli. Asynchronous programming model for the concurrent solution of the security constrained optimal power flow problem. *IEEE Transactions on Power Systems*, 9(4):2021–2027, November 1994.

43. V.C. Ramesh and S.N. Talukdar. A parallel asynchronous decomposition for online contingency planning. *IEEE Transactions on Power Systems*, 11(1):344–349, February 1996.

44. S.N. Talukdar and V.C. Ramesh. A multi-agent technique for contingency-constrained optimal power flows. *IEEE Transactions on Power Systems*, 9(2):855–861, May 1994.

45. S.-Y. Lin. A distributed state estimation for electric power systems. *IEEE Transactions on Power Systems*, 7(2):551–557, May 1992.

46. S.-Y. Lin and C.-H. Lin. An implementable distributed state estimation and distributed data processing scheme for electric power systems. *IEEE Transactions on Power Systems*, 9(2):1277–1284, May 1994.

47. D.M. Falcão, F.F. Wu, and L. Murphy. Parallel and distributed state estimation. *IEEE Transactions on Power Systems*, 10(2):724–730, May 1995.

48. D.J. Boratynska-Stadnicka, M.G. Lauby, and J.E. Van Ness. Converting an existing computer code to a hypercube computer. In *Proceedings of the IEEE Power Industry Computer Applications Conference*, Seattle, USA, May 1989.

49. M.J. Teixeira, H.J.C. Pinto, M.V.F. Pereira, and M.F. McCoy. Developing concurrent processing applications to power system planning and operations. *IEEE Transactions on Power Systems*, 5(2):659–664, May 1990.

50. N. Gubbala and C. Singh. Models and considerations for parallel implementation of Monte Carlo simulation methods for power system reliability evaluation. *IEEE Transactions on Power Systems*, 10(2):779–787, May 1995.

51. C.L.T. Borges, A.L.G.A. Coutinho, and D.M. Falcão. Power flow solution in vector computers using the Bi-CGSTAB method. In *Proceedings of the XI Congresso Brasileiro de Automática*, (in Portuguese), São Paulo, Brasil, May 1996.

52. Y. Fukuyama and H.-D. Chiang. A parallel genetic algorithm for generation expansion planning. *IEEE Transactions on Power Systems*, 11(2):955–961, May 1996.

53. R.A. Gallego, A.B. Alves, A. Monticelli, and R. Romero. Parallel simulated annealing applied to long term transmission network expansion planning. *IEEE Transactions on Power Systems*, to appear.

54. D. Brandt, R. Wachal, R. Valiquette, and R. Wierckx. Closed loop testing of a joint VAR controller using a digital real-time simulator. *IEEE Transactions on Power Systems*, 6(3):1140–1146, August 1991.

55. P. McLaren, R. Kuffel, R. Wierckx, J. Giesbrecht, and L. Arendt. A real-time digital simulator for testing relays. *IEEE Transactions on Power Delivery*, 7(1), January 1992.

56. G.T. Vuong, R. Chahine, and S. Behling. Supercomputing for power system analysis. *IEEE Computer Applications in Power*, 5(3):45–49, July 1992.

Providing Access to High Performance Computing Technologies

Jack Dongarra[1], Shirley Browne[2], and Henri Casanova[2]

[1] University of Tennessee and Oak Ridge National Laboratory
[2] University of Tennessee, Knoxville TN 37996 USA

Abstract. This paper describes two projects underway to provide users with access to high performance computing technologies. One effort, the National HPCC Software Exchange, is providing a single point of entry to a distributed collection of domain-specific repositories. These repositories collect, catalog, evaluate, and provide access to software in their specialized domains. The NHSE infrastructure allows these repositories to interoperate with each other and with the top-level NHSE interface. Another effort is the NetSolve project which is a client-server application designed to solve computational science problems over a network. Users may access NetSolve computational servers through C, Fortran, MATLAB, or World Wide Web interfaces. An interesting intersection between the two projects would be the use of the NetSolve system by a domain-specific repository to provide access to software without the need for users to download and install the software on their own systems.

1 The National HPCC Software Exchange

1.1 Overview of the NHSE

The National HPCC Software Exchange (NHSE) is an Internet-accessible resource that facilitates the exchange of software and of information among research and computational scientists involved with High Performance Computing and Communications (HPCC) [1] [3]. The NHSE facilitates the development of discipline-oriented software repositories and promotes contributions to and use of such repositories by Grand Challenge teams, as well as other members of the high performance computing community.

The expected benefits from successful deployment of the NHSE include the following:

- Faster development of better-quality software so that scientists can spend less time writing and debugging programs and more time on research problems.
- Reduction of duplication of software development effort by sharing of software.
- Reduction of time and effort spent in locating relevant software and information through the use of appropriate indexing and search mechanisms and domain-specific expert help systems.

---

[3] http://www.netlib.org/nhse/

– Reduction of duplication of effort in evaluating software by sharing software review and evaluation information.

The scope of the NHSE is software and software-related artifacts produced by and for the HPCC Program. Software-related artifacts include algorithms, specifications, designs, and software documentation. The following three types of software being made available:

– Systems software and software tools. This category includes parallel processing tools such as parallel compilers, message-passing communication subsystems, and parallel monitors and debuggers.
– Basic building blocks for accomplishing common computational and communication tasks. These building blocks will be of high quality and transportable across platforms. Building blocks are meant to be used by Grand Challenge teams and other researchers in implementing programs to solve computational problems. Use of high-quality transportable components will speed implementation, as well as increase the reliability of computed results.
– Research codes that have been developed to solve difficult computational problems. Many of these codes will have been developed to solve specific problems and thus will not be reusable as is. Rather, they will serve as proofs of concept and as models for developing general-purpose reusable software for solving broader classes of problems.

1.2 Domain-specific Repositories

The effectiveness of the NHSE will depend on discipline-oriented groups and Grand Challenge teams having ownership of domain-specific software repositories. The information and software residing in these repositories will be best maintained and kept up-to-date by the individual disciplines, rather than by centralized administration. Domain experts are the best qualified to evaluate, catalog, and organize software resources within their domain.

Netlib – Mathematical Software An example of a domain-specific repository is the Netlib mathematical software repository, which has been in existence since 1985 [2]. Netlib differs from other publicly available software distribution systems, such as Archie, in that the collection is moderated by an editorial board and the software contained in it is widely recognized to be of high quality. Netlib distributes freely-available numerical libraries such as EISPACK, LINPACK, FFTPACK, and LAPACK that have long been used as important tools in scientific computation. The Netlib collection also includes a large number of newer, less well-established codes. Software is available in all the major numerical analysis areas, including linear algebra, nonlinear equations, optimization, approximation, and differential equations. Most of the software is written in Fortran, but programs in other languages, such as C and C++, are also available. Netlib uses the Guide to Available Mathematical Software (GAMS) classification system [3] to help users quickly locate software that meets their needs.

A branch of Netlib specialized to high performance computing, called HPC-netlib, is currently under development. HPC-netlib will provide access to algorithms and software for both shared memory and distributed memory machines, as well as to information about performance of parallel numerical software on different architectures.

PTLIB – Parallel Tools Another domain-specific repository that is under development is the PTLIB parallel tools repository. PTLIB will provide access to high-quality tools in the following areas: communication libraries, data parallel language compilers, automatic parallelization tools, debuggers and performance analyzers, parallel I/O, job scheduling and resource management.

1.3 Repository Interoperation

In addition to providing access to its own software, a repository may wish to import software descriptions from other repositories and make this software available from its own interface. For example, a computational chemistry repository may wish to provide access to mathematical software and to parallel processing tools in a manner tuned to the computational chemistry discipline. A repository interoperation architecture is shown in Figure 1.

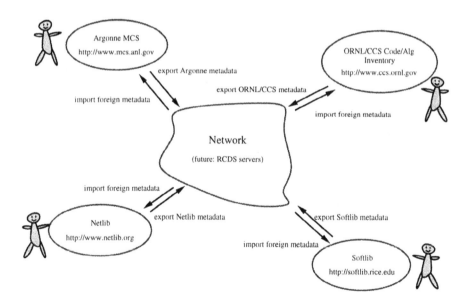

Fig. 1. Repository Interoperation Architecture

The NHSE is using the Reuse Library Interoperability Group's Basic Interoperability Data Model (BIDM) as its interoperability mechanism [4]. Participating HPCC repositories and some individual contributors have placed META and LINK tags in the headers of HTML files that describe their software resources. This information may then be picked up by other repositories and incorporated into their own software catalogs. The NHSE is developing a toolkit called Repository in a Box (RIB) that will assist repository maintainers in creating and maintaining software catalog records, in exchanging these records with other repositories (including the top-level virtual NHSE repository), and in providing a user interface to their software catalog. The Resource Cataloging and Distribution System under development at the University of Tennessee will provide a scalable substrate for repository interoperability by providing catalog and location servers that map resource names to catalog and location information.

1.4 Software Review Framework

The NHSE has designed a software review policy that enables easy access by users to information about software quality, but which is flexible enough to be used across and specialized to different disciplines. The three review levels recognized by the NHSE are the following: Unreviewed, Partially reviewed, and Reviewed. The *Unreviewed* designation means only that the software has been accepted into the owning repository and is thus within the scope of HPCC and of the discipline of that repository. The *Partially reviewed* designation means that the software has been checked by a librarian for conformance with the scope, completeness, adequate documentation, and construction guidelines. The *Reviewed* designation means that the software has been reviewed by an expert in the appropriate field, for example by an author of a review article in the electronic journal *NHSE Review* [4], and found to be of high quality. Domain-specific repositories and expert reviewers are expected to refine the NHSE software review policy by adding additional review criteria, evaluation properties, and evaluation methods and tools. The NHSE also provides for soliciting and publishing author claims and user comments about software quality. All software exported to the NHSE by its owning repository or by an individual contributor is to be tagged with its current review level and with a pointer to a review abstract which describes the software's current review status and includes pointers to supporting material.

2 The NetSolve project

An ongoing thread of research in scientific computing is the efficient solution of large problems. Various mechanisms have been developed to perform computations across diverse platforms. The most common mechanism involves software libraries. Unfortunately, the use of such libraries presents several difficulties. Some software libraries are highly optimized for only certain platforms and do not provide a convenient interface to other computer systems. Other libraries

---

[4] http://nhse.cs.rice.edu/NHSEreview/

demand considerable programming effort from the user, who may not have the time to learn the required programming techniques. While a limited number of tools have been developed to alleviate these difficulties, such tools themselves are usually available only on a limited number of computer systems. MATLAB [5] is an example of such a tool.

These considerations motivated the establishment of the NetSolve project. NetSolve is a client-server application designed to solve computational science problems over a network. A number of different interfaces have been developed to the NetSolve software so that users of C, Fortran, MATLAB, or the World Wide Web can easily use the NetSolve system. The underlying computational software can be any scientific package, thus helping to ensure good performance through choice of an appropriate package.. Moreover, NetSolve uses a load-balancing strategy to improve the use of the computational resources available. Some other systems are currently being developed to achieve somewhat similar goals. Among them are the Network based Information Library for high performance computing (Ninf) [6] project which is very comparable to NetSolve in its way of operation, and the Remote Computation System (RCS) [7] which is a remote procedure call facility for providing uniform access to a variety of supercomputers.

We introduce the NetSolve system, its architecture and the concepts on which it is based. We then describe how NetSolve can be used to solve complex scientific problems.

2.1 The NetSolve System

The NetSolve system is a set of loosely connected machines. By *loosely* connected, we mean that these machines can be on the same local network or on an international network. Moreover, the NetSolve system can be running in a *heterogeneous* environment, which means that machines with different data formats can be in the system at the same time.

The current implementation sees the system as a completely connected graph without any hierarchical structure. This initial implementation was adopted for simplicity and is viable for now. Our current idea of the *NetSolve world* is of a set of independent NetSolve systems in different locations, possibly providing different services. A user can then contact the system he wishes, depending on the task he wants to have performed and on his own location. In order to manage efficiently a pool of hosts scattered on a large-scale network, future implementations might provide greater structure (e.g., a tree structure), which will limit and group large-range communications.

Figure 2 shows the global conceptual picture of the NetSolve system. In this figure, a NetSolve client send a request to the NetSolve agent. The agent chooses the "best" NetSolve resource according to the size and nature of the problem to be solved.

Several instances of the NetSolve agent can exist on the network. A good strategy is to have an instance of the agent on each local network where there are NetSolve clients. Of course, this is not mandatory; indeed, one may have only a single instance of the agent per NetSolve system.

Every host in the NetSolve system runs a NetSolve *computational* server (also called a *resource*, as shown in Figure 2). The NetSolve resources have access to scientific packages (libraries or stand-alone systems).

An important aspect of this server-based system is that each instance of the agent has its own *view* of the system. Therefore, some instances may be aware of more details than others, depending on their locations. But eventually, the system reaches a stable state in which every instance possesses all the available information on the system.

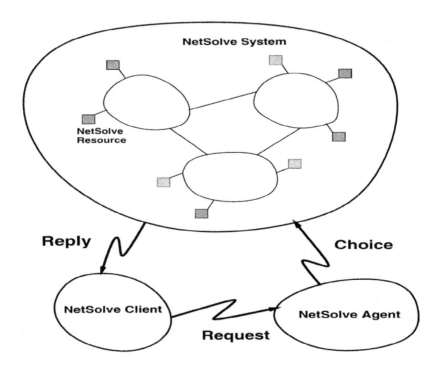

Fig. 2. The NetSolve System

Communication within the NetSolve system is achieved through the TCP/IP socket layer and heterogeneous environments are supported thanks to the XDR protocol [8].

2.2 Problem Specification

To keep NetSolve as general as possible, we needed to find a formal way of describing a problem. Such a description must be carefully chosen, since it will affect the ability to interface NetSolve with arbitrary software.

A problem is defined as a 3-tuple: $< name, inputs, outputs >$, where

- *name* is a character string containing the name of the problem
- *inputs* is a list of input objects
- *outputs* is a list of output objects

An object is itself described as follows: $< object, data >$, where *object* can be 'MATRIX', 'VECTOR' or 'SCALAR' and *data* can be any of the standard FORTRAN data types. This description has proved to be sufficient to interface NetSolve with numerous software packages. The NetSolve administrator can then not only choose the best platform on which to install NetSolve, but also select the best packages available on the chosen platform.

The current installation of NetSolve at the University of Tennessee uses the BLAS [9], [10], [11], LAPACK [12], ItPack [13], LINPACK [14] and FitPack [15]. These packages are available on a large number of platforms and are freely distributed. The use of ScaLAPACK [16] on massively parallel processors would be a way to use the power of high-performance parallel machines via NetSolve.

2.3 Client Interfaces

One of the main goals of NetSolve is to provide the user with a large number of interfaces and to keep them as simple as possible.

The MATLAB Interface :

We developed a MATLAB interface which provides interactive access to the NetSolve system. Interactive interfaces offer several advantages. First, they are easy to use because they completely free the user from any code writing. Second, the user still can exploit the power of software libraries. Third, they provide good performance by capitalizing on standard tools such as MATLAB. Let us assume, for instance, that MATLAB is installed on one machine on the local network. It is possible to use NetSolve via the MATLAB interface on this machine and in fact use the computational power of another more powerful machine where MATLAB is not available.

Within MATLAB, NetSolve may be used in two ways. It is possible to call NetSolve in a *blocking* or *nonblocking* fashion. Here is an example of the MATLAB interface to solve an linear system computation using the blocking call:

```
>> a = rand(100); b = rand(100,1);
>> x = netsolve('ax=b',a,b)
```

This MATLAB script first creates a random 100×100 matrix, *a*, and a vector *b* of length 100. The call to the **netsolve** function returns with the solution. This call manages all the NetSolve protocol, and the computation may be executed on a remote host.

Here is the same computation performed in a nonblocking fashion:

```
>> a = rand(100); b = rand(100,1);
>> request = netsolve_nb('send','ax=b',a,b)
>> x = netsolve_nb('probe',request)
      Not Ready Yet
>> x = netsolve_nb('wait',request)
```

Here, the first call to netsolve_nb() sends a request to the NetSolve agent and returns immediately with a request identifier. One can then either *probe* for the request or *wait* for it. Probing always returns immediately, either signaling that the result is not available yet or, if available, stores the result in the user data space. Waiting blocks until the result is available and then store it in the user data space. This approach allows user-level parallelism and communication/computation overlapping (see Section 2.4).

Other functions are provided, for example, to obtain information on the problems available or on the status of pending requests.

C and FORTRAN interfaces :

In addition to the MATLAB interface, we have developed two programming interfaces, one for Fortran and one for C. Unlike the interactive interfaces, programming interfaces require some programming effort from the user. But again, with a view to simplicity, the NetSolve libraries contain only a few routines, and their use has been made as straightforward as possible. As in MATLAB, the user can call NetSolve *asynchronously*.

A very attractive feature of these interfaces is that NetSolve preserves the original calling sequence of the underlying numerical software. It is then almost immediate to convert a code to NetSolve, as shown in the short FORTRAN example below :

```
C      Linear system solve : Call to LAPACK

       call DGESV(N,1,A,MAX,IPIV,B,MAX,INFO)

C      Linear system solve : Call to NetSolve

       call FNSOLVE('DGESV',NSINFO,
                    N,1,A,MAX,IPIV,B,MAX,INFO)
```

2.4 Performance

One of the challenges in designing NetSolve was to combine ease of use and excellence of performance. Several factors ensure good performance without increasing the amount of work required of the user. In addition to the availability of diverse scientific packages (as discussed in a preceding section), these factors include load balancing and the use of simultaneous resources.

Load Balancing :

Load balancing is one of the most attractive features of the NetSolve project. NetSolve performs computations over a network containing a large number of machines with different characteristics, and one of these machines is the most suitable for a given problem, meaning the one yielding the shortest response time. NetSolve provides the user with a "best effort" to find this *best* resource.

As seen on figure 2, a NetSolve client sends a request to an instance of the NetSolve agent. This instance of the agent has some knowledge about the computational resources in the system. Hopefully this knowledge is not too out of date (which is ensured by a set of protocols we do not have space to describe here) and, for each resource M, allows a fairly accurate computation of :

- T_n : the time to send the data to M and receive the result over the network, and
- T_c : the time to perform the computation on M.

All the details about the protocols involved in this computations are given in [17]. The whole idea behind this scheme is that it would be too inefficient to have the agent compute exact values for T_n and T_c for each incoming request. Instead, we prefer to have a quick estimate, which might not be as accurate.

Simultaneous resources :

Using the nonblocking interfaces to NetSolve, the user can design a Net-Solve application that has some parallelism. Indeed, it is possible to send asynchronously several requests to NetSolve. The load balancing strategy described above insures that these problems will be solved on different machines, in parallel. The client has then just to wait for the results to come back.

Here is another strength of NetSolve : as soon as a new resource is started, it takes part in the system, and can be used. Therefore, without modifying his code or knowing in fact anything about the servers, a user can see the performance of his application greatly improved.

2.5 Fault Tolerance

Fault tolerance is an important issue in any loosely connected distributed system like NetSolve. The failure of one or more components of the system should not cause any catastrophic failure. Moreover, the number of side effects generated by such a failure should be as low as possible, and the system should minimize the drop in performance. We tried to make NetSolve as fault tolerant as possible.

A first aspect of fault-tolerance in NetSolve takes place at the server level. It is possible to stop a NetSolve server (resource or instance of the agent) at any time, and restart it safely at any time. In fact, every NetSolve server is an independent entity. This insures that the NetSolve system will remain coherent after any kind of network/machine problem. In the installation of NetSolve at the University of Tennessee, the whole system is managed by a 'cron' job, and servers are restarted automatically after machines go down for back-ups for instance.

Another aspect of fault tolerance is that it should minimize the side effects of failures. To this end, we designed the client-server protocol as follows. When the NetSolve agent receives a request for a problem to be solved, it sends back a list of computational servers sorted from the most to the least suitable one. The client tries all the servers in sequence until one accepts the problem. This strategy allows the client to avoid sending multiple requests to the agent for the same problem if some of the computational servers are stopped. If at the end of the list no server has been able to answer, the client asks another list from the agent. Since it has reported all the encountered failures, it will receive a different list.

Once the connection has been established with a computational server, there still is no guarantee that the problem will be solved. The computational process on the remote host may die for some reason. In that case, the failure is detected by the client, and the problem is sent to another available computational server. This process is transparent to the user but, of course, lengthens the execution time. The problem is migrated between the possible computational servers until it is solved or no server remains.

3   Conclusions and Future Work

The NHSE is providing a means for the HPCC community to share software and information and thus broaden and accelerate the use of high performance computing technologies in scientific and engineering applications. By supplying the tools and mechanisms for HPCC repositories to interoperate, the NHSE is enabling different HPCC agencies and research groups to leverage each others efforts. During the next year, the NHSE will be bringing online several new domain-specific repositories as well as promoting the review and evaluation of software in these domains.

The NetSolve project is still at an early development stage and there is room for improvement at the interface level as well as at the conceptual level. At the interface level, we are thinking of providing a Java interface to NetSolve. At the conceptual level, the load-balancing strategy must be improved in order to change the "best guess" into a "best choice" as much as possible. The challenge is to come close to a best choice without flooding the network. The danger is to waste more time computing this best choice than the computation would have taken in the case of a best guess only. All these improvements are intended to combine ease of use, generality and performance, the main purposes of the NetSolve project.

We plan to investigate extending the NHSE Repository in a Box toolkit with a remote execution facility based on NetSolve. This facility would allow repository maintainers to provide remote access to software, instead of having users download and install the software on their own systems. We will also investigate how to provide server safe execution environments for user code so that users may upload functions for execution on a remote server. This capability is important for software packages that require user-defined functions to be provided as input.

References

1. Shirley Browne, Jack Dongarra, Stan Green, Keith Moore, Tom Rowan, Reed Wade, Geoffrey Fox, Ken Hawick, Ken Kennedy, Jim Pool, Rick Stevens, Robert Olsen, and Terry Disz. The National HPCC Software Exchange. *IEEE Computational Science and Engineering*, 2(2):62–69, 1995.
2. Jack J. Dongarra and Eric Grosse. Distribution of mathematical software via electronic mail. *Communications of the ACM*, 30(5):403–407, May 1987.
3. Ronald F. Boisvert, Sally E. Howe, and David K. Kahaner. GAMS: A framework for the management of scientific software. *ACM Transactions on Mathematical Software*, 11(4):313–355, December 1985.
4. Shirley Browne, Jack Dongarra, Kay Hohn, and Tim Niesen. Software repository interoperability. Technical Report CS-96-329, University of Tennessee, 1996.
5. Inc The Math Works. *MATLAB Reference Guide*. 1992.
6. *Ninf : Network based Information Library for Globally High Performance Computing*. Proc. of Parallel Object-Oriented Methods and Applications (POOMA), Santa Fe, 1996.
7. W. Gander P. Arbenz and M. Oettli. The remote computational system. *Lecture Note in Computer Science, High-Performance Computation and Network*, 1067:662–667, 1996.
8. Sun Microsystems, Inc. XDR: External Data Representation Standard. RFC 1014, Sun Microsystems, Inc., June 1987.
9. D. Kincaid C. Lawson, R. Hanson and F. Krogh. Basic linear algebra subprograms for fortran usage. *ACM Transactions on Mathematical Software*, 5:308–325, 1979.
10. S. Hammarling J. Dongarra, J. Du Croz and R. Hanson. An extended set of fortran basic linear algebra subprograms. *ACM Transactions on Mathematical Software*, 14(1):1–32, 1988.
11. I. Duff J. Dongarra, J. Du Croz and S. Hammarling. A set of level 3 basic linear algebra subprograms. *ACM Transactions on Mathematical Software*, 16(1):1–17, 1990.
12. E. Anderson, Z. Bai, C. Bischof, J. Demmel, J. Dongarra, J. Du Croz, A. Greenbaum, S. Hammarling, A. McKenney, S. Ostrouchov, and D. Sorensen. *LAPACK Users' Guide*. SIAM Philadelphia, Pennsylvania, 2 edition, 1995.
13. David M. Young David R. Kincaid, John R. Respess and Roger G. Grimes. Itpack 2c: A fortran package for solving large sparse linear systems by adaptive accelerated iterative methods. Technical report, University of Texas at Austin, Boeing Computer Services Company, 1996.
14. C. B. Moler J. J. Dongarra, J. R. Bunch and G. W. Stewart. *LINPACK Users' Guide*. SIAM Press, 1979.
15. A. Cline. Scalar- and planar-valued curve fitting using splines under tension. *Communications of the ACM*, 17:218–220, 1974.
16. J. Dongarra and D. Walker. Software libraries for linear algebra computations on high performance computers. *SIAM Review*, 37(2):151–180, 1995.
17. *NetSolve: A Network Server for Solving Computational Science Problems*. To appear in Proc. of Supercomputing '96, Pittsburgh, 1996.

Scan-Directional Architectures

Dinu Coltuc[1] and Jean-Marie Becker[2]

[1] Research Institute for Electrical Engineering,
Splaiul Unirii 313, Bucharest, Romania
[2] Ecole Supérieure de Chimie Physique Electronique de Lyon,
43, bd du 11 Novembre 1918, BP 2077 - 69616 Villeurbanne cedex - France

Abstract. A considerable number of algorithms, of interest for image processing and general matrix analysis problems, operate on certain pre-defined streams of data. This paper presents an architectural framework to tackle these algorithms, namely parallel/pipeline architectures builded with "scan-directional" memories. The paper analyzes the realization of such memory modules for simultaneous fast access on various scan-directions, e.g., lines, columns, diagonals, Peano-curve, bit-reversed, etc. Some of the most important treatments (matrix transposition, multi-plication, inversion, Schur complement, orthogonal transforms, linear, non-linear image filtering, etc.) fit into the scan-directional frame. Their implementation is investigated.

1 Introduction

Algorithms can roughly be divided into two main categories : data oriented or address oriented. This paper is concerned with the second class, of great interest for matrix oriented problems and notably for image/signal processing applications. A well-known example of a wholly address oriented algorithm is the FFT where 1) either the input or the output data needs bit-reversal ordering; 2) "butterflies" are computed for fixed positions of the data. In the sequel, such a predefined data-scanning order will be called a scan pattern. Usually, the scan patterns are derived from the "geometry" of the data set, e.g., rows, columns. Different algorithms may need different scan patterns. For such algorithms, the implementation efficiency depends not only on the processing speed, but also on the access time on the required scan patterns. This paper presents a scan-directional approach for the implementation of address oriented algorithms - parallel (or pipeline) architectures developed around multiple scan pattern memory modules.

2 Problem Setting

The scan directional approach copes with the implementation of address oriented algorithms. At a general level the proposed architectures are composed of memory modules and processing devices. Each memory module provides high speed access on predefined streams of data (scan patterns) to feed on several process-ing devices. More precisely, a problem is considered to fit the scan directional approach when:

1. data processing order is fully known and it is independent of data values;
2. different kinds of scan patterns are required;
3. high speed data streams are needed.

Requirements 1 and 2 are algorithm-specific. Requirement 3, which seems to be application-specific, deserves some comments. It should be understood in the context of the memory module bandwidth : namely the memory module is supposed to be implemented with much slower memory chips than the access time required for its input/output. In fact, the main aspect of this approach is to provide predefined high speed streams of data by using slow memory chips. Even if nowadays memory chips become faster and faster the problem of memory bandwidth is and will remain an important issue. To give some arguments, we mention that: once, the processing speed increases as well; twice, the need of parallelism demands that multiple data streams are simultaneously available.

To summarize, the proposed scan directional approach provides general archi-tectures to be tailored to each specific application. The core of such architectures consists of memory modules, called multiple scan pattern memory modules, that provide very high bandwidth on several data streams. While the memory module architecture is general, the mapping of the data into the memory chips strongly depends on the scan patterns at hand (application specific). Moreover, the very existence of the mapping should be investigated. An important result in our ap-proach is the existence of a memory mapping for any 2 scan patterns. For more than 2 scan patterns no such existence theorem can be given and the mapping, if any, should be established by construction. A lot of work on memory mappings has been already done in [1], [2]. The aim of this paper is to provide a unified approach on this subject and to point out the architectural framework suitable for many applications (most of them classical) in the field of matrix computation and signal/image processing.

3 Multiple Scan Pattern Memory Modules: Architecture

Let memory modules be organized as square arrays of size $N \times N$. Each data element is specified by its location, i.e., the ordered pair (x, y), $0 \leq x, y < N$, where x, y are the row and the column position, respectively. If n memory chips (banks) are used for the implementation of the memory module, one must determine an explicit mapping:

$$m : [0, N-1] \times [0, N-1] \rightarrow [0, n-1] \qquad (1)$$

Interleaved memory systems are often used for increasing the effective mem-ory bandwidth (Fig. 1). Thus, n data are simultaneously read from n memory chips, then loaded in a parallel/serial converter and shifted serially. The serial

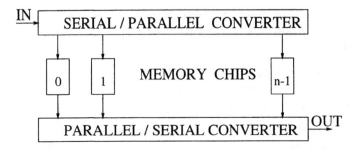

Fig. 1. Interleaved memory.

access time can be n times faster than the access time of the memory chips. The larger n, the faster the access. Let T_s be the fast serial access time of the module and T_a be the access time of the memory chips. If $nT_s > rT_a$, up to r fast data streams can be simultaneously provided; if $r = 2$ for example, it means that the memory module can provide two accesses in the same memory cycle, e.g., simultaneous reading and writing. Interleaved memories classicaly provide fast access on a single scan pattern. The mapping of the data on the memory chips is very simple: groups of n successive data in the scan direction should be consecutively stored onto the n memory chips. A convenient mapping obeying the natural ordering of the scan direction can be:

$$m(x, y) = y \bmod n \tag{2}$$

Fast access on several scan patterns requires some modifications of the classical interleaved memory (Fig. 2):

– programmable interconnections are inserted between the memory chips and the parallel/serial and serial/parallel converters;
– each memory chip should have some distinct address lines.

Thus, it is possible to access groups of data (from different addresses) in parallel and to change their order inside any group. The multiple scan patterns mapping generalizes the single scan pattern mapping: *for each scan pattern*, groups of n adjacent data should be stored, one to one, onto the n memory chips. The construction of the mapping is the main problem of the multiple scan pattern memory modules.

Comments:

– The serial access should not be understood as bit-serial. In other words, if the data to be stored is represented on p bits (e.g., 8 bits for characters, 16 bits for integers, 32 or 64 bits for reals) and the memory chips are q bits wide (e.g., $q = 1$ in Fig. 2), p/q slices as in Fig. 2 are tied together to form a memory module. Thus, p bits data are read or written at a time instant.

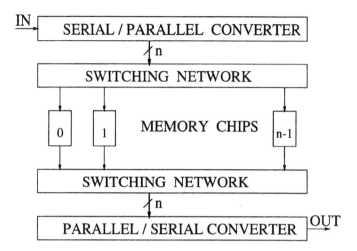

Fig. 2. Multiple scan pattern memory module.

– The multiple scan pattern memory modules provide high speed streams of data only for the specified scan patterns; otherwise, the random access time is the access time of the memory chips (for instance, if $n = 16$, the scan pattern can be delivered with more than one order of magnitude faster than a random scan).

4 Multiple Scan Pattern Memory Modules: Mapping

The existence of the mapping and the construction problems are analyzed in the framework of combinatorial theory using some results about *systems of distinct representatives* [4].

Definition 1. Let $M(S) = \{S_1, S_2, ..., S_k\}$ be a finite collection of subsets of a given set S. A set $\{s_1, s_2, ..., s_k\}$ of k distinct elements of S, such that $s_i \in S_i$ for $i = 1, ..., k$, is called a system of distinct representatives for $M(S)$ (abbreviated as SDR).

P. Hall's theorem [5] gives a necessary and sufficient condition for the existence of a SDR:

Theorem $M(S) = \{S_1, S_2, ..., S_k\}$ *has a SDR if and only if for each* $r = 1, 2, ..., k$, *the union of any selection of* r *sets from* $M(S)$ *contains at least* r *distinct elements of* S.

Definition 2. Let $M^j(S) = \{S_1^j, S_2^j, ..., S_k^j\}$, $j = 1, 2, ... q$ where the S_i^j are subsets of a common set S. A set $\{s_1, s_2, ..., s_k\}$ of k distinct elements of S such that, after some reindexing of $M^j(S)$, $s_i \in S_i^1 \cap S_i^2 ... \cap S_i^q$ for $i = 1, ..., k$, is called a system of common representatives for $M(S)$ (abbreviated as SCR).

Let us now consider the memory module as the set $S = [0, N-1] \times [0, N-1]$.

Definition 3. A scan pattern is an application $SP : [0, t] \rightarrow S$, $t \leq N^2$.

Definition 4. A complete scan pattern is an injection $SP : [0, N^2 - 1] \rightarrow S$.

Let us suppose that a complete scan pattern SP is implemented by a memory module as shown in Fig. 2. Since n data have to be simultaneously accessed, the memory module induces a partition of the data in the scan pattern in groups of n consecutive elements, namely the elements simultaneously read or written in one access of the memory chips. We consider each group as a subset of samples $S_i = \{s_{i1}, ..., s_{in}\}$, where $s_{ij} = (x_p, y_p)$ and $p = ni+j$. The scan pattern becomes the finite family of subsets $SP = \{S_1, S_2, ..., S_k\}$, $k = N^2/n$. These observations help to understand why SDRs have been introduced. The data to be mapped onto a memory chip is exactly a SDR for the collection of sets $M(S)$ induced by SP on S. Therefore a necessary and sufficient condition for the existence of the mapping is that n simultaneously distinct SDRs do exist. The mapping is realized by the effective choice of the n distinct SDRs.

Let $SP^1, SP^2, ..., SP^p$ be p scan patterns, where $SP^j = \{S_1^j, S_2^j, ..., S_k^j\}$. Similarly, the mapping of the p scan patterns is assured if n distinct SCRs do exist for the p induced collections. Each SCR is a SDR for each of the p scan patterns and n distinct SCRs mean for each scan pattern n distinct SDRs and; hence, the mapping is assured.

The existence of the mapping for two complete scan patterns is an immediate consequence of the following result, obtained by D. Koning [6]:

Theorem 5. *Let M be a non empty set, $M = A_1 \cup A_2 ... \cup A_k = B_1 \cup B_2 ... \cup B_k$, where $A_i \cap A_j = B_i \cap B_j = \emptyset$ for any i, j ($i \neq j$) and where all the A_i and B_i have the same number n of elements. Then sets B_i may be reindexed so that $A_i \cap B_i \neq \emptyset$, for any i.*

Let A and B be the two scan patterns. The conditions stated in the above theorem are met. After reindexing the B classes, let $a_1^1 \in A_1 \cap B_1, ..., a_k^1 \in A_k \cap B_k$. The set $\{a_1^1, a_2^1, ..., a_k^1\}$ represents a SDR for A and B scans. We store the SDR into memory chip 1. After taking out the samples of the SDR, we again apply the theorem to these two partitions into classes of $n-1$ elements and so on, prooving the existence of n distinct SDRs.

No existence theorem can be given for more than two scan patterns; the mapping may or may not exist depending on the scan patterns at hand. Even if the mapping does not exist, a memory module that provides the required scan patterns can be built by chaining two scan patterns memory modules. Let $p > 2$ be the number of scans. The fast access to the p scans is obtained by pipelining or parallelizing 2-scan pattern memory blocks. For a pipeline, let us consider the $p - 1$ fast access memories on respectively (SP^1, SP^2), ..., (SP^j, SP^{j+1}), ..., (SP^{p-1}, SP^p) scans. By chaining the memory blocks, each scan pattern is available. Modification of the stored data implies that the entire pipe has to be rewritten, which means $p - 1$ "steps". For a parallel architecture $(p - 1)$ 2-scan fast access memories on respectively (SP^1, SP^2), ..., (SP^1, SP^j), ..., (SP^1, SP^p) scans, have to be "tied" together. Any scan pattern is instantaneously available.

0	1	2	3	4	5	6	7
1	2	3	4	5	6	7	0
2	3	4	5	6	7	0	1
3	4	5	6	7	0	1	2
4	5	6	7	0	1	2	3
5	6	7	0	1	2	3	4
6	7	0	1	2	3	4	5
7	0	1	2	3	4	5	6

Fig. 3. Mapping for fast access on rows and columns.

5 Mapping Examples - 2, 3 and 4 Scan Patterns

Given a certain number of scan patterns, if a mapping does exist, the problem one is faced with is its construction. Once the mapping is established, the memory addressing and the configuration of the switching networks are completely determined. We give now some examples of mappings for memories with 2, 3 and 4 scan patterns. The selected scan patterns have some geometrical significance which will help for deriving the mappings by elementary mathematical considerations. The interested reader can find complete proofs in the References.

5.1 2 Scan Patterns: Rows and Columns (RC)

The memory with fast access on rows and columns is the first example of multiple scan pattern memory module presented in the literature [8] , [9]. As shown in the previous section, in the case of two complete scan patterns, mappings always exist.

The mapping can be derived by using the combinatorial formalism or by some elementary mathematics; we shall follow the second approach [1]. The raster scan requires that groups of n consecutive data on rows be simultaneously accessed. This is equivalent to the requirement that groups of n data starting at a position which is a multiple of n be mapped on different chips. Fast access on columns determines a similar condition on groups of data in vertical position. These observations suggest a memory partition into $n \times n$ blocks such that the row and column coordinates of the first element in each block are multiple of n. For the scan patterns under consideration the accesses for the traversal of such a block involve only data within the block. Since the blocks are identical, the construction of a mapping for one block solves the mapping problem for the entire memory. Furthermore, the raster scan and the column scan are symmetric to each other. The mapping for the raster scan memory presented in section 3 is $m(x) = x \bmod n$. Since we look for a symmetric mapping, it is natural to extend

Fig. 4. Peano scan.

the above formula for two variables:

$$m(x, y) = (x + y) \bmod n \qquad (3)$$

According to the mapping, the datum located at the intersection of line x and column y is stored onto the memory chip $m(x, y)$. The mapping is merely a circular shift for the groups of n pixels on rows and columns. The mapping diagram for a block of size $n = 8$ is shown in Fig. 3. The mapping is valid since groups of n data either on rows or columns are mapped onto different circuits.

5.2 3 Scan Patterns; Rows, Columns and Peano-scan (RCP)

This example considers a third scan, namely the Peano scan. Peano [2] has shown that there exist continuous curves that pass through all the points of a two (or more) dimensional continuous region. Further results were obtained by Hilbert [7]. The names given to such curves are space filling curves or Peano curves. Peano scan is a discrete series of points generated by Hilbert's method for a finite and discrete case. It represents in fact a linear traversal of a 2-dimensional grid. An example of a 32×32 Peano scan is shown in Fig. 4. Peano curves are self-similar in the sense that the $2^k \times 2^k$ Peano curve is composed by connected Peano curves of size $2^p \times 2^p$, $1 \leq p < k$, with similar shape. The $2^p \times 2^p$ curves are identical up to a symmetry. For Peano scans, this property means that the traversal of the entire image is done by a recursive scanning of square zones. Let N and n be powers of 2. The entire $N \times N$ Peano scan is done by scanning

0	1	2	3	4	5	6	7
4	5	6	7	0	1	2	3
2	3	4	5	6	7	0	1
6	7	0	1	2	3	4	5
1	2	3	4	5	6	7	0
5	6	7	0	1	2	3	4
3	4	5	6	7	0	1	2
7	0	1	2	3	4	5	6

Fig. 5. Mapping for fast access on rows, columns and Peano scan.

$n \times n$ zones. The Peano scanning of these zones is similar up to a symmetry. The Peano scan satisfies the condition for partition in $2^p \times 2^p$ blocks, as seen in the previous example. The existence of the mapping for a block guarantees the existence of the mapping for the entire image. Let us consider a mapping that generalizes (3):

$$m(x, y) = (\phi(x) + y) \bmod n \qquad (4)$$

where $\phi(x)$ is a function to be determined. While fast access on the raster scan is obviously satisfied by the mapping (since $m(x, y)$ takes n distinct values when $0 \leq y < n$), fast access on columns and Peano scan demands an appropriate selection for ϕ. This problem has been analyzed in [2], and two solutions have been proposed. Here is, without proof, the closed form formula of the second solution:

$$m(x, y) = (x^* + y) \bmod n, \qquad (5)$$

where x^* is the bit-reversal of x. A mapping example for an 8×8 block when $n = 8$ is shown in Fig. 5.

5.3 4 Scan Patterns; Rows, Columns and Diagonals (RC2D)

Due to the scan pattern intrinsic geometry, a particular mapping was built using a circular permutation with a fixed amount m for every line [1]. For row access, it is obvious that every group of n samples is mapped on a specific chip. The n neighbors on columns and diagonals, starting with the column y, are mapped respectively onto:

$$(y + xm) \bmod n \qquad (6)$$

$$(y + x(m + 1)) \bmod n, \qquad (7)$$

$$(y + x(m - 1)) \bmod n \qquad (8)$$

0	1	2	3	4	0	1	2	3	4
2	3	4	0	1	2	3	4	0	1
4	0	1	2	3	4	0	1	2	3
1	2	3	4	0	1	2	3	4	0
3	4	0	1	2	3	4	0	1	2
0	1	2	3	4	0	1	2	3	4
2	3	4	0	1	2	3	4	0	1
4	0	1	2	3	4	0	1	2	3
1	2	3	4	0	1	2	3	4	0
3	4	0	1	2	3	4	0	1	2

Fig. 6. Mapping for fast access on rows, columns and diagonals.

where x, y range from 0 to $n-1$. The condition for $xk \bmod n$ to span n different values, when x ranges from 0 to $n-1$, is that k and n be mutually prime. This means that m, $m+1$ and $m-1$ have to be prime to n. The most convenient choice for n seems to be a prime number, e.g. 5, 7, 11. (If n is a power of two this mapping fails because at least one out of 2 consecutive numbers is even, e.g. $m-1$ and m). A mapping example for $n = 5$ and $m = 2$ is shown in Fig. 7. Four blocks are presented in order to make visible the mapping for diagonals.

5.4 4 Scan Patterns; Rows, Columns, Bit Reversed on Rows and Bit Reversed on Columns (RC2B)

The mapping rule has been determined in [1]:

$$m(x, y) = \left(\sum_{j=0}^{d} \left(\left[\frac{x}{n^j} \right] + \left[\frac{y}{n^j} \right] \right) \right) \bmod n; d = [\log_n N] \qquad (9)$$

In this case n is a power of two. The mapping for $N = 16$ and $n = 4$ is shown in Fig. 9.

6 Scan-Directional Problems

As stated before, several problems fit into this class. The general scheme of a scan-directional architecture is presented in Fig. 8. It consists of k memory modules, $M_1, M_2, \ldots, M_k$, each one having up to m fast data buses linked to processing modules, P_i, which, according to the problems at hand, can range from simple multiplier-accumulators to FFT modules. Small buffers must be inserted between the memories and the processing units for data synchronization.

0	1	2	3	1	2	3	0	2	3	0	1	3	0	1	2
1	2	3	0	2	3	0	1	3	0	1	2	0	1	2	3
2	3	0	1	3	0	1	2	0	1	2	3	1	2	3	0
3	0	1	2	0	1	2	3	1	2	3	0	2	3	0	1
1	2	3	0	2	3	0	1	3	0	1	2	0	1	2	3
2	3	0	1	3	0	1	2	0	1	2	3	1	2	3	0
3	0	1	2	0	1	2	3	1	2	3	0	2	3	0	1
0	1	2	3	1	2	3	0	2	3	0	1	3	0	1	2
2	3	0	1	3	0	1	2	0	1	2	3	1	2	3	0
3	0	1	2	0	1	2	3	1	2	3	0	2	3	0	1
0	1	2	3	1	2	3	0	2	3	0	1	3	0	1	2
1	2	3	0	2	3	0	1	3	0	1	2	0	1	2	3
3	0	1	2	0	1	2	3	1	2	3	0	2	3	0	1
0	1	2	3	1	2	3	0	2	3	0	1	3	0	1	2
1	2	3	0	2	3	0	1	3	0	1	2	0	1	2	3
2	3	0	1	3	0	1	2	0	1	2	3	1	2	3	0

Fig. 7. Mapping for fast access on rows, columns and bit reversed on rows and columns.

The general scheme should be tailored to fit specific applications, i.e., number of memory modules, number of input/output ports for the memory modules and, of course, specific processing devices. The need to choose a scan directional architecture, when the algorithms are address oriented, is obviously determined by speed constraints. By a proper selection of the number n of memory chips (banks), the memory throughput can be adjusted to fit the processing devices at hand (P_i). If the processing modules are k times slower than needed and the algorithm is parallelizable, as discussed in Sect. 3, by increasing the number of memory chips, the memory module can deliver k streams of data on the specified scan patterns. Obviously, the speed is achieved with the cost of extra hardware to implement such memory modules (switching networks, address bus multipliers, etc.). Next, we point out some possible applications.

6.1 Matrix Transposition

Any of the above proposed memory modules provides fast access on columns. Thus, the effective transposition of a matrix is not necessary. However, if desired, the transposition can be made in place by block transpositions. Thus, the content

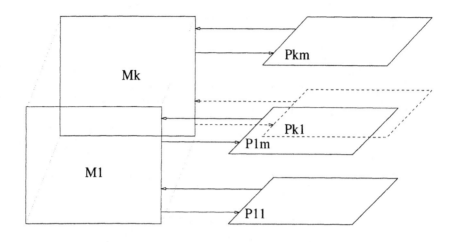

Fig. 8. Scan directional architecture.

of each block is read on rows and written back on columns; if the size of such a block is $m \times m$, a m^2 size buffer should be used.

6.2 Matrix Multiplication

The generic coefficient of two matrices, A and B, c_{ik} is the dot product of row i of A and column k of B,i.e., $c_{ik} = \sum_{j=1}^{N} a_{ij} b_{jk}$. This computation demands a multiplier-accumulator device fed on with data streams on the RC scan patterns. The operation can be performed in place by using a small buffer of size N (for the current row). The computation time of each c_{ik} is exactly the time NT_s needed to access a row of A. Thus, the entire operation is done in $N^3 T_s$ times. By simultaneously accessing m columns and computing in parallel m dot products, matrix multiplication is performed in $N^3 T_s/m$ times. In this case, the RC memory module should provide fast access on m scan patterns which are connected, by fast buses, to m processing devices $P_1, \ldots P_m$ (multiplier-accumulators).

6.3 Matrix Inversion

We present the fast implementation of a recursive matrix inversion algorithm that takes advantage of the scan-directional approach. The inverse is computed in place. It demands multiplications of matrices by scalars, tensor products and matrix additions. The main idea of the algorithm is to recursively compute the matrix U, the vectors V and W and the scalar x such that:

$$\begin{pmatrix} A & B \\ C^t & d \end{pmatrix} \begin{pmatrix} U & V \\ W^t & x \end{pmatrix} = \begin{pmatrix} I & 0 \\ 0 & 1 \end{pmatrix} \tag{10}$$

U, V, W, x are obtained through an 8 steps recursive process we are going to describe now:

1. Invert $A \rightarrow A^{-1} \Rightarrow (n-1)^3$ multiplications;
2. Compute $P = A^{-1}B \Rightarrow (n-1)^2$ multiplications;
3. Compute $Q = C^t A^{-1} \Rightarrow (n-1)^2$ multiplications;
4. Compute $r = d - C^t P \Rightarrow (n-1)$ multiplications;
5. Invert $r \rightarrow x \Rightarrow 1$ multiplication;
6. Compute $V = -xP \Rightarrow n-1$ multiplications;
7. Compute $W^t = -xQ \Rightarrow n-1$ multiplications;
8. Compute $U = A^{-1} - PW^t \Rightarrow (n-1)^2$ multiplications;

One can notice that the total count of multiplications is n^3. The algorithm described above can be regarded in two ways:

− as a recursive proof that its complexity is $O(n^3)$, i.e., by no means, worse than classical invertion algorithms;
− inviting to take advantage of its data oriented feature to be implemented by RC2D memory modules.

A by-product of the matrix inversion algorithm is Schur's complement $D - C^t A^{-1} B$.

6.4   Orthogonal Transforms

The access on rows and columns (RC memories) is needed for 2D transform computations (DFT, Walsh-Hadamard, etc.). The RC2B memory is appropriate for the computation of 2D FFT and FFT-like transforms.

6.5   Image Processing

The fast access on RC is suitable for the computation of separable 2D linear filters. We shall briefly discuss the implementation of a moving average on an image $f(i, j)$ on a window size $(2m+1) \times (2m+1)$. Each output sample, $g(i, j)$, is computed as the sum of the samples within the window filter divided by the number of samples within the window:

$$g(i, j) = \frac{1}{(2m+1)^2} \sum_{p=-m}^{m} \sum_{q=-m}^{m} f(i+p, j+q) \tag{11}$$

At first sight, one needs for each output value to access and sum $(2m+1)^2$ input samples. But the computation can be factorized by computing first a linear moving average on rows for a 1D window filter:

$$h(i, j) = \frac{1}{(2m+1)} \sum_{q=-m}^{m} f(i, j+q) \tag{12}$$

and further, another moving average on columns:

$$g(i,j) = \frac{1}{(2m+1)} \sum_{p=-m}^{m} h(i+p,j) \tag{13}$$

Thus, the 2D filter in a $(2m+1) \times (2m+1)$ window size can be computed in two video frame times, by chaining two linear 1D filters of size $2m+1$. A single processing device is used (adder and a small look-up table). As discussed above, by using an RC memory module that allows fast access on $m = 2$ scan patterns, the moving average filter can be implemented in real time, i.e., 1 video frame. Obviously, two processing modules, P_1, P_2, should work in parallel, each processing module working on a half of the image (e.g., upper-half and lower-half during the moving average on rows - in half video frame times; left-half and right-half during the 1D filtering on columns - until the end of the video frame. The moving average filter is a particular case of small size kernel convolution where the kernel is separable. Another classical example is the smoothing with the 2D operator:

$$\begin{pmatrix} 1 & 2 & 1 \\ 2 & 4 & 2 \\ 1 & 2 & 1 \end{pmatrix} \tag{14}$$

which can be done by smoothing on rows and further on columns with the 1D operator:

$$\begin{pmatrix} 1 & 2 & 1 \end{pmatrix} \tag{15}$$

Memory modules with fast access on RC2D can be used for morphological operations by octagonal structuring elements. If segments B_1, B_2, B_3, B_4 (oriented on $0°$, $45°$, $90°$ and $135°$, respectively) are structuring elements, their composition is an octagonal structuring element B:

$$B = B_1 \oplus B_2 \oplus B_3 \oplus B_4 \tag{16}$$

where $\oplus$ denotes morphological dilation. Thus, dilations (erosions) by B can be computed by chaining dilations (erosions) by B_1, B_2, B_3, B_4, [10]. The processing modules should compute logical OR, AND operations for binary dilations and erosions; max/min for the graylevel ones.

Peano scan is of interest in image processing since it achieves a good compromise between 2D and 1D windows, i.e. points that are nearby in the 2D image are nearby in their ordering along the scan and conversely. Peano scan has already been applied in image compression, color quantization, halftoning and filtering. For a fast implementation of such operations, the RCP memory is very valuable.

7 Conclusions

The paper has investigated the fast implementation of algorithms that are characterized by a certain traversal of the data. The multiple scan pattern memories eliminate data bottleneck for several applications. This approach leads to real time or near real time implementations for some basic operations of interest in image/signal processing and general matrix analysis problems.

References

1. D. Coltuc, V. Buzuloiu, "Mapping rules for multiple fast-scan-patterns memory modules", *SPIE Proceedings of the 4-th Conference on Optics, RO-MOPTO'94*, vol. 2541, pp. 285 - 288, 1995.

2. D. Coltuc, I. Pitas, "Memory Mappings for Fast Peano Scan", *ECCTD'95 - European Conference on Circuit Theory and Design*, Istanbul, Turkey, pp. 1019-1022, 1995.

3. M. Cosnard, D. Trystam, *Algorithmes et architectures paralleles*, InterEditions, Paris, 1993.

4. H. Mann, J. Ryser, *Systems of Distinct Representatives*, Amer. Math. Monthly, 1953, vol. 60, pp. 397-401.

5. P. Hall, *On Representatives of Subsets*, J. London Math. Soc., 1953, vol. 10, pp. 26-30.

6. D. Koning, *Theorie der Endlichen und Unendlichen Graphen*, Chelsea, New York, 1950.

7. D. Hilbert, "Ueber die Stetige Abbildung einer linie auf ein Flaechenstueck", *Mathematische Annalen*, vol. 38, pp. 459-460, 1881.

8. D. L. Ostapko, *A mapping and memory chip hardware which provides symmetric reading/writing of horizontal and vertical lines*, I.B.M. J. Res. Develop, vol. 28 no.4, 1984, pp. 393-398.

9. D. Coltuc, *Bidimensional memory module*, Patent O.S.I.M. Romania, no. 89118, 1984.

10. J. Serra, *Image Analysis and Mathematical Morphology*, Academic Press, 1982.

Markov Chain Based Management of Large Scale Distributed Computations of Earthen Dam Leakages

Zdzisław Onderka and Robert Schaefer, Institute of Computer Science,
Jagiellonian University, Nawojki 11, 30-072 Cracow, Poland

Summary: The detailed Markov chain model of the dynamics of the heterogeneous computer network is presented. The stochastic control problem of optimal task distribution in computer network is formulated. Some control strategies based on the stochastic forecast and deterministic rules are presented. The system mentioned above is applied to the solution of mixed FE/FD scheme coming from the earthen dam stability analysis. Numerical tests performed on the network including different workstations, scalar LAN servers and power vector unit exhibit the behavior of different strategies for various size of data to be processed.

Engineering Problem Under Consideration

The problem of water percolation through a soil resulting from local differences of the piezometric pressure distribution is classical and its various numerical models have been investigated by numerous authors (see e.g. [12] and references therein). It is one of the leading phenomena in the physics of soils and the question of its full recognition and an adequate description plays an important role in civil engineering, which follows from the fact that cohesive and organic soils cover a significant part of building sites or agricultural terrains.

The evaluation of the filtration yield is also essential for the design as well as for the exploitation of the drainage systems or earthern dams, where the determination of the seal leakage through the protective screen plays a crucial role. The quantitative filtration model is also required in the description of the consolidation phenomenon appearing for example, in the massive earthern structure body or in the soil underlaying an arbitrary structure foundation.

The paper considers the problem of determining the piezometric height distribution $w(\tau, x)$ describing the filtration of water through the cohesive porous medium for a small filtration velocity $v = (v_1, v_2, v_3)$. The function $w(\tau, x)$ solves the intial-boundary problem

$$\beta p(\tau, x)\frac{\partial w}{\partial \tau} = \sum_{i=1}^{3} \frac{\partial}{\partial x_i} v_i(\tau, x, Dw) + Q(\tau, x) + G(\tau, x) \quad (\tau, x) \in (0, T] \times \mathcal{D}, \quad (1)$$

$$w(\tau, x) = w_b(\tau, x) \quad (\tau, x) \in (0, T] \times \partial\mathcal{D}_1, \tag{2}$$

$$\sum_{i=1}^{3} n_i\, v_i(\tau, x, Dw) = q(\tau, x) \quad (\tau, x) \in (0, T] \times \partial\mathcal{D}_2,$$

$$w(0, x) = w_0(x) \quad x \in \mathcal{D}. \tag{3}$$

In the above equations $\mathcal{D}$ denotes the domain of filtration, T is the time of filtration, $n = (n_1, n_2, n_3)$ is the external unit vector orthogonal to the part $\partial\mathcal{D}_2$ of the boundary of $\mathcal{D}$, $\beta > 0$ is a water compressibility. Functions Q, G, q, p, w_b, w_0 represent: the yield of sources, the volume strain velocity of the skeleton, the boundary flux, the soil porosity, the piezometric height on $\partial\mathcal{D}_1$ and the initial height distribution respectively.

Assume (cf. [1]) that $v_i(\tau, x, Dw) = \varphi(\tau, x, |Dw|)\frac{\partial w}{\partial x_i}$, where $Dw =$grad w, $(\cdot|\cdot)$ is a scalar product in $\mathbf{R}^3$, $|Dw|^2 = (Dw|Dw)$ and

$$\varphi(\tau, x, r) = \begin{cases} M(1 - \frac{S_0}{r}(1 - \exp(-\frac{\theta r}{S_0}))), & \text{for } r > E \\ [\frac{M}{E^2}(S_0 - (S_0 + \theta E)\exp(-\frac{\theta E}{S_0}))]r \\ +M(1 - \frac{2S_0}{E} + (\frac{2S_0}{E} + \theta)\exp(-\frac{\theta E}{S_0})), & \text{for } E \geq r \geq 0, \end{cases} \tag{4}$$

All parameters in (4) are positive. A constant E depends on the floating point arithmetic accuracy and on the features of the current soil pattern. Parameters M, S_0, θ may depend on (τ, x). They are supposed to be so regular that φ is C^1.

The nonlinear system (1)-(4) can be solved using mixed Finite Element - Finite Difference scheme ([11]) of the form

$$C_h^l \frac{h^{l+1} - h^{l-1}}{\Delta_\tau} + B_h^l h^l + \frac{1}{2} DB_h^l (h^{l+1} - 2h^l + h^{l-1}) = f_h^l \quad l = 0, \ldots, k, \tag{5}$$

where $h^l = h(l\Delta_\tau), (l\Delta_\tau \in [0, T]$ for $l = 1, 2 \ldots, k)$ is the discrete solution approximating w with respect to the basis of the proper FE space which can be established after the FE mesh had been generated, C_h^l is the capacity matrix and B_h^l, DB_h^l are the discrete divergence operator and its derivative respectively, computed in the l-th time step. Each solution of system (5) starts from double initial condition (h^{-1}, h^0) and goes consecutively for $l = 1, 2, \ldots, k$. The conditioning of the mathematical model (1)-(4) and the convergence of (5) were justified in [10,11].

Domain Decomposition for Parallel Computation of System Matrices

Coefficients of matrices $C_h^l, DB_h^l, B_h^l h^l, f_h^l$ representing the discrete weak solution of parabolic PDE (1)-(4) have the form of integrals over the whole domain $\mathcal{D}$, and have to be computed in each time step $l = 1, 2, \ldots, k$ due to the lack of linearity.

Large complexity of the single coefficient computing as well as the large frequency of their repetition make them a significant part of the whole solution which need huge CPU and RAM resources. As usually in FEM, above integrals can be computed as a sum of integrals over particular elements.

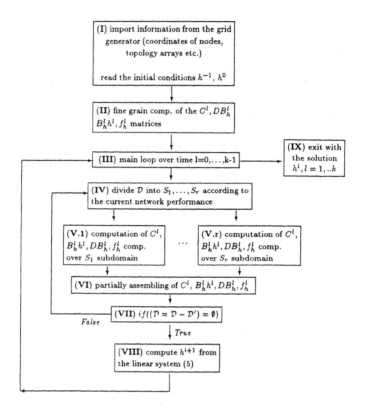

Figure 1: Flow diagram of domain decomposition

Taking into account the computation complexity involved with a single element and average computer performance we assign one, two or more elements to one task. Later in our investigation it will be considered as *elementary* one. The above mentioned assignment may be called *fine graining*.

According to the associativity rule, the sum may be divided into several groups containing the integrals over elements which belong to the arbitrarily chosen sub-domains $S_1, S_2, ... S_r \subset \mathcal{D}$. Components of the sum described above belonging to different groups may be computed independently because of the separated data requirements.

In case of huge FE meshes the initial task distribution should be changed during the integration process, in order to obtain optimal speedup. The integration over elements belonging to each subdomain S_l may be suspended keeping the partial results of matrix assembly. The remaining parts of each subdomain S_l' on which the computational process has to be continued, can be joined into $\mathcal{D}' = \bigcup_{l=1}^{r} S_l'$ and arbitrarily divided once more, according to the current network performance.

The features described above constitute a typical situation in which available computer resources as well as the computational process can be managed using a refined repetitive optimal control algorithm. The general management policy which will be undertaken is presented in the following flow diagram in Figure 1.

Tasks $(\mathbf{V}.1), \ldots, (\mathbf{V}.r)$ are suitable for execution on different scalar nodes (e.g. workstations). The linear system solution $(\mathbf{VIII})$ is dedicated to the vector architecture powerful unit.

Such solution seems to be cost-effective because:

- only the tiring linear system solution is dedicated to the vector unit which is best suited to solving linear algebra applications;

- it is difficult to vectorize the matrix coefficients computation and it is easy to parallelize. So, this computation can be executed on the system of workstation which reduce the expensive use of supercomputer.

Such complicated task distribution scheme needs to be carefully managed in order to obtain the best speedup. It forces us to introduce a stochastic model which can describe the behaviour of the computer environment.

Heterogeneous Network Model

The heterogeneous computer network can be understood as a set $H = \{M_0, \ldots, M_m\}$, of $m + 1$ distinct machines. Each machine may communicate with any other one. Machines may have different load and CPU utilization, may vary in number of incoming and outcoming packages, in throughput or may have a different architecture (sequential, dataflow, vector, parallel or other computing units). Moreover, parameters of performance may vary in time.

Let us assume, that our program $\mathbf{A}$ consists of sequential and distributed part. Let the task t_0 be the serial part (master task) and assume that the distributed part consists of N tasks (slaves) $V_D = \{t_1, \ldots, t_N\}$. Model of execution of program $\mathbf{A}$ on H assumes that each machine $M_j, j = 1, \ldots, m$ may execute at most one task of $\mathbf{A}$ at a time. In this case, we say that the computer network is *nondecomposable*. Additionally, we assume that each task is executed until completion by exactly one machine without intermediate communication or synchronization with other tasks i.e. tasks are constrained to be *atomic*. Model of program $\mathbf{A}$ is the well-known task graph $G_A = (V, E), V = V_D \cup \{t_0\}$ - directed, acyclic [1].

For a program $\mathbf{A}$ consisting of N tasks that are independent of each other and a network of m machines, there are m^N possible mappings from V_D to H. The problem of choosing an optimal mapping (in the sense of the best speedup [1,13]) is NP-complete. The heterogeneous network problem for the case $m = 2$ was solved in [2] using network flow algorithm. Bokhari [3] solved this problem for arbitrary m and N when the task graph is a tree. Related results are derived in [4,5,6,13]. In [6] several assigment procedures are described with their time complexity, but most of these heuristic schemes are *work-greedy*.

Let $F : V \to H$ is a mapping of tasks to machines, such that

$$V_D = \bigcup_{i=1}^{m} F^{-1}(M_i), \ F^{-1}(M_j) \cap F^{-1}(M_l) = \emptyset, \ j \neq l, \ \sum_i card(F^{-1}(M_i)) = N$$
$$F^{-1}(M_0) = t_0, \tag{6}$$

If $m < N$, F has to map multiple tasks to at least one machine. Usually this assumption for nondecomposable heterogeneous network is forced for augmenting task graph to a *scheduled task graph* [2].

In case of dynamic task distribution F is composed of the sequence of functions $\{F_n\}_{n=0,1,...}$ i.e.

$$\bigcup_n F_n^{-1}(M_i) = F^{-1}(M_i) \quad \forall M_i \in H \tag{7}$$

$$F_n^{-1}(M_j) \cap F_n^{-1}(M_l) = \emptyset, \quad j \neq l$$

$$\bigcup_{i=1}^{m} \bigcup_n F_n^{-1}(M_i) = N$$

In the rest of this paper we assume that tasks (slaves) are identical, F maps N slave tasks to all m machines ($m < N$) and master task t_0 to machine M_0, G_A is also a scheduled task graph and the network H is heterogeneous.

The Workload Model

Let us consider two physical performance parameters, $load_{M_j}$ and net_{M_j} which express the average number of jobs in the load queue and the number of incoming packets, for example, ethernet, per second. Later in this paper we will call them the additional machine load and additional communication load respectively. The tests show that these two parameters are the main influence on the slowing down of the application execution in our situation. Then the vector $\{s_{M_1}, \ldots, s_{M_m}\}$, $s_{M_j} = (load_{M_j}, net_{M_j})$, $j = 1, \ldots, m$ has principal impact on the workload of H.

We will consider only finite number of intervals for $load_{M_j}$ and net_{M_j} which corresponds to the discrete states of workload.

Let us define the slowing down coefficient η^M for machine $M \in H$ by the formula:

$$\overline{T}_t^M = (1 + \eta^M)T_t^M \tag{8}$$

where T_t^M and $\overline{T}_t^M$ are the times required to execute the pattern task $t \in V_D$ on the machine M without and with the additional load respectively.

Let introduce the ordered set of tresholds $\Gamma = (\eta(0), \ldots, \eta(K))$, $\eta(i) < \eta(i+1)$, $i = 0, \ldots, K-1$, $\eta(i) \in \mathcal{R}_+$, $i = 0, \ldots, K$. We define, that the machine $M \in H$ is in the state of workload $i \in \{0, \ldots, K-1\}$ if its current value of the slowing down parameter $\eta^M \in [\eta(i), \eta(i+1))$, $i = 0, \ldots, K-1$ and is in the state K if $\eta^M \in [\eta(K), \infty)$. So, if we assume the same set of tresholds for each $M \in H$ we obtain the uniform space of states $S = \{0, 1, \ldots, K\}$. It means that each machine $M \in H$ is in this same state of workload $i \in S$ if its slowing down ratio η^M is in this same range i.e. $\eta^M \in [\eta(i), \eta(i+1))$. Now we can easily define a bijection (numbering) $\eta : S \ni i \rightarrow \eta(i) \in \Gamma$.

We assume that there exist time intervals $\Delta_n = [\tau_n, \tau_{n+1})$, $n = 0, 1, \ldots$ for which average values of workload parameters represent in a satisfactory way the mean performance of each machine in Δ_n. Therefore, we can utilize only these average values in our stochastic model.

Let us consider for each M_j:

- a probability space $(\Omega, \mathfrak{S}, P_j)$, where Ω (sample space) is a set of events that machine admits some average value of the ratio of slowing down in the time interval Δ_n for some n, identical for each M_j. In consequence each event corresponds to an unique state from S. $\mathfrak{S}$ is a σ-algebra on Ω and P are the probabilistic measures which vary for $n = 0, 1, \ldots$ and $j = 1, \ldots, m$.

- a sequence of random variables $X_0^{M_j}(\omega), X_1^{M_j}(\omega), \ldots, \omega \in \Omega,\ j = 1, \ldots, m$ related to the time intervals $\Delta_n, n = 0, 1, \ldots$, defined over the same space Ω and taking values in the state space S. This sequence forms a discrete stochastic process (chain) which describes dynamic behavior of load of $M_j \in H$ in time. In this paper we assume that this is the first degree Markov chain [13,16].

- we assume that the random variables $X_i^{M_j}$ and $X_l^{M_r}$ are independent for $j \neq r$ and arbitrary i, l, i.e. there is no cross correlations in the computer network.

Although we will be working with both l_{M_j} and c_{M_j}, we can simplify the future formal description omitting communication load c_{M_j}, because of their identical formal description.

Let $p_{ij}(n) = P(X_{n+1}(\omega) = j | X_n(\omega) = i)$ identify the probability that process is in state j at the step $n + 1$ provided that it is in state i at the step n. These probabilities form the one-step transition probabilities matrix $P(n) = \{p_{ij}(n)\}$ of the chain. Note that, on summing the probabilities $p_{ij}(n)$ over all possible states j in the state space S, the result is equal to unity.

Having the stochastic processes $\{X_n^{M_j}\}_{n=0,1,\ldots}$ which describes dynamic behavior of load of $M_j \in H$ for $j = 1, \ldots, m$ and bijection η, we may define the new stochastic processes $\{\overline{T}_t^{M_j}(n)\}_{n=0,1,\ldots}$ for $M_j \in H, j = 1, \ldots, m$ describing dynamic behavior of time required to execute the single task t on machine M_j. Moreover,

$$\overline{T}_t^{M_j}(n) = (1 + \eta(X_n^{M_j})) \cdot T_t^{M_j} \quad \forall j \in \{1, \ldots, m\},\ n = 0, 1, \ldots \qquad (9)$$

The Control Problem of Dynamic Task Distribution

The sequence $\{F_n\}_{n=\mu,\mu+1,\ldots}$ belongs to the set of admissible policies U_μ if probability that $\overline{T}_t^{M_j}(n) \cdot card(F_n^{-1}(M_j)) < \Delta_n$ is greater then zero. The quantity $\mu \geq 0$ denotes the starting step for each policy from U_μ. Let $c_M(n, u)$ be an *immediate cost function*, which is involved with single step $n = \mu, \mu + 1, \ldots$ of realized policy $u \in U_\mu$ for the machine $M \in H$. One of the siplest case of $c_M(n, u)$ is the elapsed time of tasks execution at the step n. Let us assume, that $\forall u \in U_\mu$, and any μ, and $n \geq \mu$, and any initial probability distribution $\beta(M)$ of the state of machine $M \in H$, there exists the expected value $E_{\beta(M)}(c_M(n, u))$ (see [8]).

Next, we define the cost criteria which will appear in the control problem. Let $C_M^Z(\mu, \beta(M), u)$ be the *finite horizon cost* related to the policy $u \in U_\mu$ provided in the time period $[\mu, Z]$. It is defined for any machine $M \in H$ and any initial distribution $\beta(M)$ at the starting step μ as:

$$C_M^Z(\mu, \beta(M), u) = \sum_{n=\mu}^{Z} E_{\beta(M)}(c_M(n, u)) \tag{10}$$

The *finite horizon cost for network H* is defined as:

$$C^Z(\mu, \beta, u) = max_{M \in H}\{C_M^Z(\mu, \beta(M), u)\} \tag{11}$$

where β is now the system $\{\beta(M_1), \ldots, \beta(M_m)\}$.

We define the control problem **CP** as:

Find a policy $u \in U_\mu$ that minimizes $C^Z(\mu, \beta, u)$ over the whole U_μ.

Additionally, we will consider the *a'posteriori* cost function $\overline{C}^Z(\mu, \beta, u)$ defined as:

$$\overline{C}^Z(\mu, \beta, u) = \sum_{n=\mu}^{Z} max_{M \in H}\{c_M(n, u)\} \tag{12}$$

It could be understood (in a simple case) as real total elapsed time of program execution under the policy u.

One of the simplest policies utilizes the immediate cost function $c_M(n, u)$ which is equal to the time of execution of the $F_n^{-1}(M)$ set of tasks in the n-th time interval. For such policies we can define the *a'posteriori* value of the *speedup* (13):

$$\overline{S}_A^Z = \frac{min_{M \in H}\{T_A^M\}}{T^{net} + \overline{C}^Z(\mu, \beta, u)} \tag{13}$$

where T_A^M is the *a'posteriori* execution time of program **A** on machine $M \in H$ and T^{net} is the *a'posteriori* communication time.

The Tasks Distribution Policies

We will study further only the **CP** problem with the simplified cost function described above. In other words we will consider policies which are leading only to the fastest execution of the program **A**. They generally fall in two groups: the group of deterministic policies and the group of stochastic policies $u \in U_\mu$ based on the Markov model described in previous section.

We used the following policies from the first group [15]:

- *single task* - in each step master task distribute m tasks; a single task to a single machine so, $card(F_n^{-1}(M)) \leq 1, \forall n = \mu, \mu + 1, \ldots$, $\forall M \in H$ and therefore, $card(F^{-1}(M_j)) - card(F^{-1}(M_i)) \leq 1$, $i, j = 1, \ldots, m$;

- *multiple task* - all N tasks are divided into m equal groups of tasks and a single group is distributed to a single machine. Similarly, $card(F_n^{-1}(M)) \leq N/m, \forall n = \mu, \mu+1, \ldots$, $\forall M \in H$ and also $card(F^{-1}(M_j)) - card(F^{-1}(M_i)) \leq 1$. If $\mathbf{mod}(N, m) \neq 0$ then the rest of tasks are distributed like in *single task* policy;

- *dynamic single distribution* - a single task is executed by each machine at a time and the policy is *work-greedy* i.e. the policy does not let a machine idle when there is a task the machine could execute. This policy is the same as the one proposed in [17] by C.P. Kruskal to allocate independent task on a MIMD parallel computer. In [17] is described the estimation of total execution time for policy which allocate K task batchs at a time to each processor;

- *stationary* - the whole set of tasks (of $\mathbf{A}$) is divided into m seperate subsets with cardinals proportional to the current state of load of the machine i.e. for each $M_j \in H$

$$\left| \frac{card(F^{-1}(M_j))}{card(V_D)} - \varphi_j \right| \tag{14}$$

is minimal, where $\varphi_j = \frac{\lambda_j}{\Lambda}$, $\sum_{j=1}^m \varphi_j = 1$. The quantity $\lambda_j = 1/\overline{T}_t^{M_j}$ is the power coefficient for $M_j \in H, j = 1, \ldots, m$ and $\Lambda = \sum_{j=1}^m \lambda_j$ is the power coefficient for all $M \in H$, where $\overline{T}_t^{M_j}$ is the current value of time required to execute a single task t on M.

In the group of stochastic policies the distribution of tasks is based on the current state and the independent knowledge about the process of varying the load of each machine $M_j \in H$.

In order to execute the second group of policies, we introduce the following classes of agents: *state agents* - that monitors the load in time and prepare the state forecast for each machine in H; a *decision making agent*, which is involved in establishing task allocation policy i.e. prepare action.

A *state agent* on each machine M_j generates the probability distributions in time intervals $\Delta_n, n = 0, 1, \ldots$. Denote a state probability distribution as $\Pi(n)$, $\Pi_i(n) = P(X_n = i)$ and $\Pi(\mu) = \beta(M)$ as the initial distribution at the starting step μ. If the machine is in state $l \in \{1 \ldots, K\}$ at the step μ, the initial distribution $\Pi_i(\mu) = 1$ if $i = l$ and 0 otherwise. The state probabilities $\Pi(n)$ can be evaluated recursively: $\Pi_i(n) = \sum_{j \in S} p_{ji}(n-1) \cdot \Pi_j(n-1)$, $n > \mu$ (see [13,16]).

Having the vector of state probabilities at each step, the *state agent* involved with machine M_j can evaluate the state value at each time step, and then can prepare forecast of execution time for one task.

The *decision making agent* cooperates with the *state agents* and makes a decision based on actual and forecasted average state of each machine in H (or forecasted computation time of task).

In case of the dynamic, stochastic policies we are looking for a sequence of functions $\{F_n\}_{n=\mu,\mu+1,\ldots}$ each with feature (7) which belongs to U_μ and which maximizes the foreseen speedup (see [1, 15]). We suggest the following algorithm:

1. $n = \mu$;

   $V(n) = V_D$;

2. $B = \sum_{M_j \in H} \left\lfloor \frac{\Delta_n}{E_{\beta(M_j)}(\overline{T}_t^{M_j})} \right\rfloor$

3. $if(\ B > card(V)\ )$

    $\{$ $for(\ each\ \ M_j \in H\ )$

        Let $F_n^{-1}(M_j) \subset V(n)$ such that $card(F_n^{-1}(M_j) = \left\lfloor \dfrac{\Delta_n}{E_{\beta(M_j)}(\overline{T}_t^{M_j})} \right\rfloor$;

    $V(n+1) = V(n) \setminus \bigcup_j F_n^{-1}(M_j);$

    $n = n+1;$

    $goto\ 2;$

    $\}$

    $else/*last\ step*/$

    $\{$ $for(\ each\ \ M_j \in H\ )$

        Let $\lambda_j = \dfrac{1}{E_{\beta(M_j)}(\overline{T}_t^{M_j})};$

    $\Lambda = \sum_{M_j} \lambda_j;$

    $for(\ each\ \ M_j \in H\ )$

    $\varphi_j = \frac{\lambda_j}{\Lambda};$

    $for(\ each\, M_j \in H\ )$

        Let $W_j = F_n^{-1}(M_j)$ such that the set $\{|\frac{card(F_n^{-1}(M_j))}{card(V_n)} - \varphi_j|\}$

        is minimal over $\{W_j\}_{M_j \in H\ and\ \bigcup_j W_j = V_n};$

    $\}$

This policy leads to *Constrained Markov Decision Process* [7,8] in which we minimize the maximum over network H of a cost function (that is the time of execution) which is associated with the used policy and the initial state.

Let the initial state of a machine M be $s = j$ and the initial distribution is $\Pi(\mu) = \beta(M)$. The expected execution time of single task on M in each time step $n = \mu, \mu + 1, \ldots$ is as follows:

$$E_{\beta(M)}\left(\overline{T}_t^M(n)\right) = \sum_{j=1}^{K}(\Pi_j(n) \cdot (1 + \eta(j))T_t^M) \tag{15}$$

$$for\ n = \mu\ \ E_{\beta(M)}\left(\overline{T}_t^M(\mu)\right) = (1 + \eta(j))T_t^M$$

and the expected time of execution of $k_i(n) = card(W_i(n))$ tasks is equal to:

$$E_{\beta(M)}\left(k_i(n) \cdot \overline{T}_t^M(n)\right) = k_i(n) \cdot E_{\beta(M)}\left(\overline{T}_t^M(n)\right) \tag{16}$$

Then, the finite horizon cost for machine M (10) is as follows:

$$C_M^Z(\mu, \beta(M), u) = \sum_{n=\mu}^{Z} (k_i(n) \cdot \sum_{j=1}^{K} (\Pi_j(n) \cdot (1 + \eta(j))T_t^M)) \tag{17}$$

$$for\ n = \mu \quad C^Z(\beta(M), u) = k_i(\mu) \cdot (\Pi_j(\mu) \cdot (1 + \eta(j))T_t^M))$$

and the finite horizon cost for network H (11) is:

$$C^Z(\mu, \beta(M), u) = max_{M \in H} \{\sum_{n=\mu}^{Z} (k_i(n) \cdot \sum_{j=1}^{K} (\Pi_j(n) \cdot (1 + \eta(j))T_t^M))\} \tag{18}$$

We set λ_j using the following statistics:

- $1/E_{\beta(M)} \left(\overline{T}_t^{M_j}(n)\right)$ (see equations (15)),

- $1/x_p^{M_j}$ where $x_p^{M_j}$ is the quantile of $\overline{T}_t^{M_j}(n)$ computed under the given risk level $1 - p$,

- $1/(\alpha \cdot \overline{T}_t^{M_j}(n) + (1 - \alpha) \cdot x_p^{M_j})$, where $T_t^{M_j}(n)$ is the current value of stochastic process and $\alpha \in [0, 1] \cap \mathbf{R}$ which defines the *combined inertial - Markovian policy*.

in the first implemented group of dynamic Markovian policies.

Numerical Tests

In the case of the described (in first part of this paper) engineering problem, $V_D(n)$ is the set of tasks which compute coefficients of matrices $C_h^l, DB_h^l, B_h^l h^l, f_h^l$ over the subdomain S_i, $i = 1, \ldots, r$ in blocks (**V.1**),...,(**V.r**) (see Figure 1). Master task t_0 control the main loop for $l = 0, \ldots, k - 1$ (**III**), realize *fine grain computations* (**II**) and make partial assembly of the computed matrix coefficients (**VI**). Moreover, the master task cooperate with *decision making agent* and distribute tasks according to prepared decision. The decision processes are computed in block (**II**). The *decision making agent* can realize an arbitrary policy which may or not require the cooperation with the *state agents*. Applying policies described earlier to our problem, there are some simplifications of flow diagram (Figure 1):

- in case of the single task distribution the subdomains $S_i, i = 1, \ldots, r$ contain only single FE mesh element,

- in case of multiple task distribution the whole domain $\mathcal{D}$ is divided into equal subdomains, so $\mathcal{D} = \mathcal{D}'$ if $\mathrm{mod}(N, m) = 0$ otherwise we assign at most one additional task to one machine without introducing a significant unbalance and loop (**IV**), ..., (**VII**) is performed once.

- in case of *work-greedy* policy there is no synchronization in (**VI**) and division into subdomains is the same like in case of single task distribution policy,

100 tasks	fastest machine in H	single distribution	multiple task policy	work-greedy policy
time [s]	215.5	144.4	78.3	64.6

200 tasks	fastest machine in H	single distribution	multiple task policy	work-greedy policy
time [s]	431.4	289.3	156.5	129

Table 1: Tables with results without additional load

100 tasks	fastest machine in H	single distribution	multiple task policy	work-greedy policy
time [s]	278.4	183.6	143.3	119.7

200 tasks	fastest machine in H	single distribution	multiple task policy	work-greedy policy
time [s]	543.5	326.3	269.5	238.3

Table 2: Tables with results under medium additional load

- in case of stationary policy the loop (**IV**), ..., (**VII**) will execute only once because the whole domain $\mathcal{D}$ is divided once into subdomains according to current network performance, so $\mathcal{D} = \mathcal{D}'$,

- in case of any stochastic policy block (**IV**) is executed according to forecast of network performance. In each time step of task distribution the set $V_D(n)$ is based on the rest part of $\mathcal{D}$ (computed in block (**VII**)). The flow diagram (Figure 1) is best suited to this kind of policies based on the described Markov model.

Several numerical tests which applied described deterministic policies as well as policies involving Markov model identification forecast with decision making, were provided for the engineering problem under consideration (5). The network H consists of one SPARC 4 (as master - machine M_0), two SPARC stations ELC, SUN 490 and a CONVEX 3200 using 2 processors as vector nodes.

Tables 1 and 2 show the example results for deterministic policies for the application **A** consisting of 100 and 200 tasks each with complexity about 2015360 floating point, integer and conditional operations. In the first column there is the number of tasks, in the second one there is the time of execution for our application on the fastest machine in the network H described above. Third, fourth, and fifth columns contain the execution times for the *single distribution policy*, and the *multiple task policy* and the *work-greedy policy* respectively. Table 1 shows execution times without additional load on the machines in H and Table 2 presents execution times with medium additional load.

In case of Markov model of workload we assume that the uniform space of load states is $S = \{0, 1, 2\}$ for each machine in H (i.e. $K = 2$). The state $s = 0$ is the

60

state without additional load and $s = 2$ is the state with the greatest additional load.

Initial observations of the network behaviour lead us to the tresholds of sloving down parameters presented below, so the bijection η is defined as follows:

$$\eta(s) = \begin{cases} 0 & \text{if } s = 0 \\ 0.4 & \text{if } s = 1 \\ 0.8 & \text{if } s = 2 \end{cases} \tag{19}$$

It means for example, that if measured load for machine M by the *state agent* is such that M is in the state 2 and $T_t^M = 2.1$ sec, then the foreseen time of execution of single task $\overline{T}_t^M$ is at least 1.8*2.1 sec.

During the process of identification of Markov chain we assume that it is not stationary and periodic with period equal to 24 hours and time intervals $\Delta_n = 1$ hour, $n = 0, \ldots, 23$ and it is stationary seperately in each Δ_n, $n = 0, \ldots, 23$.

The program for network performance parameters investigation (based on PVM software system) monitors *the percentage of the CPU time, the average number of jobs in the load queue, the number of sending and receiving ethernet packages* and *the network collisions* (see the description of the *rstat* Unix function) for each machine in H during 24 hours period.

The results were that the most suitable performance parameters describing machine workload and network workload were *the load of jobs queue* and *the number of sending and receiving ethernet packets*.

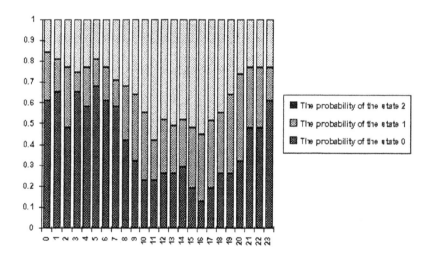

Figure 2: The probabilities of each state in the period of 24 hours

So, we obtained 24 one-step matrices of transition probabilities between the states of load and between the states of network for each $M \in H$.

number of tasks in the subset	80	40	14	2	1
No. of subsets of the same type	60	80	120	80	160

Table 3: The type and the number of the elementary tasks for coarse grain problem

The Figure 2 shows typical values of probability of each state for each step (one hour) in the 24 hour period obtained for LAN server SUN 490. Each value is the total probability of the state in which the machine will be in the next step on condition that the machine is in state 0 or 1 or 2 (in the current step).

The *state agent* and the *decision making agent* are implemented using PVM software system. Each *state agent* works on each slave machine $M \in H - \{M_0\}$ and periodically (1 hour) sends the machine state to the *decision making agent* which works on master machine M_0 and prepares decision about tasks distribution. Agents communicate with *decision making agent* by sending messages as well as *decision making agent* and the application **A**.

The next tests were performed in order to show the influence of the process granularity on the control results. The first group of tests consisted in distributed computing of 10000 identical *elementary* tasks as in previous examples for three different policies. So, these results are related to the fine granularity of the application.

The second group of tests were applied to the coarse grain application. The same set of *elementary* tasks (10000) were partitioned into several subsets of different size (cardinal number) - see Table 3. Each of these groups were managed as new *elementary* coarse grain tasks.

Results for *work-greedy*, stationary and Markov based policies are shown in Table 4 and 5 for the first and the second group of tests respectively. The expected value statistic $\lambda_j = 1/E_{\beta(M)}(\overline{T}_t^{M_j}(n))$ was applied to estimate the computational power in the stochastic policy mentioned above.

The Table 4 exhibit that the *work-gready* policy is the the best and the stochastic policy is slightly better then the stationary one in the case of fine grain application. On the other hand, in the case of coarse grain application (see Table 5) stochastic policy dominates over *work-gready* and stationary.

The dominance of stochastic policy in case of the coarse grain application may be explain by:

- the stochastic policy provides the better estimation of computational power of $M \in H$ with respect to the whole execution time of application than the stationary one;

- the *work-gready* policy does not use any information about the current and foreseen computational power explicitly. It may lead to significant fault in case of task distribution of different size.

work-greedy policy	stationary policy	stochastic policy
11931.5 [sec]	12646.2 [sec]	12217.8 [sec]

Table 4: Execution times for fine grain problem

work-greedy policy	stationary policy	stochastic policy
13616.3 [sec]	15362.9 [sec]	12760.3 [sec]

Table 5: Execution times for coarse grain problem

Future Work

The described numerical problem (5) is a part of whole CAD computation technology. We have investigated several policies of task distribution and tested them on the subproblem dealing with the computation of matrix coefficients. Future work will be involved with the management of the complete process of the solution of the CAD problems which is composed of the main following stages:

1. *Solid Decomposition* - is a sequential part of computation CAD technology (see [18]). During this stage the division of the whole domain $\mathcal{D}$ into subdomains is prepared. The most essential restriction of this division is an effective free RAM memory of each machine $M \in H$;

2. *Mesh Generation, Mesh Adoption* - these are processes strongly scalable and fine grain (see [19]). These computations could be distributed but this is not necessary because of little time and memory complexity of all this process;

3. *Non Linear Algebraic Solver* which consists of two subproblems: *computations of matrix coefficients* and *linear algebraic solver* - the first subproblem is fine grain and well scalable. It consist of a great number of identical small tasks with low time and memory complexity so, it is well suited to the distributed computation. The second subproblem is the most time consuming part of whole CAD computational technology. It consist of only several different tasks with high time and memory complexity, then this stage is rather weakly scalable and coarse grain. This is due to the previous division of the whole domain $\mathcal{D}$. Moreover, the execution time of this CAD stage is considerably enlarged due to the loop over time (Figure 1 - **(III)** main loop) in which the described subproblems are computed. It will be important to investigate the proper algorithm which will fit the forecast time of execution of each task to the one or several steps of Markov chain in this case;

4. *Postprocessing* and *a'posteriori* Error Estimation.

CAD stages described above could be distributed in different ways, depending on scalability and the kind of granularity. During task distribution different policies could be utilized: stationary policies as well as stochastic ones with different kind of repetition and forecast. For example, the *matrix coefficients computation* could be distributed using the *dynamic single distribution* policy and the *linear algebraic solver computations* could be distributed using *stochastic policy*. The total supervising system for distributed CAD processing, involving object oriented implementation is in progress (see [20]).

An additional region of further investigation is the sensitivity analysis of the proposed stochastic control policies. It will be studied with respect to the fault of the stochasti process estimation and fault of state forecast.

Conclusions

1. The performed tests (on a network consisting of 6 machines of above type) revealed that *work-greedy* policies are the best in the group of stationary policies with respect to our engineering problem.

2. The stochastic policies are better suited to the coarse granularity problem of tasks distribution than the static ones in case of huge computations for which the foreseen sequential time of execution requires several hours. Results are compared in the Tables 4 and 5.

3. The proposed policies as well as the forecast technologies are simplest from the set of stochastic control policies [8]. The future refinement will focus on the *feedback* policies in case of enormously large computations.

4. The presented approach dedicates the most time-consuming part of computations (linear system solution) to the computer with vector architecture. In the future work the described managing system will be used to supervise both the distribution of tasks coming from parallel mesh and matrix coefficients generation as well as from the PCG-SBS [14] parallel linear solver.

5. In case of described CAE (*Computer Aided Engineering*) application the real speedup which can be obtained by the proposed method of tasks distribution is usually much greater than the theoretically computed one concerning only CPU operations and inter task communication (see [1] for detail definition of the speedup). A single computer RAM is usually not large enough for storing the matrices appearing in (5) if a large scale problems have to be computed with a high accuracy. Therefore, the real time of the sequential execution of such computation is greater than theoretical one, because it is enlarged by a swapping time (time overhead of frequent disk transmitions).

6. The described policy of the stochastic task distribution can be utilized not only for CAE parallel programs but also for other scalable large distributed applications.

References

[1] V. Donaldson, F. Berman, R. Paturi, *Program Speedup in Heterogeneous Computing Network*, Journal of Parallel and Distributed Computing 21, 316-332 (1994).

[2] H.S. Stone, *Multiprocessor scheduling with the aid of network flow algorithms*, IEEE Trans. Software Engrg. SE-3, 1 (Jan 1977), 85-93.

[3] S.H. Bokhari, *A shortest tree algorithm for optimal assignments across space and time in a distributed processor system*. IEEE Trans. Software Engrg. SE-7, 6 (Nov. 1981), 583-589.

[4] S. White, A. Alund, V.S. Sunderam, *Performance of the NAS Parallel Benchmarks on PVM-Based Networks*, Journal of Parallel and Distributed Computing 26, 61-71 (1995).

[5] D.M. Nicol, D.R. O'Hallaron, *Improved algorithms for mapping pipelined and parallel computations*. IEEE Trans. Comput. 40, 3 (Mar. 1991), 295-306.

[6] S. Manoharan, N.P. Topham, *An assessment of assignment schemes for dependency graphs*, Parallel Computing 21 (1995) 85-107.

[7] E.Altman, *Asymptotic Properties of Constrained Markov Decision Process*, INRIA, Centre Sophia Antipolis, France, Raport No.1598, 1992.

[8] E.Altman, *Constrained Markov Decision Process*, INRIA, Centre Sophia Antipolis, France, Raport No.2574, 1995

[9] R. Schaefer, *Numerical Models of the Prelinear Filtration*, Rozprawy Habilitacyjne U.J., Vol. 212, Kraków, 1991, (in Polish)

[10] R. Schaefer, S. Sędziwy, *Semivariational Numerical Model of Prelinear Filtration with the Special Emphasis on Nonlinear Sources*, CAMES, vol. 3 1996, pp. 83-96.

[11] R. Schaefer, S. Sędziwy, *Filtration in cohesive soils: modelling and solving* in: Proc. of the Conf.:Finite Elements in Fluids, New Trends and Appl.,Barcelona 1993, Vol. II., 887 – 891

[12] J. Bear, *Dynamics of Fluids in Porus Media*, Elsevier, New York, 1972

[13] K. Kant, *Introduction to Computer System Performance Evaluation*, Dep. of Comp. Science The Pensylvania Satet Univ. McGraw-Hill, Inc, 1992

[14] M. Papadrakakis, S. Bitzarakis, *Domain Decomposition PCG Methods For Serial and Parallel Processing*, Computing Systems in Engineering, 1995

[15] Z. Onderka, *Markov Chain Based Management of Large Scale Distributed CAD Computations* in preparation

[16] H. Kuszner, *Intruduction to Sochastic Control*, Holt, Reinhart and Winston, Inc., New York Copyright 1971.

[17] Clyde P. Kruskal, Alan Weiss, *Allocating Independent Subtasks on Parallel Processors*, IEEE Trans. on Software Engrg. vol. SE-11, No. 10, October 1985.

[18] Flasinski M., Schaefer R., Toporkiewicz W., *Optimal Decomposition of IE-Graph Represented Structures*, Proc. of the 7-th Int. Conf. on Engrg. Computer Graphics and Descriptive Geometry, pp. 405-409, ICECGDG'96, July 1996 Krakow, Poland.

[19] Przybylski P., *The Distributed Algorithm of an Unstractural Triangular Mesh Generator - Object Oriented Approach*, Proc. of XII Conf. on Computer Methods in Mechanics, pp. 283-284, May 1995 Warsaw-Zegrze, Poland.

[20] Krok J., Lezanski P., Orkisz J., Przybylski P., Schaefer R., *Basic Concepts of an Open Distributed System for Cooperative Design and Structure Analysis*, CAMES, vol. 3 1996, pp. 169-186.

Irregular Data-Parallel Objects in C++

Jean-Luc Dekeyser   Boris Kokoszko   Jean-Luc Levaire

Philippe Marquet

{dekeyser,kokoszko,levaire,marquet}@lifl.fr

Laboratoire d'Informatique Fondamentale de Lille
Université de Lille

Abstract. Most data-parallel languages use arrays to support parallelism. This regular data structure allows a natural development of regular parallel algorithms. The implementation of irregular algorithms requires a programming effort to project the irregular data structures onto regular structures. We first propose in this paper a classification of existing data-parallel languages. We briefly describe their irregular and dynamic aspects, and derive different levels where irregularity and dynamicity may be introduced. We propose then a new irregular and dynamic data-parallel programming model, called Idole. Finally we discuss its integration in the C++ language, and present an overview of the Idole extension of C++.

1 Irregularity and Data-Parallelism

The evolution of data-parallel languages mimics closely the evolution of sequential languages. Keeping in mind efficiency and simplicity, compilers have supported, in a first step, only regular data structures: arrays in sequential languages, vectors and matrices in parallel languages. Handling irregular data structures imposes then to the programmer an explicit management of the memory. A consequence is that irregularity will take place at the level of the algorithm (gather/scatter operations). The integration of irregular data structures in data-parallel languages can be compared to the emergence of dynamic memory management and pointers in sequential languages. While preserving the semantics of the data-parallel model, irregular data structures essentially allow the specification of interdependencies between parallel object elements, which are similar to the interdependencies expressed by pointers in sequential data structures. Let's consider the following example : in a binary tree, each node wants to compute the minimum of a value held by its two child nodes. (Parallel happy birthday: in a family and on several generations, each parent wants to know the next birthday to wish among its children.) This algorithm is effectively a data-parallel algorithm. The same algorithm is applied to a set of data of similar type. For a data-parallel language supporting irregular data structures, the translation is straightforward:

```
if (rchild and lchild exist) then
  min_child = min (rchild.val, lchild.val)
```

To write this algorithm using arrays, it is necessary to linearize the binary tree into an array (*val*) and to manage links using two index tables (*lchild (:)* and *rchild (:)*). The algorithm then consists in two gather operations:

```
where (rchild(:) and lchild (:))
  min_child(:) = min (val (rchild(:)),
                      val (lchild(:)))
```

Similarly, during a tree update (insertion, deletion), a data-parallel language with irregular data structures will provide adequate tools (e.g. dynamic virtual processor allocation, PV_alloc ()). In an array language, the programmer will have to ensure himself the management of the distributed memory.

2 Data-Parallel Language Classification

It is currently difficult to provide an exhaustive list of data-parallel languages. Some of them have been proposed by constructors, and were often recognized as standards. Apart from these "dinosaurs", a great number of languages are developed by research teams. As for sequential languages, none are unanimously accepted by the users, even if their contributions are significant. From this diversity of data-parallel languages, we propose to extract some common criteria that will be the root of a classification of data-parallel languages. We extend the classification proposal of D. Lazure [17] and obtain the following criteria.

2.1 Object Declaration

The *explicit* declaration of parallel objects in a language implies that the identification of the data-parallel code is directly achievable during the compilation phase. On the contrary, for an *implicit language* (sequential), a primary phase of automatic parallelization is required to identify objects that may be safely handled in parallel. The following criteria only concern *explicit languages*.

2.2 Virtualization

The size of data-parallel structures depends on the problem and on the algorithm used to solve it. Usually, the target machine does not offer the exact number of processors allowing a direct mapping of one element of the structure to one elementary processor. *Virtual machine languages* rely on the notion of a virtual machine, which will offer the desired number of virtual processors. It is actually up to the compiler to ensure the emulation of these virtual processors on the physical processors. *Physical machine languages* impose that the size of data-parallel structures is strictly equal to the number of physical processors. It is now up to the programmer to ensure "by hand" the virtualization. The following criteria only concern *virtual machine languages*.

2.3   Virtual Machine

A virtual machine is called an *instantiation machine* if each object declared on the machine inherits the size and the rank of the machine. All objects declared on one instantiation machine have thus the same number of elements and the same topology than the machine itself. A virtual machine is called an *alignment machine* if the objects declared on the machine may have any number of elements in the range of the size of the machine. Moreover the elements of an object may be mapped to any subset of virtual processors of the machine. For instance, on a 2D grid virtual alignment machine, you can declare an object corresponding to only one line of the grid, or also an object corresponding to the four corners of the grid.

2.4   Mapping

During the compilation phase of a virtual machine language, the compiler will perform a mapping of virtual processors to physical processors. Some languages require that the programmer explicitly states a mapping for each virtual machine (cyclic, block...). Others leave the compiler to decide the mapping to use.

2.5   Data-Parallel Semantics

In implicit languages, there is no explicit notion of parallel variables. The program only accesses to their elements through an indexing mechanism (e.g. A[i]). In explicit languages, parallel objects are seen in their whole. A program can then express a computation involving two parallel variables directly using their names (e.g. A + B). For an instantiation machine, the semantics of the computation is unique as objects are defined on the whole machine: perform the computation on every virtual processors. For an alignment machine, there exists two different interpretations, as objects are not necessarily defined on the whole virtual machine. The difference between the two interpretations lies in the method used to map the two objects one on each other, in order to perform a computation element by element. The first interpretation, called *semantics of the virtual processor*, uses the virtual machine as a referential to realize the mapping between the two objects. The more natural approach is to consider the intersection of the two sets of virtual processors where each object is defined. Only virtual processors where both objects are declared will perform the computation. In this way, the semantics is always well-defined. Moreover, the compiler will not have to generate implicit communications because all objects of the same virtual processor are usually located on the same physical processor. The other interpretation, called *semantics of the index*, linearizes the two objects (or considers them as arrays when possible), and directly maps the corresponding arrays one on each other. The referential used to perform the mapping between the objects is then the referential of the objects themselves. The problem here is that the two objects should have the same topology, or at least the same number of elements when linearized. This leads to a semantics which is not always well defined. Moreover, the mapping of each object on the virtual machine is not necessarily the same. This forces the compiler to perform some extra tests and eventually generate communications between physical processors.

This classification is illustrated with some typical languages in Figure 1.

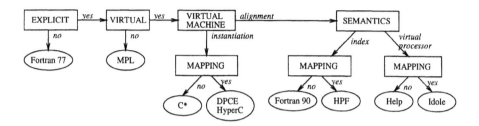

Fig. 1. Data-parallel language classification.

2.6 Content of the Paper

From our classification, we identify three levels where dynamic and irregular properties of data-parallel structures may be introduced (section 3). The Idole programming model is described in section 4. Idole is a virtual machine language with alignment, and supports the semantics of the virtual processor. Irregularity and dynamicity appear in Idole at the level of the virtual machine and at the level of objects. The integration of this model in C++ is discussed in section 5. Section 6 presents through examples the basic steps of programming in Idole. Section 7 focuses on the irregular and dynamic aspects of the Idole language.

3 Irregularity and Dynamicity in Data-Parallel Languages

In the proposed classification, no assumption has been made on the structure of parallel objects. Nevertheless, it is difficult to imagine handling irregular data structures for any of the classes. Implicit languages can handle them since they are sequential, but current parallelization techniques do not succeed in recognizing them. Similarly, physical machine languages manipulate exclusively regular data. The architecture of data-parallel machines is indeed often regular and static. In the class of virtual machine languages, one finds the expression of the irregularity and the dynamicity at different levels.

3.1 The Virtual Machine

A virtual machine is composed of a set of virtual processors. The topology of this machine is often a static multi-dimensional grid. In some languages (C*[13], DPCE [11], HyperC [16, 7], Paralaxis [4, 3]), irregular and/or dynamic topology constructions facilitate the development of data-parallel algorithms. As a rule, objects associated with an irregular virtual machine inherit its irregularity (and in some cases its dynamicity).

3.2 The Objects

In instantiation virtual machines, all the objects have the rank of the virtual machine. In this case, irregularity at the object level makes no sense. In the case of alignment virtual machines (Help [9, 17]), objects do not necessarily cover the whole virtual machine. The allocation domain of an object on this machine can be regular, when it may be described directly using an array subscript notation (for example multi-dimensional arrays). It is irregular when its description is impossible using such a notation (for example elements on the perimeter of a 2-D domain).

3.3 The Mapping

The lack of irregularity at the object level as well as at the virtual machine level leads languages to propose dynamic and irregular mapping functions (HPF-2 [14, 15, 20], Vienna Fortran [6, 21], Vienna Fortran 90 [1], PST [19]). This allows, by a judicious gathering of array elements, to ensure the physical locality of elements according to the irregular addressing of the algorithm. However, in this case, the programmer still has to construct its algorithm using regular structures. Only the irregular mapping will allow to achieve better performance in the execution of the algorithm. Such a situation requires a particular effort from the programmer.

This classification has been presented in [10]. Many irregular and dynamic language constructions illustrating these different levels are detailed.

4 Idole Programming Model

The Idole language supports a virtual alignment machine with a virtual processor semantics. Such a machine is called a collection in the language. The Idole programming model integrates as a basic principle the dynamicity and the irregularity of parallel objects. The expression of the irregularity is found at two levels: the virtual machine and the objects themselves.

This twofold irregularity is justified by the following. The structure of a collection is common to all objects of the collection. A transformation of this structure have identical consequences on all these objects. For instance, when deleting a virtual processor of the collection, all objects that own an occurrence on this virtual processor also undergo a deletion. One can thus modify the structure of the virtual machine according to the algorithm. Such a structure modification represents a change in the application domain of the data-parallel algorithm. Let's consider stack algorithms [12, 8]. In this kind of algorithm, the data structure is a distributed stack. The same algorithm is applied on each element of the stack. Each element can generate another element or can be deleted. For example, in particle dynamics experimentation, a particle is tracked step by step until it stopped. On each step, a particle can generate a child particle which is pushed on the stack for further computation. This algorithm is well suited to data-parallel implementation but it relies a data-structure which evolved in an irregular way.

Inside the collection, the alignment machine authorizes the creation of objects not allocated on the whole collection. Unlike Help or HPF that only manipulate multi-dimensional objects, Idole accepts objects allocated on any subset of virtual processors. No assumption is made on the structure of this set. As in Help, Idole objects may vary in rank and size. The irregularity at this level allows to limit, for a particular phase of the algorithm, the range of the data-parallel processing. This set may actually characterize both an active domain of computation as well as an allocation domain for parallel objects.

5 Idole Language: A C++ Extension

The Idole language is derived from the C++ language. The object-oriented technology provides two main features: the encapsulation of functions in object methods, and the refinement of object properties through the inheritance mechanism. We have chosen to take advantage of these characteristics in the data-parallel section of the language.

The direct integration of the data-parallel programming model in C++ has already been studied by Lickly and Hatcher [18]. Their opinion is that such an integration would not be satisfying. Indeed, C++ does not allow to express data-parallel computations, without adding a special mechanism. Neither does it provide a way to define parallel control structures. These both aspects have been included in Idole. We modify in a common way the C++ semantics of computation for data-parallel objects: element operations are performed in parallel on each element when applied to a data-parallel object. Syntactically, Idole also provides data-parallel control structures, mainly the **where** statement and the other usual structures. Idole specific extensions are collection classes, virtual processor classes, shapes, and finally data-parallel objects.

Each Idole extensions will be illustrated using adaptative mesh algorithm [5]. In this example, the nodes concentration is not the same on the whole mesh. Moreover, adaptative mesh technique relies on refinement of the mesh (*cf.,* Figure 2) to concentrate on points where error contributions are large. By using those irregular mesh, number of points needed for the computation are reduced to accurately capture region where an high calculus precision is required.

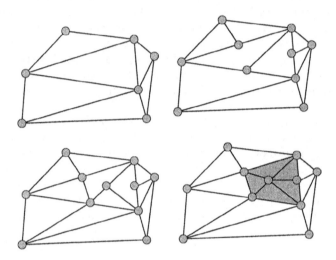

Fig. 2. An Adaptative mesh (Refinement by bisection of the longest side)

5.1 Collection Classes

Idole collections correspond to the virtual machines of the programming model, providing thus the way to express data-parallelism. Collections are involved as an extra specifier in the declarations of data parallel objects, in addition to the element class specifier. We extend C++ to support this new mechanism. Idole introduces classes of collection with the keyword `collection_class`, and adds a new predefined class `bag`. This collection class corresponds to unstructured sets of processors, and implements basic attributes (size) and methods on collections. The mechanism of inheritance is extended to collection classes. For convenience, Idole provides also the collection class `grid`, derived from `bag`, which corresponds to multidimensional grids.

It would have been convenient to introduce collections just as specific classes in the language. They actually act as classes in the declaration of parallel objects. But in the model, the dynamicity at the virtual machine level dictates that collections should have some modifiable attributes. The only way to provide this feature in C++ is to consider collections as objects themselves. Thus, a collection is an object, instance of a collection class. Irregularity and dynamicity at the virtual machine level are specified by the programmer using predefined primitives

(See section 7). Note that a modification of the structure of a collection may result in modifications on objects defined on that collection.

In case of adaptative mesh, we define a collection to handle the whole mesh structure. Each node of the mesh is represented by a virtual processor. These processors will be interconnected to describe the mesh structure.

5.2 Virtual Processors

The virtual processor semantics implies an explicit communication mechanism between virtual processors of a virtual machine. Communication links between virtual processors will be identified in Idole through communication ports holding processor references (entries). This set of communication links defines the virtual machine topology. This topology may vary at runtime, as the value of a communication port is dynamic. Communication ports are declared at the level of the virtual processors.

Processor classes are derived from the predefined **processor** class. An instance of this class is a virtual processor without predefined communication ports. It provides methods to declare communication ports (these are only allocated at the processor instantiation) and methods to connect a communication port to a processor of any type. The mechanism of inheritance on processor classes allows to expand already defined processor classes.

Collections are built using processor objects. Conceptually, processors are created at the instantiation of collections. One processor handles (and performs computations on) one element (if present) of each DPO in the collection. A collection can be composed of different types of processor. Moreover, virtual processor can be dynamically allocated. A collection class can also be defined in a generic manner to support any kind of virtual processor: the notion of C++ template has been extended to collection classes.

When the refinement of a region is required, this region is divided in two sub-regions by bisection of the longest side. First, we create a new virtual processor to get a new node using dynamic allocation function. Then we update the values of the communication ports to reflect the new structure of the mesh.

5.3 Shapes

Processing domains or DPO allocation domains can be defined on collections. These domains have the type **shape**. A shape refers a subset of the virtual processors of a given collection. A shape is built by addition and deletion of virtual processors. Specific block description operations are available for **grid** collection shapes. Set operations (union and intersection) are also defined on shapes.

In our example, in last refinement step of Figure 2, we hilight the region where high precision is required. We define a shape by adding all the virtual processors of this region. Using this shape, we can limit the computation domain to this region and allocate DPO to focus on this domain.

5.4 Data-Parallel Objects

A DPO is defined by specifying both a collection (or a shape) and an element class. The element class is a predefined or user-defined C++ class. Methods of the element class are implicitly extended to the DPO: a method of element called on a DPO is applied to every elements of the DPO. Global operations on the DPO are defined in the collection class in a special section named **dpo**. In particular, it is possible to define DPO methods involving inter-processor communications, reshaping, etc.

In our example, DPO elements are data required to describe the physical properties of each node of the mesh. DPOs can be defined on the whole mesh or only on a shape. For example, we can declare a DPO on the shape defined in the section 5.3. This DPO will hold extra data required for more acurrate calculus.

6 Idole Programming Method

Figure 3 describes the Idole programming methodology.

6.1 Processor Class Definition

The first step consists in declaring the processor classes. The macro **DECL_PROCESSOR** declares a new class **processor4** of processors, each of which having four communication ports. The macros **CONNECTION** allow the programmer to name these communication ports. We use macros here in order to simplify the syntax.

```
DECL_PROCESSOR(processor4, processor, 4)
  CONNECTION(north,0)
  CONNECTION(south,1)
  CONNECTION(west,2)
  CONNECTION(east,3)
END_PROCESSOR
```

6.2 Collection Class Definition

The second step consists in declaring the collection classes. We define below a 2D grid class **2d_grid**, derived from the predefined class **grid**. The processor class

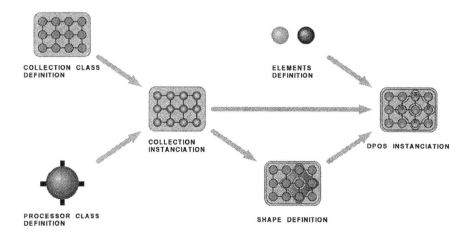

Fig. 3. Diagram of a Idole programming model.

on which collections will be built is left as a template. It will be specified at the collection instantiation. In the constructor of the collection class, the constructor of the predefined class **grid** is called. The body of the **2d_grid** constructor will also usually contain an initialization phase of the communication ports of virtual processors. Communication ports may be referenced here using their indices, as the processor class may be unknown at that time. Every collections are dynamic in Idole. We specify here a method **add_new_line** to add a new line of processors in collections. Finally, we define a method of DPO **move** in the **dpo** section, which shifts a DPO inside the grid.

```
template<class P>
collection_class 2d_grid : grid<P,2> {
  2d_grid(int x_size, int y_size)
     : grid<P,2>(x_size, y_size) {...}
  ...
  void add_new_line(void) {...}
  ...
dpo:
  ...
  void move(int x_offset, int y_offset) {...}
} ;
```

6.3 Collection Instantiation

We can now instantiate a collection from a collection class and a processor class. We define a 4 × 3 2D grid collection named **a_grid**, which derives from the collection class **2d_grid** and is built using **processor4** type processors.

```
2d_grid<processor4> a_grid(4,3) ;
```

6.4 Shape Definition

The optional fourth step is to define some useful domains on the collection a_grid. We define here a shape a_domain, and initializes it by adding virtual processors which satisfy a given condition.

```
shape<a_grid> a_domain ;
...
where(a_condition_is_true_on_a_grid)
  a_domain.add() ;
```

6.5 Elements Definition

The fifth step is to reuse already defined element classes, or to define your own element classes. Theses classes are classic C++ class, e.g. a class of complex numbers.

```
class complex {...} ;
```

6.6 DPO Instantiation

We can now define parallel variables, the DPOs, by specifying their collections (or shapes) and their element types. Complete DPOs are defined on the whole collection whereas incomplete DPOs are only defined on a domain of the collection.

```
a_grid<int>        a_complete_dpo ;
a_domain<complex>  an_incomplete_dpo ;
```

The final step consists in writing the code of the algorithm itself. The next section presents some sample programs written in Idole.

7 Irregular and Dynamic Programming in Idole

This section emphases on the dynamic and irregular aspects of the Idole programming language.

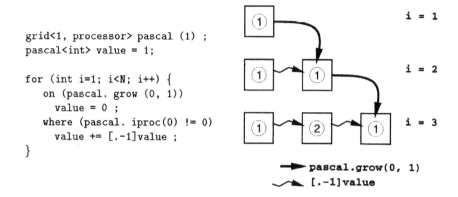

```
grid<1, processor> pascal (1) ;
pascal<int> value = 1;

for (int i=1; i<N; i++) {
   on (pascal. grow (0, 1))
      value = 0 ;
   where (pascal. iproc(0) != 0)
      value += [.-1]value ;
}
```

Fig. 4. Dynamic manipulation of regular collections in Idole.
The collection *pascal* is a regular collection of rank 1 and size 1, derived from the predefined collection grid. This collection is going to evolve during the algorithm. An integer variable *value* is defined on this collection. The constructor on is used to limit the current *shape* (the current computation domain). The primitive grow realizes the expansion of the collection *pascal* (dimension 0 by 1). This results in a shape made of all the newly created processors. The method iproc returns a parallel value representing the virtual processor indices in the collection (dimension 0). The left indexing of the variable *value* performs a communication to get value from the left neighbor.

7.1 Collection grid

Collection **grid** provides a multi-dimensional indexing of virtual processors. These indices allow to reference virtual processors. Regular communications on the **grid** use connections between ports. Irregular communications between disconnected virtual processors are realized using the indices. The dynamicity of the **grid** collections is restricted to the modification of the size of the grid. Virtual processors on the edge of the domain can be added or deleted by transform operations. The example of Figure 4 illustrates these primitives.

7.2 Collection bag

A collection **bag** is built by virtual processor additions. This operation is sequential (one creates a virtual processor) or parallel (a subset of virtual processors each create simultaneously a new virtual processor). Port connections allow the creation of a collection with any topology. The structure of the collection depends on these connections. Only regular communications (through the same communication port) are supported. A dynamic and irregular example of virtual processor handling is presented in Figure 5.

```
DECL_PROCESSOR (tree_node, processor, 3)
   CONNECTION (parent, 0)
   CONNECTION (rchild, 1)
   CONNECTION (lchild, 2)
END_PROCESSOR

bag<tree_node> mytree ;

add_rchild (void) {
   mytree <tree_node_entry<mytree>> child ;
   mytree <tree_node_entry<mytree>> self
       = (tree_node_entry<mytree>) mytree.getvpid() ;

   where(child = fork (mytree, tree_node)) {
     self-> rchild () = child ;
   } elsewhere {
     self = (tree_node_entry<mytree>) mytree.getvpid() ;
     self-> parent () = (tree_node_entry<mytree>) mytree.getpvpid() ;
     self-> lchild () = NULLPROC ;
     self-> rchild () = NULLPROC ;
   }
}
```

Fig. 5. Dynamic manipulation of virtual processors in Idole.
The processor class *tree_node* is derived from the predefined class **processor**. A processor *tree_node* offers 3 communications ports. These ports are actually defined as methods of the class *tree_node*. The macro-definition DECL_PROCESSOR also defines the generic type *tree_node_entry*, which are references (entries) to processors of that new class. The bag collection *mytree* is built using *tree_node* processors. The *add_rchild* method adds a right child to all active processors. *child* is a DPO of reference to *tree_node* processors; it is defined on the collection *mytree*. The DPO *self* is initialized from the getvpid () method of bag. getvpid () returns an entry (a reference) to the current processor. The fork () constructor performs the parallel creation of *tree_node* processors. In the first alternative of the where, the parent processors connect to their child. In the second alternative, the child processors init themselves and connect to their parent (the getpvpid () method returns a reference to the parent processor).

7.3   Idole Objects

Idole proposes the manipulation of dynamic and irregular parallel objects. This dynamicity covers the size and the position of objects. Idole introduces irregularity in the shape of incomplete objects. The allocation domain of objects can be any subset of virtual processors of the collection; this is supported for **grid** collections as well as for **bag** collections. An illustration of these features is given in Figure 6.

```
grid<1, processor> pascal(N) ;
pascal[0:0]<int> value = 1 ;
shape<pascal> oldshape ;

for (int i=1; i<N; i++) {
  oldshape = value. shape ;
  value. shape. add (pascal[i]) ;
  value[i] = 1 ;
  on (oldshape)
    where (pascal. iproc(0) != 0)
      value += [.-1]value ;
}
```

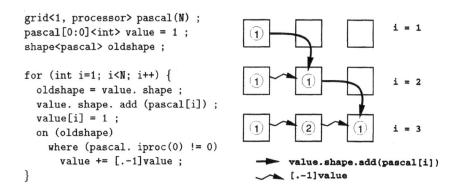

Fig. 6. Dynamic object manipulation in Idole.
pascal is a regular one-dimensional collection of size *N*. The parallel object *value* is initially defined on only the first processor of this collection. The shape *oldshape* is initialized with the shape of the object *value*. This object is then successively altered by the addition of one processor (method add).

8 Conclusion

We identified three levels to express the irregularity and the dynamicity in data-parallel languages. At the level of the virtual machine, the irregularity and dynamicity are expressed for all the objects associated to this virtual machine. At the level of the object, when the irregularity and the dynamicity are its own, they characterize the application domain of the algorithm for a given phase. Finally at the level of mapping functions, an irregular distribution of a regular data structure ensures a good load balancing on the physical machine.

We presented the Idole programming model. Idole supports the irregularity and dynamicity at the two first levels. The Idole model is well suited to an implementation in an object-oriented language. However C++ does not allow a direct implementation of the model. We have then chosen to define the Idole language as a C++ extension. Idole allows the definition of programmable virtual machines, the handling of parallel variables and the usage of data-parallel control structures.

References

1. Siegfried Benkner. *Vienna Fortran 90 and its Compilation*. PhD thesis, University of Vienna, September 1994.
2. François Bodin and Dennis Gannon. Distributed pC++: Basic ideas for an object oriented parallel language. Technical report, Indiana University, Bloomington, IN, 1993.

3. Thomas Braünl. Parallaxis-III: A langage for structured data-parallel programming. Technical report, Computer Science Dept., Univ. Stuttgart, Germany, September 1995.

4. Thomas Bräunl. Structured SIMD programming in PARALLAXIS. *Structured Programming*, 10(3):121–132, July 1989.

5. Eric J. Bylaska, Scott R. Kohn, Scott B. Baden, Alan Edelman, Ryiochi Kawai, Maria E. G. Ong, and John H. Weare. Scalable parallel numerical methods and software tools for material design. In *7th SIAM Conf. on Parallel Proc. for Scientific Computing*, San Francisco, February 1995.

6. Barbara M. Chapman, P. Mehrotra, and Hans Zima. Programing in Vienna Fortran. *Scientific Programming*, 1(1):31–50, Fall 1992.

7. Philippe Clermont and Nicolas Paris. Handling and optimizing unstructured computations in data-parallel languages. In *Proc. of the New Frontiers Workshop on Future Directions of Massively Parallel Processing*, pages 28–35, McLean, VA, October 1992.

8. Jean-Luc Dekeyser, Cyril Fonlupt, and Philippe Marquet. Analysis of synchronous dynamic load balancing algorithms. In *Parallel Computing: State-of-the Art Perspective (ParCo'95)*, volume 11 of *Advances in Parallel Computing*, pages 455–462, Gent, Belgium, September 1995. Elsevier Science Publishers.

9. Jean-Luc Dekeyser, Dominique Lazure, and Philippe Marquet. A geometrical data-parallel language. *ACM Sigplan Notices*, 29(4):31–40, April 1994.

10. Jean-Luc Dekeyser and Philippe Marquet. Supporting irregular and dynamic computations in data-parallel languages. In *Spring School on Data Parallelism*, pages 197–219, Les Ménuires, France, March 1996. Lectures Notes in Computer Science Tutorials Series.

11. DPCE Subcommittee. Data-parallel C extensions. Technical Report version 1.6 X3J11/94-080, Numerical C Extensions Group of X3J11, 1994.

12. Cyril Fonlupt. *Distribution Dynamique de Données sur Machines SIMD*. Thèse de doctorat (PhD Thesis), Laboratoire d'Informatique Fondamentale de Lille, Université de Lille 1, December 1994. (In French).

13. James L. Frankel. C* language reference manual. Technical report, Thinking Machines Corporation, Cambridge, MA, May 1991.

14. High Performance Fortran Forum. High Performance Fortran language specification. *Scientific Programming*, 2(1-2):1–170, 1993.

15. High Performance Fortran Forum. HPF-2 scope of activities and motivating applications, November 1994. version 0.8.

16. HyperParallel Technologies, Palaiseau, France. *HyperC Documentation Kit*, 1993.

17. Dominique Lazure. *Programmation Géométrique à Parallélisme de Données — Modèle, Langage et Compilation*. Thèse de doctorat (PhD Thesis), Laboratoire d'Informatique Fondamentale de Lille, Université de Lille 1, January 1995. (In French).

18. Daniel J. Lickly and Philip J. Hatcher. C++ and massively parallel computers. *Scientific Programming*, Winter 1993.

19. Andreas Müller and Roland Rühl. Extending High Performance Fortran for the support of unstructured computations. In *Proc. Int'l Conf. on Supercomputing*, Barcelona, Spain, July 1995.

20. Rob Schreiber. Evolution and future of High Performance Fortran. In *Spring School on Data Parallelism*, Les Ménuires, France, March 1996. Lectures Notes in Computer Science.

21. Hans Zima, Peter Brezany, Barbara Chapman, Piyush Mehrotra, and Andreas Schwald. Vienna Fortran—a language specification, version 1.1. Technical Report TR 92-4, ACPC, University of Vienna, 1992.

Control and Data Flow Analysis
for Parallel Program Debugging

Dieter Kranzlmüller, Andre Christanell, Siegfried Grabner, Jens Volkert

GUP-Linz, Johannes Kepler University Linz

Altenbergerstr. 69

A-4040 Linz, Austria/Europe

phone: ++43 732 2468 9499

fax: ++43 732 2468 9496

email: kranzlmueller@gup.uni-linz.ac.at

Abstract

Parallel program analysis for error detection requires support of mighty tools. Another demand to these tools is usability. The monitoring and debugging environment MAD provides a solution for the domain of distributed memory computers. It consists of several modules which can be applied to improve the reliability of programs for message passing systems.

Following the monitoring of an initial execution, the user can inspect a visual representation of the program flow, the event graph. Detection of simple errors and race conditions is possible. Through control and data flow analysis the origin for faulty behavior can be located. Based on a graphical representation the user is directed from the occurrence of the error to the line of code responsible for this behavior.

1 Introduction

Debugging of programs is a difficult task in the software life-cycle. The problems are further increased, if the target system is a parallel machine. Then programs consist of several concurrently executing and communicating tasks. A solution to this situation comes with the increasing number of tools, that provide support for all means of parallel debugging. These tools can be categorized as follows [AgCh 96]:

- Integrated environments providing both debugging and performance tuning
- Tools for debugging only
- Tools for performance monitoring only

Well-known examples for tools applied to performance analysis are *Paragraph* [HeEt 91] and *Paradyn* [Mill 95]. Tools for error detection are *HeNCE* [Begu 93] and *p2d2* [Hood 96]. A good overview of many debugging and tuning approaches can be found in [Hond 95] and [PaNe 93].

The *M*onitoring *A*nd *D*ebugging environment *MAD* is another toolset in this area. It combines both tasks, error detection and performance analysis by providing several modules that perform specific activities of these tasks.

The target architectures of MAD are distributed memory machines that use explicit message passing for communication between processes. A first implementation was built for the nCUBE 2, a MIMD machine with hypercube topology. With the availability of a standard message passing interface [MPI 94], the degree of portability between platforms has been increased. Thus, MAD can now be applied to any machine that uses MPI.

During the design phase of MAD, two goals have been established - functionality and usability. To reach a high degree of functionality and to preserve extensibility, the whole environment is based on a modular approach. Each of the features is included in specialized modules, which communicate via a fixed interface. This has the advantage, that certain modules can easily be exchanged and new modules can be added to the system.

In terms of usability a good degree is reached with the combination of all tools under a graphical user interface, instead of a textual interface. While textual representations might be sufficient for sequential debugging, their use in parallel debugging is rather limited. As it is stated in [Panc 96], "Text... is inadequate to express the complex, multidimensional relationships of an executing parallel program". The graphical interface of MAD is based on the ET++ application framework [Wein 88], which is running on the X-Window system of many UNIX workstations.

In addition to "standard functionality" like monitoring and visualization of communication, the MAD environment provides some extended features. One of these features is the control and data flow analysis. It has been introduced recently and is discussed in this paper.

The paper is organized as follows. First we give an introduction to the errors, that are possible in parallel programs. Afterwards a brief overview of the MAD components, the tools that perform the debugging activities, is given. Section 4 presents the control- and data flow analysis tool CDFA, while section 5 concentrates on the backtracking of variables. Afterwards the current approach is summarized and concluded with an outlook on future developments in this project.

2 Error Classification

When debugging parallel programs, many different bugs are observable. Sometimes a program will break off before the desired end of the program or produce incorrect results. Another kind of error is a "deadlock", where the program does not finish its task, because some processes lock each other. Other failures include the "stampede effect", where processes cannot be stopped and stampede over the evidence of an error, and the "irreproducibility effect". The latter appears, when race conditions are included in a program. Then the program is nondeterministic and might lead to different behavior in successive program runs, even with the same input.

All the bugs in parallel programs can be divided into the following two classes of errors:

- local errors
- global errors

The difference between these errors is the range of their consequences. Local errors reside completely on one single process and appear within one block of code. The observable faulty behavior and the reason for it does not interfere with any other process before the occurrence of the error, neither through communication nor synchronization. Because of the locality of such an error, traditional debugging methods as already established in sequential debuggers can be used to localize and correct them.

The more difficult kind of errors in parallel programs are global errors. Although the observable behavior might appear only on one single process, they influence other processes in the parallel system. The influenced processes are all communication partners of the process with the faulty program code. In many cases the faulty behavior and the origin of the error are on different processes. Thus, detection of global errors is more complicated than detection of local errors.

Figure 1 shows a diagram of the two different kinds of errors. The local error occurs on process 2, together with the reason for the faulty behavior. The range between error and origin can be investigated with a traditional debugger.

The global error in figure 1 manifests itself on process 1, but originates from process 0. In this case, a parallel debugger with capabilities for detecting this kind of errors has to be used. Nevertheless, after following the global error across the communication to process 0, again a traditional debugger might be useful.

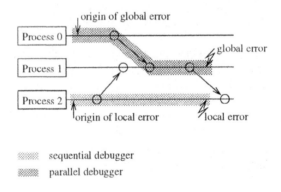

Fig. 1. Error Classification

Global errors, that are relatively easy to recognize, are wrong communication structures and wrong data in transmitted messages. The worst kind of global errors are called race conditions [NeMi 92]. Races are typical cases of unintended nondeterminism [ChMi 91]. Therefore it is possible, that a program containing a race condition produces correct results most of the time. Sporadically however, incorrect results might be observed. The reasons are manifold, ranging from different states of the caches to unsynchronized clocks and different processor speeds. The problem is, that this kind of bugs is difficult to detect and dedicated functionality has to be provided.

3 Architecture of the MAD Environment

This section describes the architecture of the MAD environment. The basic parts of MAD as well as any other debugging tool are [Hond 95]

- a monitor and
- an inspector.

3.1 EMU - Event Monitoring Utility

The monitor, which generates the data for all modules of the MAD environment, is the *Event Monitoring Utility EMU*. After instrumentation of the source code, it observes the execution of a program and generates tracedata at each interesting event. In message passing programs the most interesting events are communication events like send, receive and test [GrVo 93].

The problem with any monitor in the field of parallel program analysis is the intrusion on the running program. The delay introduced by the monitoring activities may change the behavior of the program ("probe effect" [Gait 86]). The monitor EMU tries to improve this situation through configuration options. These options allow the user to control the monitoring influence by means of time delay, memory consumption and generated data [GrKr 94].

An example configuration option is the amount of data generated at interesting events. The data that is needed for the analysis, depends on what analysis has to be performed. For example, if the user is only interested in analysis of correct ordering of the events, logical clocks as proposed by [Fidg 88] are sufficient. On the other hand, if the user needs data for performance analysis, exhaustive timings have to be obtained during the program execution.

The architecture of EMU is not static. New modules are included every now and then and the degree of complexity in using EMU is already rather high. Therefore a user interface has to be provided. It is visualized in figure 2. The subwindow contains configuration options, which are easy to understand even for inexperienced users. For usage of the whole configuration set, the item "Advanced Monitoring" has to be selected.

3.2 ATEMPT - A Tool for Event Manipulation

The tracefiles of EMU are analyzed post-mortem by the inspection tools of the environment. The most important inspection tool in MAD is *ATEMPT - A Tool for Event ManiPulaTion* [Grab 95a]. It generates a global communication graph with the data stored in the tracefiles and includes functionality for investigating the monitored program based on this graphical representation.

The global communication graph, also called event graph, consists of vertices, which represent the occurrence of an event, and of arcs, which represent communication or sequential program flow between the events. An example for an event graph is already visualized in figure 1.

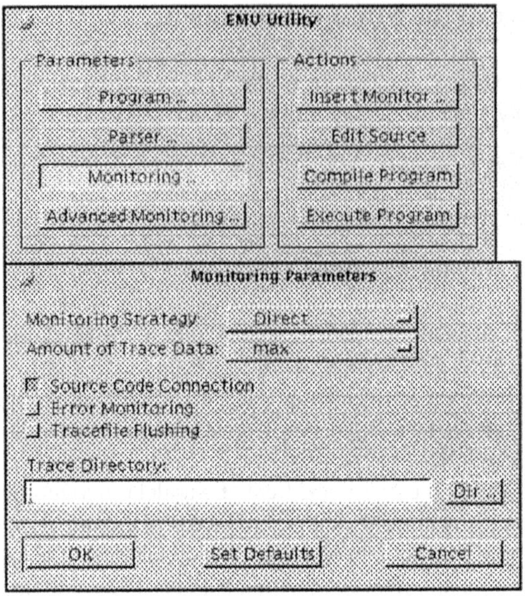

Fig. 2. EMU User interface

Based on the event graph some automatic analysis can be performed. As a result automatic visualization of simple errors is possible. Such errors are isolated events (send event without receive event or vice versa) and events with different message length at sender and receiver. An example is visualized in figure 3. It shows an event graph for a parallel householder algorithm on 8 nodes. The automatic detected errors are highlighted with different colors indicating different errors (remark: annotations have been attached for printing purposes).

Additionally to the error detection, performance analysis features are included in ATEMPT. An example is shown in figure 4. Again it contains an event graph, but this time with actual timings, which have been measured during execution. The time scale is included at the bottom of the window. The (blue) bars in front of receive events indicate blocking time, with the length of the bars representing delay through blocking. Additional information can be provided by statistical overviews of the execution, containing computation time, communication time, and monitoring overhead.

3.3 Race Condition Detection

Another feature of MAD, which is based on ATEMPT, is the *race condition detection* mechanism [Kran 95]. At race conditions the program flow or the resulting data depends on the actual ordering of the occurrence of two or more events [GrVo 93]. In message passing programs this can happen e.g. at receives where the origin or the type of the message to be received is undefined. There, any message would be accepted, thus any message ordering can occur.

Automatic visualization of errors:
- isolated events
- events with different
  message length

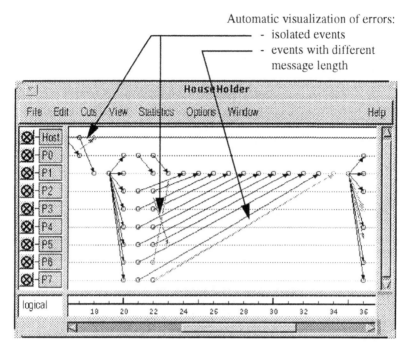

Fig. 3. Error detection with ATEMPT

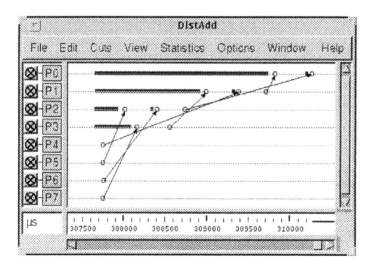

Fig. 4. Performance Analysis with ATEMPT

The race condition detection mechanism of ATEMPT is based on two steps. Firstly the user is provided with *race condition candidates*. These are events which might be reordered during different executions. With this information the user can carry out two activities [GrVo 96]:

- Change the program code to remove unintended race condition candidates, e.g. by using additional message types to distinguish racing messages.

- Exchange the race condition candidates graphically based on the event graph and initiate a trace driven simulation with PARASIT to verify the results of different message orderings.

3.4 PARASIT - Parallel Simulation Tool

The difference between the monitored behavior and the behavior defined by event manipulation is verified in the second step, the trace driven simulation. The original idea of using a trace driven simulation is based on the debugging cycle.

The debugging cycle defines the activity of running a program several times under control of a debugger and applying different methods of analysis to detect errors. This requires an *equivalent execution* [Leu 90], where a program can be run successively with the same ordering of interprocess communication events. This requirement is no problem for deterministic programs, where each execution of the program with the same input set results in the same event order and the same output.

However, in nondeterministic programs, e.g. programs with race conditions, the problem with an equivalent execution is inherent. Therefore a mechanism has to be provided, that preserves an equivalent execution compared to a monitored execution of a program.

This task is carried out with the *PARAllel SImulation Tool PARASIT*, which performs trace driven simulation [MiCh 88]. It follows the program flow defined by the source code and forces communication to take place as ordered in the specified event graph. Therefore a debugging cycle can be applied to any program run, deterministic or nondeterministic, because equivalent execution is preserved by PARASIT.

Furthermore the results of the event manipulation can be verified. In this case the replay of PARASIT is used to compare the altered execution to the original execution for detecting the race conditions [Grab 95b]. This is achieved by performing an equivalent execution of the program up to the place where event manipulation took place. Afterwards, as the behavior of the program is unknown after the exchange, the program has to be executed without control of PARASIT.

4 Control and Data Flow Analysis

The methods described in section 2 are useful for many activities of software repair. Additional analysis can be performed, if other methods of visual representation or additional analysis techniques are integrated in the environment. This section describes, how a control and data flow analysis tool can support debugging.

Event graph in
ATEMPT
Filebrowser

Variable inspection
Communication
event inspection

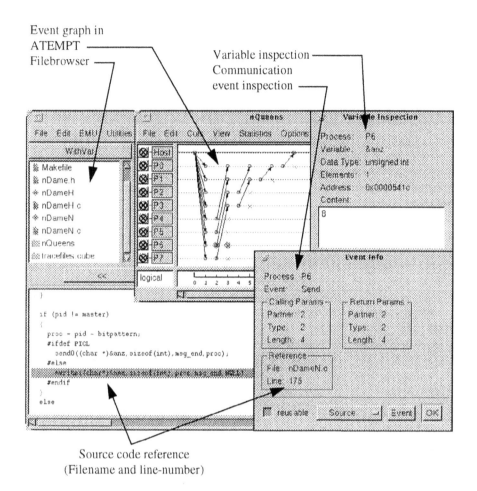

Source code reference
(Filename and line-number)

Fig. 5. Source Code Connection in Filebrowser

In program analysis for debugging two tasks have to be carried out:

- Detection of faulty behavior.
- Location of code responsible for faulty behavior.

The first activity, detection of faulty behavior, is based on the observed behavior of the program. The representations of this behavior can be based on various forms of visualization. Of course, the event graph is only one possibility to provide meaningful information to the user. Another idea incorporated in MAD is the visualization of program behavior based on the underlying hardware architecture. This is achieved with the *MU*ltiprocessor *C*lass *H*ierarchy *MUCH* which includes hardware models and visualizations for the nCUBE 2 multiprocessor and the CONVEX SPP virtual shared memory computer. Additional information on this module is available in [Kran 96].

The second activity of debugging is the location of code, that is responsible for the faulty behavior. Sometimes, this code is the statement, that is associated with the observed event, sometimes the faulty behavior has to be backtracked to the original lines of code, that are really responsible.

4.1 Source Code Connection

A first extension to solve the location problem comes from *source code connection*. When selecting and inspecting events in the event graph display, the user is automatically provided with the original line of code. This is achieved with a file browser, that highlights the statements responsible for the selected events. An example is visualized in figure 5.

4.2 Function Call-Graph and Control Flow Graph

A similar connection is established with the Control- and Data Flow Analyzer *CDFA*. In CDFA the target program is analyzed based on the underlying source and connected to the program flow observed with EMU.

When starting CDFA the source code analysis is performed. If the program uses the SPMD (single program multiple data) programming model, the analysis is similar to the sequential counterpart, because only one source code has to be analyzed. If it is MPMD (multiple program multiple data), the analyzes has to be performed for each of the contributing programs.

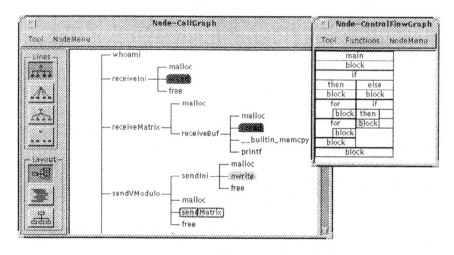

Fig. 6. Function Call Graph and Control Flow Graph

The results of these analysis steps are data structures that contains control and data flow information about the program. The difference between sequential code analysis and parallel code analysis is based on the communication functions. These communication functions are special building blocks of the algorithm, because the parallel analyses, especially the detection of global errors, is based on their occurrence in the source code.

The visual representation within CDFA is either a *function call-graph* or a *control flow graph*. A function call-graph represents all the functions of a program, whereas a control flow graph represents the control flow within a specified function. These two representations are static and therefore only based on the source code.

An example of this representation is visualized in figure 6. The left window shows a function call-graph, the right window shows the control flow graph for one of the functions. The blocks, that are marked with different colors, are blocks containing communication functions. Depending on the kind of communication function used in the selected code part, three different colors are used to represent the three different kinds of communication functions (send, receive, and test).

Additionally the display of the two graphs can be manipulated to hide information that is not needed. Therefore the user can select certain blocks and collapse segments of the tree in the function call-graph or basic blocks of the control flow graph. The part that represented the structure of the program under the selected element, is then removed from the display. The collapsed elements are visualized with special boxes (e.g. procedure SendMatrix in function call-graph of figure 6).

A connection similar to the source code connection can be obtained, if trace data is available. Again, the user inspects an event in the event graph display of ATEMPT, which opens the event inspection window. Additionally, the currently selected block of code is marked in the graphical representations that are visible in CDFA. This connection is very useful, if the user inspects a complicated program or a program, that was developed by someone else. The graphical representation allows the user better insight and understanding of the program.

5 Variable Backtracking

Another benefit of CDFA comes from the variable backtracking functionality, which is also based on a combination of underlying data structure and tracefiles. While the results of a bug - the erroneous behavior - is visible and noticeable to the user, the origin of the error is only detectable through following the program flow back in time. This can be very time-consuming and difficult, if the program flow is complex or the error must be tracked via communication channels.

The method used in CDFA is known from sequential debugging as "program slicing" [Weis 84]. The idea is to cut the program in small slices until only interesting blocks of code are remaining. The question is, why should the debugging user see all the source code, when the error is only in a small subset of statements.

The slicing in CDFA can be done in both ways, either textually based on the source code in the filebrowser, or graphically based on the visualizations in CDFA. The procedure is as follows:

Firstly the user selects a variable X to be tracked. This can be any variable in the source code, even if no data about this variable is included in the tracefiles. Then one can decide, whether to investigate the definition set or the use set of variable X.

The definition set of a variable is, where this variable was written to during the execution of the code. These are all places, where X was on the left side of an assignment. On the contrary the use set of variable are all the places, where X was accessed with a read. This means, that all statements or blocks are selected, where X is on the right side of an assignment.

Example:
Consider the following source code:

```
1:  for(i=0;i<=maxcol;i++) {
2:      if(ProcID == P0) {
3:          piv = i;
4:          for(row=i+1;row<maxrow;row++)
5:              if(mat[i][row] > mat[i][piv]) pivot = row;
6:          send(P1, pivot);
7:          }
8:      else
9:          receive(P0, pivot);
10:     for(col=current;col<maxcol;col++)
11:         mat[col][row] = mat[col][row]/pivot;
12:     }
```

In case the user selects variable row as variable to be investigated, the definition set of row is as follows:

```
4:              for(row=i+1;row<maxrow;row++)
```

The use set of row is

```
4:              for(row=i+1;row<maxrow;row++)
5:                  if(mat[i][row] > mat[i][piv]) pivot = row;
11:             mat[col][row] = mat[col][row]/pivot;
```

The same results are achieved with CDFA and the filebrowser, where only the statements contained in the selected sets are highlighted.

So far, all the slicing is equal to the sequential counterpart. The difference in parallel programs appears, when the user selects a variable, that depends on a message transferred from another process. In this case, two new statements have to be considered, namely the send and receive of the underlying message passing interface. The following connection is used:

```
receive (message X) <-- send (message X)
```

Similar to an assignment, the variable X is used on the sender side (only read access) and defined on the receiver side (write access to the message buffer). This means, that for parallel programs the definition and use sets of variables have to be tracked across communication channels. Furthermore, as messages need not contain the same variable names at both sides of the communication (sometimes not even the same data type), recursive tracking has to be possible. Therefore a variable X at the receiver part of a message might as well be a variable Y at the sender side.

Again, depending on the example code above, the user retrieves the following code sets, when selecting variable pivot as target for the variable backtracking (only definition set):

On process P0:

```
5:                    if(mat[i][row] > mat[i][piv]) pivot = row;
```

On process P1:

```
9:                    receive(P0, pivot);
```

This information can also be presented graphically, when initiating flowback analysis in CDFA. Then the control flow graph for the processes is generated and the history-path of the variable is highlighted, showing two or more control flow graphs of processes. By clicking on the building blocks of the control flow, the user is again connected to the original statement in the source code.

Figure 7 provides a screenshot of the CDFA module for variable backtracking. it shows the control flow for two corresponding (communicating) processes. The blocks including access to the inspected variable are shaded.

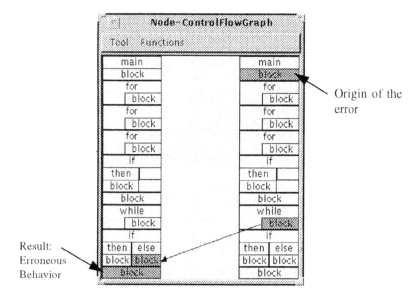

Fig. 7. Variable backtracking in parallel code with CDFA

6 Conclusion

This paper presents the debugging toolset MAD for detecting errors in parallel programs. Observation of an initial program execution is performed by the monitoring tool EMU, which generates tracefiles for each occurring event. Primary investigations can be performed with ATEMPT, which provides automatic features for detecting simple errors and communication bottlenecks. After detecting these errors the user can hunt race conditions with the event manipulation feature and the trace driven simulator PARASIT.

An extension to the toolset comes from CDFA. The idea of CDFA is to provide control- and data flow inspection mechanisms, which allow the user to backtrack suspicious variable assignments to their origin, the statements in the source code. Furthermore visual representations of function call-graph and control flow graph increase the user's understanding for the overall program structure. Thus the idea of incorporating a control and data flow analyzer in a debugging environment can present further insight in the observed program behavior.

At present we are working on integrating dynamic slicing mechanisms [Kamk 96] into the trace driven simulator, which work in combination with CDFA. Additionally of major interest is the usability of CDFA for verification of race condition candidates. With a combination of the race detection mechanism and the control and data flow information it shall be possible to reduce the set of possible racing messages to the more appropriate race condition candidates. First results in this area are already very promising.

Acknowledgment

A lot of work described in this paper was implemented by various students doing their diploma thesis at GUP Linz. These were P. Brandstetter, D. Müller-Wipperfürth, and R. Schall.

Remark

More information about MAD, ATEMPT, and CDFA as well as binaries are available via World Wide Web at the URL:

http://www.gup.uni-linz.ac.at:8001/research/debugging

References

[AgCh 96]   J. K. Aggarwal, P. Chillakanti, "Software for parallel computing - a perspective", in: A. Y. Zomaya (Ed.), "Parallel Computing: Paradigms and Applications", Intl. Thomson Computer Press, London, pp. 376-393 (1996).

[ChMi91]   J.-D. Choi, S.L. Min, "RACE FRONTIER: Reproducing Data Races in Parallel-Program Debugging", Proc. 3rd ACM SIGPLAN Symposium on Principles & Practice of Parallel Programming PPOPP, Williamsburg, Virginia, pp. 145 - 154 (April 1991).

[Fidg 88]   C.J. Fidge, "Partial orders for parallel debugging", Proc. of Workshop on Parallel and Distributed Debugging, ACM, pp. 183 - 194 (1988).

[Gait 86]   J. Gait, "A Probe Effect in Concurrent Programs", IEEE Software - Practise and Experience, Vol. 16(3), pp. 225-233 (March 1986).

[Grab 95a]   S. Grabner, D. Kranzlmüller, J. Volkert, "Debugging Parallel Programs using ATEMPT", Proc. HPCN Europe 95 Conference, Milano, Italy, pp. 235-240 (May 1995).

[Grab 95b]   S. Grabner, D. Kranzlmüller, J. Volkert, "Debugging of Concurrent Processes", Proc. 3rd Euromicro PDP Workshop, Sanremo, Italy, pp. 547-554 (Jan. 1995).

[GrKr 94]   S. Grabner, D. Kranzlmüller, "A Comparison of Monitoring Strategies for Distributed Memory Computers", Proc. of CONPAR 94-VAPP VI, Linz, Austria, pp. 66-75 (Sept. 1994).

[GrVo 93]   S. Grabner, J. Volkert, "Debugging Parallel Programs using Event Graph Manipulation", VECPAR 93 Conference, in: Computing Systems in Engineering, Elsevier, Vol. 6, Nos. 4/5, pp. 443-450 (1995).

[GrVo 96]   S. Grabner, J. Volkert, "Debugging Distributed Memory Programs Using Communication Graph Manipulation", Proc. of HPCS 96 Conference, Montreal, Canada (June 1996)

[Hond 95]   A. Hondroudakis, "Performance Analysis Tools for Parallel Programs", Version 1.0.1, Edinburgh Parallel Computing Centre, The University of Edinburgh, available at: http://www.epcc.ed.ac.uk/epcc-tec/documents.html (July 1995).

[HeEt 91]   M. T. Heath, J. A. Etheridge, "ParaGraph: A Tool for Visualizing Performance of Parallel Programs", Technical Report Oak Ridge National Laboratories (1993).

[Hood 96]   R. Hood, "The p2d2 Project: Building a Portable Distributed Debugger", Proc. SIGMETRICS Symposium on Parallel and Distributed Tools, SPDT 96, Philadelphia, PA, pp. 127-136 (May 1996).

[Kamk 96]   M. Kamkar, P. Krajina, P. Fritzson, "Dynamic Slicing of Parallel Message-Passing Programs", Proc. 4th Euromicro PDP Workshop, Braga, Portugal, pp. 170-177 (Jan. 1996).

[Kran 95]   D. Kranzlmüller, S. Grabner, J. Volkert, "Race Condition Detection with the MAD Environment", Proc. of PART'95, Second Australasian Conference on Parallel and Real-Time Systems, Fremantle, Western Australia (Sept. 1995).

[Kran 96]   D. Kranzlmüller, R. Koppler, Ch. Holzner, S. Grabner, J. Volkert, "Parallel Program Visualization with MUCH", ACPC'96, Klagenfurt, Austria, (Sept. 1996).

[Leu 90]   E. Leu, A. Schiper, A. Zramdini, "Execution Replay on Distributed Memory Architectures", Proc. of 2nd IEEE Symp. on Parallel & Distr. Processing, Dallas, TX, pp. 106-112 (Dec. 1990).

[MPI 95]   Message Passing Interface Forum, "MPI Standard Version 1.1", at ftp://ftp.mcs.anl.gov/pub/mpi/mpi-1.jun95/mpi-report.ps, (June 1995).

[Mill 95]   B.P. Miller, J.K. Hollingsworth, M.D. Callaghan, "The Paradyn Parallel Performance Measurement Tools and PVM", in: J. Dongarra, B. Tourancheau (Eds.), "Environments and Tools for Parallel Scientific Computing", SIAM Press (1994).

[MiCh 88]   B.P. Miller, J.-D. Choi, "A Mechanism for Efficient Debugging of Parallel Programs", Proc. of the SIGPLAN/SIGOPS Workshop on Parallel & Distr. Debugging, Madison, Wisconsin, pp. 141-150 (May 1988).

[NeMi 92]   R.H.B. Netzer, B.P.Miller, "What are Race Conditions? - Some Issues and Formalizations", ACM Letters on Progr. Languages and Systems, Vol. 1(1) (1992).

[Panc 96]   C. M. Pancake, "Visualization techniques for parallel debugging and performance-tuning tools", in: A. Y. Zomaya (Ed.), "Parallel Computing: Paradigms and Applications", Intl. Thomson Computer Press, London, pp. 376-393 (1996).

[PaNe 93]   C.M. Pancake, R.H.B. Netzer, "A Bibliography of Parallel Debuggers, 1993 Edition", Proc. of the ACM/ONR Workshop on Parallel and Distributed Debugging, San Diego, pp. 169-186 (May 1993).

[Wein 88]   A. Weinand, E. Gamma, R. Marty, "ET++ - An Object Oriented Application Framework in C++", OOPSLA'88 Conference Proceedings, SIGPLAN Notices, Vol. 23 (11), San Diego, pp. 168-182 (Sept. 1988).

[Weis 84]   M. Weiser, "Program Slicing", IEEE Transactions on Software Engineering, Vol. SE-10, No. 4, pp. 352-357 (July 1984).

ProHos-1 - A Vector Processor for the Efficient Estimation of Higher-Order Moments

J. C. Alves, A. Puga, L. Corte-Real and J. S. Matos

FEUP - Faculdade de Engenharia da Universidade do Porto/INESC, Praça da República 93, 4007 Porto CODEX, PORTUGAL

Abstract. Higher-order statistics (HOS) are a powerful analysis tool in digital signal processing. The most difficult task to use it effectively is the estimation of higher-order moments of sampled data taken from real systems. For applications that require real-time processing, the performance achieved by common microprocessors or digital signal processors is not good enough to carry out the large number of calculations needed for their estimation. This paper presents ProHos-1, an experimental vector processor for the estimation of the higher-order moments up to the fourth-order. The processor's architecture exploits the structure of the algorithm, to process in parallel four vectors of the input data in a pipelined fashion, executing the equivalent to 11 operations in each clock cycle. The design of dedicated control circuits led to high clock rate and small hardware complexity, thus suitable for implementation as an *ASIC* (Application Specific Integrated Circuit).

1 Introduction

Higher-order statistics (HOS—higher-order cumulants and higher-order spectra) are powerful tools for the analysis and classification of signals, particularly when their source exhibit non-linearities or non-gaussianities [1]. For example, in opposition to second-order statistics (autocorrelation and power spectrum), the higher-order counterparts are not phase blind. This characteristic enables the identification of non-minimal phase systems using the third and fourth-order cumulants.

The application of HOS in various engineering fields requires their efficient computation, specially when real-time processing is required. Examples are the application in pattern recognition [2] and real-time imaging systems [3]. Under certain characteristics of the signal to process (zero mean), the harder task to estimate higher-order cumulants is the evaluation of the higher-order moments of the sampled data. The 2nd, 3rd and 4th-order cumulants of a real one-dimensional stationary process are defined as:

$$C_2^y(\tau_1) = m_2(\tau_1) \tag{1}$$

This work is supported by Junta Nacional de Investigação Científica e Tecnológica (JNICT), contract no. PBIC/TIT/2489/95 — HAREOS

$$C_3^y(\tau_1, \tau_2) = m_3(\tau_1, \tau_2) \tag{2}$$

$$C_4^y(\tau_1, \tau_2, \tau_3) = m_4(\tau_1, \tau_2, \tau_3) - m_2(\tau_1) \cdot m_2(\tau_2 - \tau_3) -$$
$$m_2(\tau_2) \cdot m_2(\tau_3 - \tau_1) - m_2(\tau_3) \cdot m_2(\tau_1 - \tau_2) \tag{3}$$

where $m_2(\tau_1)$, $m_3(\tau_1, \tau_2)$ and $m_4(\tau_1, \tau_2, \tau_3)$ are the 2nd, 3rd and 4th order moments, defined as follows:

$$m_2(\tau_1) = E[x(k) \cdot x(k + \tau_1)] \tag{4}$$

$$m_3(\tau_1, \tau_2) = E[x(k) \cdot x(k + \tau_1) \cdot x(k + \tau_2)] \tag{5}$$

$$m_4(\tau_1, \tau_2, \tau_3) = E[x(k) \cdot x(k + \tau_1) \cdot x(k + \tau_2) \cdot x(k + \tau_3)] \tag{6}$$

In order to illustrate the need of HOS, let us consider the two following difference equations:

$$y_I(k) = x(k) - a \cdot x(k - 1) + b \cdot x(k - 2) \tag{7}$$
$$y_{II}(k) = x(k) - a \cdot x(k + 1) + b \cdot x(k + 2) \tag{8}$$

Equation (7) describes a minimum phase system, while equation (8) represents a non-minimum phase system. If we take classic methods of system identification, the two systems cannot be distinguished, because their 2nd-order cumulants are equal:

$$C_2^{y_I}(\tau_1) = C_2^{y_{II}}(\tau_1) = \begin{bmatrix} 1 + a^2 + b^2 & -a(1 + b) & b \end{bmatrix} \tag{9}$$

However, the 3rd-order cumulants for the two systems are different, allowing the distinction between them:

$$C_3^{y_I}(\tau_1, \tau_2) = \begin{bmatrix} 1 - a^3 + b^3 & -a + ba^2 & b \\ -a + ba^2 & a^2 - ab^2 & -ab \\ b & -ab & b^2 \end{bmatrix} \tag{10}$$

$$C_3^{y_{II}}(\tau_1, \tau_2) = \begin{bmatrix} 1 - a^3 + b^3 & a^2 - ab^2 & b^2 \\ a - ab^2 & -a + ba^2 & -ab \\ b^2 & -ab & b \end{bmatrix} \tag{11}$$

Previous work has been done on the development of dedicated hardware and software parallel architectures for the estimation of higher-order moments. An application specific array of processors was proposed in [4], derived from a systematic algorithm to architecture transformation. The order-recursive algorithm is transformed into a locally recursive algorithm, that is mapped into a planar triangular array of identical processing nodes. The resulting systolic architecture can compute in parallel the estimates of higher-order moments up to the 4th-order, scheduling the input data with a minimum latency and requiring communication links only between adjacent nodes. In spite of the performance attained, the hardware complexity would be very high for practical applications: to process a data block with M values, the triangular array needs $M \cdot (M+1)/2$ processing nodes, each one with four multipliers and four adders. This work was extended in [5], in order to adapt the locally recursive algorithm for implementation in the MasPar-1 parallel machine.

2 Computation of higher-order moments estimates

The estimation of higher-order moments requires a large number of arithmetic operations over long vectors of data. However, important saves in the computational effort can be achieved, taking advantage of HOS symmetries, order-recursion and adequate quantization of the data values to process.

2.1 Regions of interest

The statistics presented above exhibit certain symmetry properties [6] that reduce their non-redundant domain to:

$$\tau_3 \leq \tau_2 \leq \tau_1 \wedge \tau_1, \tau_2, \tau_3 \geq 0 \qquad (12)$$

All the other parts of their domains are reflections of that unique region. In addition, for practical applications the domain of the moment statistics is bound by a function of the size of the data vector processed, usually referred as $P2$, $P3$ and $P4$, respectively for the 2nd, 3rd and 4th order moments [6, 7, 1]:

$$P2 = \frac{L}{5}, P3 = \sqrt{\frac{L}{5}}, P4 = \sqrt[3]{\frac{L}{5}} \qquad (13)$$

Consequently, the computation of the moment components (commonly called *taps*) is reduced to the following regions of interest, respectively for the 2nd, 3rd and 4th-order moments:

$$0 \leq \tau_1 \leq P_2 \qquad (14)$$

$$0 \leq \tau_1 \leq P_3 \wedge 0 \leq \tau_2 \leq \tau_1 \qquad (15)$$

$$0 \leq \tau_1 \leq P_4 \wedge 0 \leq \tau_2 \leq \tau_1 \wedge 0 \leq \tau_3 \leq \tau_2 \qquad (16)$$

2.2 Estimation procedure

The higher-order moments of a one-dimensional vector of data can be estimated by the following algorithm:

- Segment the input data vector into M segments of L samples each.
- Normalize each segment: subtract the mean of the segment to each sample and divide it by the standard deviation of the whole data vector.

– Compute the moment estimates of each segment m, in the regions defined by (14), (15) and (16):

$$m_2^{x_m}(\tau_1) = \frac{1}{L} \sum_{l=0}^{L-1-\tau_1} x_m(l) \cdot x_m(l+\tau_1) \tag{17}$$

$$m_3^{x_m}(\tau_1, \tau_2) = \frac{1}{L} \sum_{l=0}^{L-1-\tau_1} x_m(l) \cdot x_m(l+\tau_1) \cdot x_m(l+\tau_2) \tag{18}$$

$$m_4^{x_m}(\tau_1, \tau_2, \tau_3) = \frac{1}{L} \sum_{l=0}^{L-1-\tau_1} x_m(l) \cdot x_m(l+\tau_1) \cdot x_m(l+\tau_2) \cdot x_m(l+\tau_3) \tag{19}$$

– Compute each moment tap as the mean of the corresponding taps of each segment:

$$m_2(\tau_1) = \frac{1}{M} \sum_{m=0}^{M-1} m_2^{x_m}(\tau_1) \tag{20}$$

$$m_3(\tau_1, \tau_2) = \frac{1}{M} \sum_{m=0}^{M-1} m_3^{x_m}(\tau_1, \tau_2) \tag{21}$$

$$m_4(\tau_1, \tau_2, \tau_3) = \frac{1}{M} \sum_{m=0}^{M-1} m_4^{x_m}(\tau_1, \tau_2, \tau_3) \tag{22}$$

2.3 Software implementation

The operations involved in the estimation of higher-order moments are basically the evaluation of the inner product of two, three and four data vectors with variable shifts among them. As can be seen above, the term in summation (17) is used in summation (18) and the term of (18) is used again in (19). Because of this, when the 2nd, 3rd and 4th-order moments are computed over the same data vector, order-recursion gains can be obtained by saving terms used in higher-order taps estimation. Figure 1 shows an efficient order-recursive program [4, 6] that implement the estimation algorithm. Vectors m2(), m3() and m4() are assumed initialized with 0, and the final division by the length of the data segment is not included in the algorithm.

A significant reduction of the number of multiplications is achieved by this implementation. However, results obtained by a C program running in a PC Pentium have shown non significant gain in the execution time, when compared to the code that repeats the multiplications.

```
for t1=0 to P2
  for i=0 to L-t1-1
  begin
    p2 = x[i]*x[i+t1]; m2[t1]=m2[t1]+p2;
    if t1<=P3
    begin
      for t2=0 to t1
      begin
        p3=p2*x[i+t2]; m3[t1,t2]=m3[t1,t2]+p3;
        if t1<=P4
        begin
          for t3=0 to t1
          begin
            m4[t1,t2,t3]=m4[t1,t2,t3]+p3*x[i+t3];
          end
        end
      end
    end
  end
end
```

Fig. 1. Order-recursive algorithm for the estimation of higher-order moments

2.4   Quantization error

Besides the estimation procedure itself, another main issue regarding the efficient implementation is the numerical resolution along the computational process. In fact, the choice of the width of the input data word is a very important decision, because it determines the numerical precision of the results and greatly influences the hardware complexity of the estimator implementation. In order to evaluate the influence of the quantization error of the input data on the numerical precision of the results, a set of simulations have been done, using the systems described by equations (7) and (8). Exponential distributed white noise (with mean equal to 1) was applied to each system and the 2nd and 3rd-order output cumulants were estimated. The output signal was quantized using 2 to 64 bits, their cumulants were estimated and compared with the estimates obtained without quantization error. Figure 2 presents the results obtained for data words ranging from 8 to 16 bits. It can be verified that, to achieve 99.9% precision in the 3th-order cumulant, it will be necessary to use at least 9 bit words to represent the data values, and 8 bits are enough for the 2nd-order cumulant. At the right end of the chart, it can be seen that the effects of quantization on the final results are irrelevant for both systems, when more than 16 bits are used. All the values presented in fig. 2 are the worst case of 10 independent simulations. The following measure of precision was used:

$$Precision = 100 \left(1 - \frac{\|\hat{C}^q - \hat{C}\|}{\|\hat{C}\|} \right) \%$$

where $\hat{C}^q$ is the estimate of the cumulant with quantization error and $\hat{C}$ is the estimate of the cumulant without quantization error.

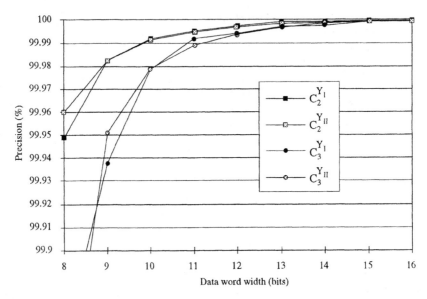

Fig. 2. Variation of the precision of the cumulant estimation with the data word width

Further work will extend this study to a wider range of systems, different number representations (either integer, fixed point or floating point) and analyze the over- and underflow events that can occur during the calculations. This will enable us to establish accurate needs for the estimator hardware, specially in what concerns to the width of the calculator datapath and format for the data representation.

3   ProHos-1 architecture

ProHos-1 is an experimental vector architecture for the efficient estimation of 2nd, 3rd and 4th order moments (fig. 3). Instead of the works referred above, that have been directed towards the implementation of the higher-moments estimator on large parallel machines, our architecture aims to achieve a good performance/cost tradeoff, to be in the included in low cost imaging systems for PC computers.

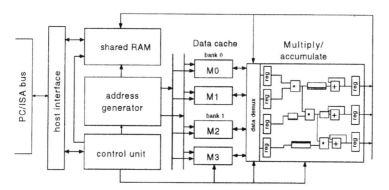

Fig. 3. ProHos-1 architecture

The processor is composed of six main parts: host interface, shared memory, address generator, data cache memory, multiply-accumulate unit and control unit, as depicted in fig. 3.

The processor connects to the PC/ISA bus, and receives the data vector to process in a shared memory, placed into a slot of the PC's memory address space. The PC start ProHos-1 by writing into an I/O port, waits for the completion of the execution, and reads the computed moment estimates placed by ProHos-1 in the shared memory.

The architecture is built around a pipelined multiply-accumulate unit (MAC) composed of three multipliers and three accumulators. By now, this unit works on 8-bit integer data values, producing 32-bit results with no overflow control. The work done so far has concentrated on the development of the memory architecture and control unit, in order to obtain a high input data throughput into the MAC unit. This is constrained principally by the limited bandwidth of the access to the RAM memory where the data vector to process is stored. To overcome this, a set of 4 fast static RAM were used as data cache memories. These memories hold one data segment and supply four values at a time to the inputs of the MAC unit, addressed by a dedicated control circuit based on binary counters.

3.1 Host interface

The host interface implements the logic necessary to interface ProHos-1 with the PC-ISA bus. This includes the address decoding, control access for the shared RAM and one 16-bit bidirectional I/O port. This port is used to control the shared RAM access and the state of ProHos-1. As a read port, it allows the host to inquire the present state of the processor (either running, stopped or accessing the shared RAM). This can be used to allow the host to read values already computed and stored in the shared RAM, while ProHos-1 is still running.

3.2 Multiply/accumulate unit

The multiply/accumulate unit (MAC) is a pipelined datapath composed by three cascaded pipelined multipliers, each one followed by one accumulator, as shown in fig. 4. This way, the summation of $x(i) \cdot x(i+\tau_1)$, $x(i) \cdot x(i+\tau_1) \cdot x(i+\tau_2)$ and $x(i) \cdot x(i+\tau_1) \cdot x(i+\tau_2) \cdot x(i+\tau_3)$ is done in parallel, producing the coefficients of the 2nd, 3rd and 4th order moments simultaneously. The dummy pipeline stages create additional delays in the data flow through the datapath, to provide each input-output path with the same number of pipe stages. This simplifies the control procedure because the results at the end of each accumulator are delayed for the same number of clock cycles with respect to the inputs. Once the unit is filled, it can provide the results of 3 multiplications and three additions in each clock cycle. Because the pipeline needs to be filled and emptied only once during the calculation of the whole data vector, the number of clock cycles needed for this is not significant and is diluted over the total processing time.

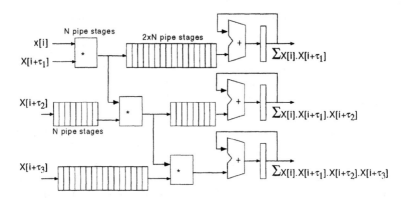

Fig. 4. Multiply-accumulate unit

3.3 Data cache memories

The data cache is made of 4 fast static RAMs organized in two banks. One bank is used to supply the four data values to the MAC unit, while the other is loaded with the next segment to process. Each memory provides two data values to the MAC unit in each clock cycle. An address multiplexer and data demultiplexer, controlled by a finite state machine running at twice the frequency of the system clock, sequences the two addresses for each memory, and loads appropriate registers at the four inputs of the MAC module. When the processing of the current segment ends, the function of the cache banks is switched and the processing of a new segment can start immediately with no need of additional clock cycles to empty and fill again the pipelined MAC unit. Figure 5 shown the address/data flow when bank 0 is used to supply data values to the MAC unit, and bank 1 is loaded with the next segment to process.

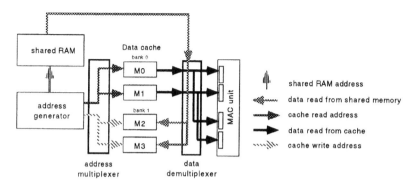

Fig. 5. Example of cache memories operation: while bank 0 is being used to provide data values to the MAC unit, bank 1 is filled with the next data segment from the shared memory

3.4 Address generation

To process a data segment, their values must be sent at a given order to the inputs of the MAC unit. To compute the 4th order moment $m_4(\tau_1, \tau_2, \tau_3)$, four data samples $x(i)$ must be sent to the MAC unit, starting at indexes 0, τ_1, τ_2 and τ_3, and running until the last value of the data segment is reached. The address generation unit produces the addresses for the cache memories, in the right order to provide to the MAC unit the appropriate sequence of data samples. This unit is composed of two sections: the domain generator and the address sequencer.

The domain generator produces the values τ_1, τ_2 and τ_3 for the unique 2nd, 3rd and 4th order moment coefficients. As shown in (13), the number of taps if $P2$ for $m_2(\tau_1)$, $P3$ for $m_3(\tau_1, \tau_2)$ and $P4$ for $m_4(\tau_1, \tau_2, \tau_3)$. However, the present implementation assumes the same value P for the three functions. In this case, the sequence of coefficients is generated by the following three nested loops:

```
for τ₁=0 to P
  for τ₂=0 to τ₁
    for τ₃=0 to τ₂
      { ... }
```

The domain generator has been obtained as a direct transformation of the loops above into dedicated hardware. The logic circuit has three binary counters, three equality comparators and a few glue logic, as shown in the left part of figure 6. The three values τ_1, τ_2 and τ_3 produced by this circuit are feed into the right section of the address generator, at the beginning of the processing of a new segment.

The sequence of addresses needed to access four data values from the cache memories is created by the right part of the circuit of figure 6. Because the samples are accessed in sequence, the address generation is made automatically by one counter and three adders. The counter generates the base address, counting from $L - 1 - \tau_1$ to zero. Each adder adds τ_1, τ_2 and τ_3 to the value produced by this counter, creating the 4 addresses that refer to the delayed data samples. These addresses are feed into the two cache memories, via the address multiplexer. When the last address is reached, a new set of τ_1, τ_2, τ_3, meanwhile computed by domain generator, is loaded and the next sequence of addresses is generated.

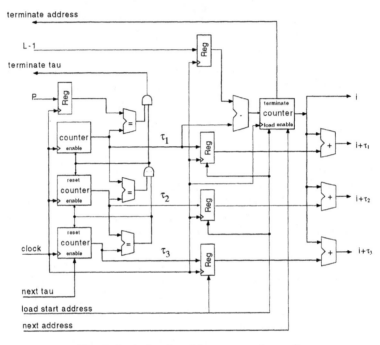

Fig. 6. Logic for the address generation unit

3.5 Control unit

The control unit is responsible for the overall operation of the processor, providing the appropriate control signals for the other modules. The controller is made of three hierarchical hardwired finite state machines: the main controller, address sequencer and address/data multiplexer. The address/data multiplexer multiplexes the cache addresses and loads the input registers of the MAC unit, synchronously with the address generation unit. This finite state machine runs at twice the frequency of the system clock, and executes two read operations from the cache in each system clock cycle. The address sequencer controls the address generation unit, synchronized with the address/data multiplexer. Finally, the main controller listens commands sent from the host, controls the other finite state machines and arbitrates the access to the shared RAM.

4 Results and conclusions

The architecture proposed in this paper was simulated using the Verilog hardware description language [8]. A mixed behavioral/structural hardware model was built, that has enable us to verify the correctness of the architecture.

Once the pipelined MAC unit is filled, the processor executes, in each clock cycle, 3 multiplications and 3 additions in the MAC unit, plus 4 additions and one comparison in the address sequencer. The global throughput is thus 11 operations per clock cycle, not counting with the additional operations performed to generate the domain of the moment functions and the address generation to load the data segments into the cache. Considering the cache memories as static RAMs with a 25ns read cycle, the processor can run with a 50ns (20MHz) system clock cycle, reaching a 220 MOPs performance.

For a data vector with 4000 8-bit data samples, ProHos-1 can compute the unique moment coefficients up to the 4th order for $P2 = P3 = P4 = 10$, in approximately 44ms. This produces a total of 285 different coefficients (220 for $m_4()$, 56 for $m_3()$ and 10 for $m_2()$), performing the equivalent to 8.6 million operations. Running a C implementation of the program of fig. 1, compiled with Microsoft-C 6.0 and optimized for speed, the same problem is solved in a PC Pentium@120 MHz in about 660ms, 15 times more than the estimated execution time for ProHos-1. This comparison does not take into account the time needed to fill the shared RAM and to read the 285 32-bit results from ProHos-1 through the PC-ISA bus. Using 16-bit I/O data transfers, the processing time is increased by approximately 2.3ms, representing 5.2% increase in the estimated processing time and reducing the overall speedup from 15× to 14.2×.

Presently a prototype processor is under development, based on Field-Programmable Gate Arrays [9] and off-the-shelf components to implement the parts that will not fit in the FPGAs. The realization of the control and address generation units in programmable logic will enable us to experiment with different sequencing strategies, with no need to redesign the physical hardware.

The use of FPGAs with fast reconfiguration will enable us to explore the reusability of the programmable hardware to implement other estimation procedures. An example is the estimation of higher-order spectra using the estimated moments.

The design of the whole system as an application specific integrated circuit (ASIC) is also planned for a near future.

References

1. C. L. Nikias and J. M. Mendel. Signal processing with higher-order spectra. *IEEE Signal Processing Magazine*, 10:10–37, 1993.
2. Michail K. Tsatsanis and Georgios B. Giannakis. Object and texture classification using higher-order statistics. *IEEE Transactions on Pattern Analysis and Machine Intelligence*, 14(7):733–750, jul 1992.

3. John M. M. Anderson and Georgios B. Giannakis. Image motion estimation algorithm using cumulants. *IEEE Transactions on Image Processing*, 4(3):346–357, mar 1995.

4. Haris M. Stellakis and Elias S. Manolakos. An array of processors for the real-time estimation of fourth- and lower-order moments. *Signal Processing*, (36):341–354, 1994.

5. John N. Kalamatianos and Elias S. Manolakos. Parallel computation of higher-order moments on the maspar-1 machine. In *Proceedings of the International Conference on Acoustics Speech and Signal Processing*, may 1995.

6. A. T. Puga and A. P. Alves. Higher-order statistics symmetry properties. In *Proceedings of 7th Portuguese Conference on Pattern Recognition- RECPAD'95*, 1995.

7. S. M. Kay. *Modern Spectral Estimation*. Prentice Hall, Englewood Cliffs, NJ, 1988.

8. D. Tomas and P. Moorby. *The verilog Hardware Description Language*. Kluwer Academic Publishers, 1991.

9. John V. Oldfield and Richard C. Dorf. *Field-Programmable gate Arrays: Reconfigurable logic for rapid prototyping and implementation of digital systems*. John Wiley & Sons, 1995.

The Use of Computational Kernels in Full and Sparse Linear Solvers, Efficient Code Design on High-Performance RISC Processors[4]

Michel J. Daydé[1] and Iain S. Duff[2,3]

[1] ENSEEIHT-IRIT, 2 rue Camichel, 31071 Toulouse Cedex, France
[2] Rutherford Appleton Laboratory, Oxfordshire, OX11 0QX, England
[3] CERFACS, 42 av. G. Coriolis, 31057 Toulouse Cedex, France

Abstract. We believe that the availability of portable and efficient serial and parallel numerical libraries that can be used as building blocks is extremely important for both simplifying application software development and improving reliability.

This is illustrated by considering the solution of full and sparse linear systems. We describe successive layers of computational kernels such as the BLAS, the sparse BLAS, blocked algorithms for factorizing full systems, direct and iterative methods for sparse linear systems.

We also show how the architecture of the today's powerful RISC processors may influence efficient code design.

1 Introduction

One of the common problems for application scientists is to exploit as efficiently as possible the hardware of high-performance computers (either serial or parallel) without totally rewriting or redesigning existing codes and algorithms. We believe that the availability of portable and efficient serial and parallel numerical libraries that can be used as building blocks is extremely important for both simplifying application software development and improving reliability.

The availability of powerful RISC processors is of major importance in today's market since they are used both in workstations and in the most recent parallel computers. They are usually more efficient than vector processors on scalar applications. The main reason for their success in the marketplace is their very good cost to performance ratio. They are used as a CPU both in workstations and in most of the current MPPs (DEC Alpha in the CRAY T3D, SPARC in the CM5 and PCI CS2, HP PA in the CONVEX EXEMPLAR, and POWER processors in the IBM SP1 and SP2). We report in Table 1 the uniprocessor performance of some current RISC processors on the double precision 100-by-100 and 1000-by-1000 LINPACK benchmarks (Dongarra, 1992). We also record their peak performance.

[4] Part of this work was funded by Conseil Régional Midi-Pyrénées under project DAE1/RECH/9308020 and by the Alliance Program from the British Council.

Computer	LINPACK 100*100	LINPACK 1000*1000	Peak performance
DEC 8400 5/300	140	411	600
IBM POWER2-990	140	254	286
HP 9000/755	41	107	200
SGI POWER Challenge	104	261	300

Table 1. Performance in MFlops of RISC processors on the double precision LINPACK benchmarks

We briefly consider the impact of the memory hierarchy on the performance of RISC architectures in Section 2. In Section 3, we show how portable and efficient serial and parallel versions of the Level 3 BLAS can be designed for RISC-based computers using code tuning techniques such as blocking, copying, and loop unrolling. In Section 4, we give an example of a blocked factorization algorithm of special interest for optimization algorithms. This block version of the modified Cholesky factorization of Eskow and Schnabel is designed to make intensive use of the Level 3 BLAS in the same way as other block algorithms. In Section 5, we indicate how BLAS for full matrices can be used within codes for the direct solution of sparse linear equations. In Section 6, we show how an efficient preconditioned conjugate gradient algorithm for symmetric, partially separable, unassembled linear systems can be designed to make use of the previously described computational kernels. We use Element-by-Element preconditioners to exploit the structure of the problems. We demonstrate how a numerical preprocessing step can dramatically improve both the numerical behaviour and the performance of the preconditioners. We describe, in Section 7, an extension of the BLAS for handling sparse matrix operations and indicate its use in the iterative solution of sparse equations. We present some concluding remarks in Section 8.

2 Impact of the Memory Hierarchy of RISC Processors on Performance

The ability of the memory to supply data to the processors at a sufficient rate is crucial on most modern computers. This necessitates a complex memory organization, where the memory is usually arranged in a hierarchical manner. The minimization of data transfers between the levels of the memory hierarchy is a key issue for performance (Gallivan, Jalby and Meier, 1987, Gallivan, Jalby, Meier and Sameh, 1988).

Most of the RISC-based architectures use a more complex memory hierarchy than is usually the case for vector processing units. This normally involves one or several levels of cache. Calculations are pipelined over independent scalar operations instead of vector operations. This is why the design of codes for

RISC processors may substantially differ from that of code for vector processors. A high reuse of the data located at the highest levels of the memory hierarchy is required for efficient codes for RISC-based architectures. The use of tuned computational kernels that can be used as building blocks is crucial for both simplifying application software development and achieving high performance.

The cache memory is used to mask the memory latency (typically the cache latency is around 1-2 clocks while it is often 10 times higher for the memory). The code performance is high so long as the cache hit ratio is close to 100%. This will happen if the data involved in the calculations can fit in the cache or if the calculations can be organized so that data can be kept in cache and efficiently reused. One of the most commonly used techniques for that purpose is called blocking and an example of this is reported in the following section. Blocking enhances spatial and temporal locality in computations. Unfortunately, blocking is not always sufficient since the cache miss ratio can be dramatically increased in quite an unpredictable way if the memory accesses use a stride greater than 1 (see Bodin and Seznec, 1994).

Some strides are often called *critical* because they generate a very high cache miss ratio (for example, when referencing cache lines that are mapped into the same physical location of the cache). These critical strides obviously depend on the cache management strategy. For example, if the cache line length is equal to four words and the cache is initially empty, the execution of the loop

```
do i=1,n,4
    temp = temp + a(i)
enddo
```

will cause a cache miss on each read of a(i), assuming that a(i) is one word.

Copying blocks of data (for example submatrices) that are heavily reused may help to improve memory and cache accesses (by avoiding critical strides for example). However, since such copying may induce a large overhead, it is not always a viable technique. We illustrate the use of copying in our blocked implementation of the BLAS in Section 3. We note that blocking and copying are also very useful in limiting the cost of memory paging.

3   The BLAS Computational Kernels

As we have previously discussed, it is very important to use standard building blocks in application codes. They are extremely useful for simplifying the design of codes while guaranteeing portability and efficiency. The building blocks for much of our work, both in the solution of sparse as well as full systems, and in more complicated areas of scientific computation, are the Basic Linear Algebra Subprograms known as the BLAS. For reasons of efficiency, we are interested in the higher level BLAS, in particular the Level 3 BLAS (Dongarra, Du Croz, Duff and Hammarling, 1990) that include kernels like the matrix-matrix multiply routine _GEMM. Indeed, in Daydé, Duff and Petitet (1994a) and Kågström, Ling and Loan (1993), it is shown how all the Level 3 BLAS routines can be designed for high performance using the _GEMM kernel. We consider the performance and the implementation of _GEMM here and show, in Section 3.2, how it can be used to design the other Level 3 BLAS kernels on one example: the symmetric rank-k update _SYRK. The effect of using BLAS in the solution of linear equations, first when the coefficient matrix is full, then when it is sparse will be discussed respectively in Sections 4 and 5.

A tuned manufacturer-supplied version of the BLAS is today available on most high-performance computers and, in cases when it is not provided, a standard Fortran implementation is available on the **netlib** electronic server (Dongarra and Grosse, 1987).

3.1 Design of a Fast _GEMM for RISC Processors

We have developed a set of Level 3 BLAS computational kernels in single and double precision for efficient implementation on RISC processors (Daydé and Duff, 1996). This version of the Level 3 BLAS is an evolution of the one described by Daydé et al. (1994a) for MIMD vector multiprocessors. They report on experiments on a range of computers (ALLIANT, CONVEX, IBM and CRAY) and demonstrate the efficiency of their approach whenever a tuned version of the matrix-matrix multiplication routine _GEMM is available.

Our basic idea for efficient implementation of the BLAS on RISC processors is to express all the Level 3 BLAS kernels in terms of subkernels on submatrices that involve either _GEMM operations or operations involving triangular submatrices. Additionally, all the calculations on blocks are performed using tuned Fortran codes with loop-unrolling. Copying is occasionally used. Of course, the relative efficiency of this approach depends on the availability of a highly tuned _GEMM kernel.

This approach is relatively independent of the computer: only the block size parameter, here called NB, and in some cases the loop-unrolling depth should be tuned according to the characteristics of the target machine. NB is determined by the size of the cache and the loop-unrolling depth from the number of scalar registers.

We describe only the blocked implementation of _GEMM in this subsection. The other kernels are designed in the same way and we consider, as an example, _SYRK in Section 3.2. Further details can be found in Daydé and Duff (1996).

_GEMM performs one of the matrix-matrix operations

$$\mathbf{C} = \alpha \ op(\mathbf{A}) \ op(\mathbf{B}) + \beta \mathbf{C},$$

where α and β are scalars, $\mathbf{A}$ and $\mathbf{B}$ are rectangular matrices of dimensions m×k and k×n, respectively, $\mathbf{C}$ is a m × n matrix, and $op(\mathbf{A})$ is $\mathbf{A}$ or $\mathbf{A}^t$.

We consider the case corresponding to op equal to "No transpose" in both cases and block the computation as:

$$\begin{pmatrix} C_{1,1} \ C_{1,2} \\ C_{2,2} \ C_{2,2} \end{pmatrix} = \alpha \begin{pmatrix} A_{1,1} \ A_{1,2} \\ A_{2,1} \ A_{2,2} \end{pmatrix} \begin{pmatrix} B_{1,1} \ B_{1,2} \\ B_{2,1} \ B_{2,2} \end{pmatrix} + \beta \begin{pmatrix} C_{1,1} \ C_{1,2} \\ C_{2,1} \ C_{2,2} \end{pmatrix}$$

_GEMM can then obviously be organized in terms of a succession of matrix-matrix multiplications on submatrices as follows:

1. $C_{1,1} \leftarrow \beta C_{1,1} + \alpha A_{1,1} B_{1,1}$ (_GEMM)
2. $C_{1,1} \leftarrow C_{1,1} + \alpha A_{1,2} B_{2,1}$ (_GEMM)
3. $C_{1,2} \leftarrow \beta C_{1,2} + \alpha A_{1,1} B_{1,2}$ (_GEMM)
4. $C_{1,2} \leftarrow C_{1,2} + \alpha A_{1,2} B_{2,2}$ (_GEMM)
5. $C_{2,1} \leftarrow \beta C_{2,1} + \alpha A_{2,1} B_{1,1}$ (_GEMM)
6. $C_{2,1} \leftarrow C_{2,1} + \alpha A_{2,2} B_{2,1}$ (_GEMM)
7. $C_{2,2} \leftarrow \beta C_{2,2} + \alpha A_{2,1} B_{1,2}$ (_GEMM)
8. $C_{2,2} \leftarrow C_{2,2} + \alpha A_{2,2} B_{2,2}$ (_GEMM)

The ordering of these eight computational steps is determined by consideration of the efficient reuse of data held in cache. We have decided to reuse the submatrices of **A** as much as possible and we perform all operations involving a submatrix of **A** before moving to another one (see Figure 1). For our simple example, it means that we perform the calculations in the order: Step 1, Step 3, Step 5, Step 7, Step 2, Step 4, Step 6, and Step 8. This approach is similar to that used by Dongarra, Mayes and Radicati di Brozolo (1991b). In practice, NB is usually chosen so that all the submatrices of **A**, **B**, and **C** required for each submultiplication fit in the largest on-chip cache. On some machines, access to off-chip caches has so low latency that we can improve performance by using a larger block size. This is true, for example, on the SGI Power Challenge. Since all the computational kernels call _GEMM, the block size NB is always determined as the most appropriate block size for _GEMM, that is, NB is the largest even integer such that

$$3(NB)^2 prec < CS,$$

where *prec* is the number of bytes corresponding the precision used (4 bytes for single precision and 8 bytes for double precision in IEEE format) and CS is the cache size in bytes. We choose an even integer to facilitate loop-unrolling. For example with a 64Kbytes cache, NB is set to 52 using 64-bit arithmetic.

Part of the double precision blocked code is shown in Figure 1. Its main features are the following:

− The multiplication of **C** by β is performed before all other calculations.
− The submatrix of **A** is multiplied by α and transposed into array AA to avoid non-unit strides because of access by rows in the innermost loops of the calculations. These are organized in such a way that AA is kept in cache as long as required.

We use two tuned Fortran codes to perform calculations on submatrices (see Figure 2): DGEMML2X2 is a tuned code for performing matrix-matrix multiplication on square matrices of even order; and DGEMML is a tuned code that includes additional tests over DGEMML2X2 to handle matrices with odd order. It is occasionally slightly less efficient than DGEMML2X2.

```
*
*      Form C := beta*C
*
       IF( BETA.EQ.ZERO )THEN
           DO 20  J = 1, N
              DO 10  I = 1, M
                 C( I, J ) = ZERO
    10        CONTINUE
    20     CONTINUE
       ELSE
           DO 40  J = 1, N
              DO 30  I = 1, M
                 C( I, J ) = BETA*C( I, J )
    30        CONTINUE
    40     CONTINUE
       END IF
*
*      Form  C := alpha*A*B + beta*C.
*
       DO 70  L = 1, K, NB
          LB = MIN( K - L + 1, NB )
          DO 60  I = 1, M, NB
             IB = MIN( M - I + 1, NB )
             DO II = I, I + IB - 1
                DO LL = L, L + LB - 1
                   AA(LL-L+1,II-I+1)=ALPHA*A(II,LL)
                ENDDO
             ENDDO
             DO 50  J = 1, N, NB
                JB = MIN( N - J + 1, NB )
*
*  Perform multiplication on submatrices
*
                IF ((MOD(IB,2).EQ.0).AND.(MOD(JB,2).EQ.0)) THEN
                    CALL DGEMML2X2(IB,JB,LB,AA,NB,B(L,J),LDB,C(I,J),LDC)
                ELSE
                    CALL DGEMML(IB,JB,LB,AA,NB,B(L,J),LDB,C(I,J),LDC)
                END IF
    50        CONTINUE
    60     CONTINUE
    70  CONTINUE
```

Fig. 1. Part of the blocked code for DGEMM

We have used two versions for all the tuned codes: the TRIADIC option for computers where triadic operations are either supported in the hardware (for example the floating-point multiply-and-add on IBM RS/6000) or are efficiently compiled, and the NOTRIADIC option for other computers. The use of triadic operations should not normally degrade the performance severely on processors that do not support these operations since efficient code generation can transform them into dyadic operations. However, in early versions of SPARC compilers, we saw that there was sometimes such a degradation. Thus we prefer to offer both options. Part of the tuned code for DGEMML2X2 using the TRIADIC options is shown in Figure 2.

```
*
*              C := alpha*A*B + C.
*
               DO 70  J = 1, N, 2
                  DO 60  I = 1, M, 2
                     T11 = C(I,J)
                     T21 = C(I+1,J)
                     T12 = C(I,J+1)
                     T22 = C(I+1,J+1)
                     DO 50  L = 1, K
                        B1 = B(L,J)
                        B2 = B(L,J+1)
                        A1 = A(L,I)
                        A2 = A(L,I+1)
                        T11 = T11 + B1*A1
                        T21 = T21 + B1*A2
                        T12 = T12 + B2*A1
                        T22 = T22 + B2*A2
50                   CONTINUE
                     C(I,J) = T11
                     C(I+1,J) = T21
                     C(I,J+1) = T12
                     C(I+1,J+1) = T22
60                CONTINUE
70             CONTINUE
```

Fig. 2. Part of the tuned code for DGEMML2X2 (TRIADIC option)

In Figures 3 and 4, we illustrate respectively the double and single precision average performance of the standard and the blocked versions of _GEMM (averaged over square matrices of order 32, 64, 96, and 128 when **A** and **B** are not transposed). We also include the peak performance of the computer and the performance of the manufacturer-supplied version when available to us. This tuned subset of the BLAS – called the RISC BLAS – is publically available via anonymous ftp at ftp.enseeiht.fr in pub/numerique/BLAS/RISC.

This blocked implementation of _GEMM gives a gain in performance of

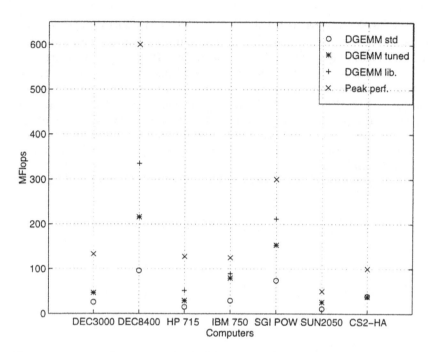

Fig. 3. Average performance of DGEMM from RISC BLAS ("No transpose", "No transpose").

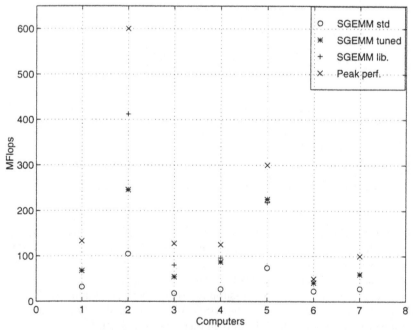

Fig. 4. Average performance of SGEMM from RISC BLAS ("No transpose", "No transpose").

greater than a factor of two compared with the standard Fortran coded version. Furthermore, we have observed that the performance is even better if the matrices are already held in the cache (which was not the case in these experiments).

3.2   Use of _GEMM in Designing the Kernel _SYRK

We now consider the use of _GEMM in designing other kernels as in Daydé et al. (1994a). We use the kernel _SYRK to illustrate this.

_SYRK performs one of the symmetric rank-k operations:

$$C = \alpha A A^t + \beta C, \text{ or } C = \alpha A^t A + \beta C$$

where α and β are scalars, C is an n × n symmetric matrix (only the upper or lower triangular parts are updated), and A is a n × k matrix in the first case and a k × n matrix in the second case.

We consider the case corresponding to $C = \alpha A A^t + \beta C$ where only upper triangular part of C is updated (that is, "Upper", and "No transpose"), and we block the computation as

$$\begin{pmatrix} C_{1,1} & C_{1,2} \\ 0 & C_{2,2} \end{pmatrix} = \alpha \begin{pmatrix} A_{1,1} & A_{1,2} \\ A_{2,1} & A_{2,2} \end{pmatrix} \begin{pmatrix} A_{1,1}^t & A_{2,1}^t \\ A_{1,2}^t & A_{2,2}^t \end{pmatrix} + \beta \begin{pmatrix} C_{1,1} & C_{1,2} \\ 0 & C_{2,2} \end{pmatrix}.$$

1. $C_{1,1} \leftarrow \beta C_{1,1} + \alpha A_{1,1} A_{1,1}^t$ (_SYRK)
2. $C_{1,1} \leftarrow C_{1,1} + \alpha A_{1,2} A_{1,2}^t$ (_SYRK)
3. $C_{1,2} \leftarrow \beta C_{1,2} + \alpha A_{1,1} A_{2,1}^t$ (_GEMM)
4. $C_{1,2} \leftarrow C_{1,2} + \alpha A_{1,2} A_{2,2}^t$ (_GEMM)
5. $C_{2,2} \leftarrow \beta C_{2,2} + \alpha A_{2,1} A_{2,1}^t$ (_SYRK)
6. $C_{2,2} \leftarrow C_{2,2} + \alpha A_{2,2} A_{2,2}^t$ (_SYRK)

The symmetric rank-k update is expressed as a sequence of _SYRK for updating the submatrices $C_{i,i}$ and _GEMM for the other blocks. The updates of the submatrices of C can be performed independently. The _GEMM updates of off-diagonal blocks can be combined. We note that, at the price of extra operations, we could perform the update of the diagonal blocks of C using _GEMM instead of _SYRK.

Part of the corresponding blocked code is shown in Figure 5. We note that it is more efficient to perform the multiplication of matrix C by β before calling _GEMM rather than performing this multiplication within _GEMM. The codes for DSYRKL2X2 and DSYRKL are designed using loop unrolling and copying.

In Figures 6 and 7, we illustrate respectively the double and single precision average performance of the standard and the blocked versions of _SYRK (averaged over square matrices of order 32, 64, 96, and 128 for the case "Upper" and "No transpose"). We also include the peak performance of the computer and the performance of the manufacturer-supplied version when available to us.

```
          DO 130, I = 1, N,NB
             NB_LIG_C=MIN(NB,N-I+1)
*
*      Multiplication of diagonal block of C
*

             IF (BETA.EQ.ZERO) THEN
                DO J = 1, NB_LIG_C
                   DO II = 1, J
                      C(II+I-1,J+I-1) = ZERO
                   ENDDO
                ENDDO
             ELSE
                DO J = 1, NB_LIG_C
                   DO II = 1, J
                      C(II+I-1,J+I-1) = BETA*C(II+I-1,J+I-1)
                   ENDDO
                ENDDO
             END IF

             DO 90, L=1,K,NB

             NB_COL_A=MIN(NB,K-L+1)

             IF ((MOD(NB_LIG_C,2).EQ.0).AND.(MOD(NB_COL_A,2).EQ.0))THEN
                   CALL DSYRKL2X2(NB_LIG_C,NB_COL_A,ALPHA,
     $                           A(I,L),LDA,ONE,C(I,I),LDC)
             ELSE
                   CALL DSYRKL(NB_LIG_C,NB_COL_A,ALPHA,
     $                           A(I,L),LDA,ONE,C(I,I),LDC)
             END IF

90           CONTINUE

             NB_COL_C=N-NB_LIG_C-I+1
             NB_COL_A=K

             CALL DGEMM('N','T',NB_LIG_C,NB_COL_C,NB_COL_A,
     $                  ALPHA,A(I,1),LDA,A(I+NB_LIG_C,1),LDA,
     $                  BETA, C(I,I+NB_LIG_C),LDC)

130      CONTINUE
```

Fig. 5. Part of the blocked code for DSYRK

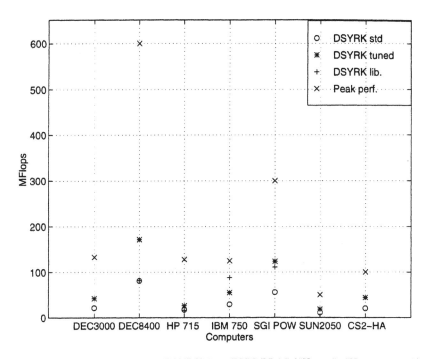

Fig. 6. Average performance of DSYRK from RISC BLAS ("Upper", "No transpose").

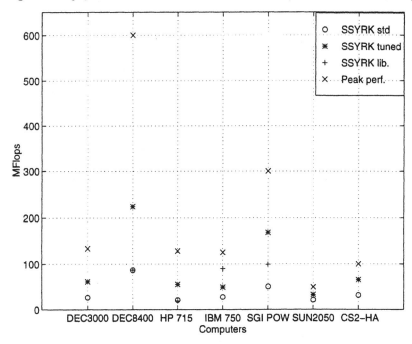

Fig. 7. Average performance of SSYRK from RISC BLAS ("Upper", "No transpose").

For this kernel, our gains over using standard BLAS are significant, usually by a factor of close to two. We consistently outperform the vendor code on the SGI by a significant amount in single precision. Our blocked code is substantially better than the vendor kernel on the DEC 8400 and would be even faster if we used the vendor-supplied _GEMM.

3.3 Parallel Versions of the BLAS

The increased granularity of higher Level BLAS also allows more efficient parallelization because of a reduction in the synchronization overheads (Daydé et al., 1994a and Amestoy, Daydé, Duff and Morère, 1995). A parallel version of the Level 3 BLAS is easily obtained using the loop-level parallelism available on most of the shared and virtual shared memory multiprocessors.

We have developed parallel versions of the Level 3 BLAS for two virtual shared memory computers: the BBN TC2000 and the KSR1. On matrices of order 1536, our SGEMM code executes at 150 Mflops on 24 processors on the BBN TC2000, and it is possible to obtain 1320 Mflops with 72 processors on the KSR1 for matrices of order 768 (Amestoy et al., 1995). The RISC BLAS are used as the tuned serial codes.

Two additional libraries have been designed in order to make a parallel BLAS available on message-passing systems: the BLACS (Basic Linear Algebra Communication Subprograms, see Dongarra and Whaley, 1995) that are used as a communication layer (on top of message passing libraries such as PVM, NX, MPI, CMMD,..), and the PBLAS (Parallel Basic Linear Algebra Subprograms, see Choi, Dongarra, Ostrouchov, Petitet, Walker and Whaley, 1995b).

We indicate, in Figure 8, the performance of the parallel matrix-matrix product from PBLAS (we consider both the single and double precision kernels, that is PSGEMM and PDGEMM respectively) on the MEIKO CS2-HA installed at CERFACS, using square matrices of order 500 and 1000. The processor is a 100 Mhz HyperSPARC of peak performance equal to 100 MFlops. We use the version of BLACS on top of the NX message passing library. The tuned serial BLAS used is the RISC BLAS.

4 Solution of Full Linear Systems

4.1 Introduction

The BLAS can be used successfully in designing codes for the solution of the linear system

$$Ax = b. \tag{1}$$

when A is full.

The LAPACK library (Anderson, Bai, Bischof, Demmel, Dongarra, DuCroz, Greenbaum, Hammarling, McKenney, Ostrouchov and Sorensen, 1992) uses block algorithms as much as possible to take advantage of the higher level of

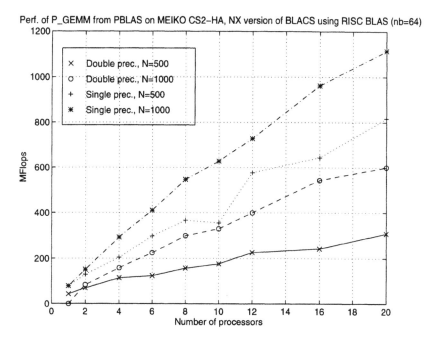

Fig. 8. Performance of parallel matrix-matrix product from PBLAS on MEIKO CS2-HA

BLAS. LAPACK contains subroutines to solve systems of linear equations, linear least-squares problems, eigenvalue problems, and singular value problems. It is designed to give high efficiency on vector processors, RISC-based computers, and shared memory multiprocessors.

4.2 Use of Parallel BLAS

We have used parallel versions of the BLAS mentioned in Section 3.3 to exploit, in a transparent manner, parallelism in codes from the LAPACK Library on shared and virtual shared memory computers (see Daydé and Duff, 1991 and Amestoy et al., 1995).

The ScaLAPACK library (Choi, Demmel, Dhillon, Dongarra, Ostrouchov, Petitet, Stanley, Walker and Whaley, 1995a) is an extension of LAPACK for distributed memory computers. It illustrates the importance of building blocks for reusing existing software as much as possible in the development of portable and efficient codes. ScaLAPACK is based on the BLAS, LAPACK, PBLAS and BLACS libraries. We show, in Figure 9, the performance of the LU, Cholesky and QR factorizations from ScaLAPACK version 1.0 using both single and double precision on relatively small matrices of order 1500 on the MEIKO CS2-HA at CERFACS.

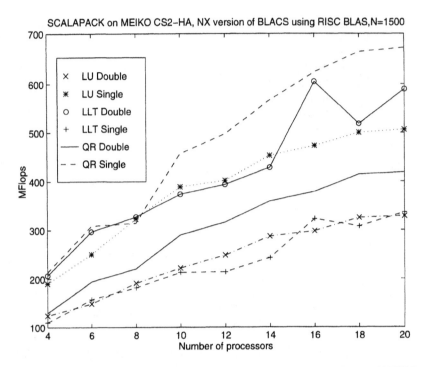

Fig. 9. Performance of full matrix factorizations from ScaLAPACK on MEIKO CS2-HA

4.3 Blocked Eskow-Schnabel Modified Cholesky Factorization

The modified Cholesky factorization modifies an indefinite matrix to obtain a Cholesky factorization of a nearby positive definite matrix. It is an important computational kernel in optimization. It was introduced by Gill and Murray (1974) and improved by Gill, Murray and Wright (1981) and more recently by Eskow and Schnabel (1991b, 1991a). It is particularly useful in optimization algorithms –both in unconstrained and constrained optimization (Gill et al., 1981)– to generate a descent direction when the Hessian matrix is not sufficiently positive-definite. It is also used in some trust region techniques (see Dennis and Schnabel, 1983, Schnabel, Koontz and Weiss, 1985), in the LANCELOT package (Conn, Gould and Toint, 1992), and in sparse preconditioners (Schlick, 1993). It has also been used to build Element-by-Element preconditioners for large scale optimization applications (Daydé, L'Excellent and Gould, 1994b, 1995).

We refer here to the Eskow-Schnabel Modified Cholesky Factorization. Let $\boldsymbol{A}$ be a symmetric n-by-n, not necessarily positive-definite matrix, then the Eskow-Schnabel factorization computes:

$$\boldsymbol{P}^T(\boldsymbol{A} + \boldsymbol{E})\boldsymbol{P} = \boldsymbol{L}\boldsymbol{L}^T$$

where $\boldsymbol{P}$ is a permutation matrix and $\boldsymbol{E}$ is the n-by-n zero matrix if $\boldsymbol{A}$ is sufficiently positive definite. Otherwise E is a non-negative diagonal matrix chosen so that $A + E$ is sufficiently positive-definite. There is no need to know *a priori* if A is positive-definite and the matrix E is determined during the factorization process. The Eskow-Schnabel Modified Cholesky Factorization exhibits better properties in terms of computational costs and an *a priori* upper bound on $\|E\|_\infty$ than those described by Gill and Murray (1974) and Gill et al. (1981).

The Eskow-Schnabel Modified Cholesky Factorization. The Eskow-Schnabel Modified Cholesky factorization is based on a simple idea, namely that once an approximation of the most negative eigenvalue of A is determined (using the Gerschgorin circle theorem bounds), it is easy to compute E as a diagonal matrix that shifts the most negative eigenvalue towards a sufficiently positive value. Of course, it is desirable that the eigenvalues be shifted towards the positive values by a quantity that is not substantially greater than the most negative eigenvalue of A.

The algorithm for the Modified Cholesky Factorization is organized in two phases. The first phase corresponds to a standard Cholesky factorization. The algorithm switches to a second phase to perform a modified factorization when the matrix is detected to be not sufficiently positive-definite. Its main characteristic is that diagonal pivoting is used based on the maximum diagonal element. When the matrix is found to be not positive-definite (some diagonal element becomes non-positive), an approximation of the most negative eigenvalue of A is computed using the Gerschgorin Circle Theorem bounds in order to determine the amount to add to the diagonal of A. It is exactly the Cholesky factorization if A is sufficiently positive-definite, and in that case the number of flops of the modified factorization is the same as standard Cholesky ($\frac{n^3}{3}$ flops, plus order of n operations to precalculate the new diagonal entries). When A is not sufficiently positive-definite, the cost of this factorization is at most $2n^2$ additions and $\frac{n^2}{2}$ multiplications greater than the standard Cholesky factorization.

The blocked version of the modified Cholesky factorization performs exactly the same factorization as the original non-blocked one. It is designed in the same way as other block algorithms (Anderson et al., 1992, Daydé and Duff, 1989, Daydé and Duff, 1991, and Dongarra, Duff, Sorensen and van der Vorst, 1991a). It is organized using block columns of the matrix. At the k th step, the block column k is factorized using Level 1 and Level 2 BLAS with an unblocked algorithm derived from the standard Eskow-Schnabel factorization. The subsequent updates are effected using Level 3 BLAS. Of course the diagonal pivoting and update of Gerschgorin bounds complicate things since this implies that the diagonal values and the bound estimates must be updated after each column factorization. Also, the switch from Phase 1 to Phase 2 may imply stopping the factorization of a block column to perform all updates resulting from the last steps in Phase 1 before moving to Phase 2. More details can be found in Daydé (1996).

Numerical Experiments. We report on experiments on 3 types of full matrices built using the **genmat** procedure available in the software from Eskow and Schnabel. This generates a random matrix that has eigenvalues in a range of prescribed values: matrix of type 1 has eigenvalues in the range -1. to 100, matrix of type 2 has eigenvalues in the range -10000. to -1, and matrix of type 3 has eigenvalues in the range 1. to 10.

We show, in Table 2, the performance of the Level 1 and Level 2 BLAS version of the Modified Cholesky factorization (called **dsymd2**) and the performance of the block version (called **dsymdf**) on two RISC workstations: the HP 715/64 and the IBM RS/6000-750. We give the performance achieved using different block sizes (16, 32, and 64) and for matrices of order 100 and 500. We also give the peak performance in MFlops of each computer in parentheses. We use the manufacturer-supplied BLAS on these computers. We observe that the blocked factorization is at least twice as fast as the unblocked one. The codes described here are available using ftp anonymous to ftp.enseeiht.fr in directory pub/numerique/MODCHOL.

Computer	Order	Type	dsymd2	dsymdf		
				16	32	64
HP 715/64	100	1	8.3	16.7	16.7	33.3
		2	8.3	11.1	16.7	33.3
(128 MFlops)		3	8.3	16.7	16.7	16.7
	500	1	4.2	17.7	18.4	14.9
		2	4.2	17.6	18.3	14.8
		3	4.2	17.8	18.5	14.9
IBM RS/6000-750	100	1	16.7	33.3	16.7	33.3
		2	16.7	33.3	33.3	33.3
(125 MFlops)		3	16.7	33.3	33.3	16.7
	500	1	19.3	46.8	43.9	46.8
		2	20.0	44.8	45.8	41.3
		3	19.2	47.3	48.4	43.8

Table 2. Performance in MFlops of the double precision Modified Cholesky Factorization.

5 Use of Level 3 BLAS in the Direct Solution of Sparse Linear Systems

We refer to the studies of Amestoy and Duff (Amestoy and Duff, 1989, Amestoy, 1991, Amestoy and Duff, 1993, Amestoy et al., 1995) on the sparse LU factorization of square matrices on shared memory multiprocessors. They use a multifrontal approach for the factorization. Further background on this approach can be obtained from the original papers by Duff and Reid (1983, 1984).

In a multifrontal method, the sparse factorization proceeds by a sequence of factorizations on small full matrices, called frontal matrices. The ordering for the sequence of computations and the frontal matrices are determined by a computational tree, called assembly tree, where each node represents a full matrix factorization and each edge the transfer of data from child to parent node. This assembly tree is determined from the sparsity pattern of the matrix and from a reordering that reduces the fill-in during the numerical factorization (such as the minimum degree ordering that we use here). During the numerical factorization, eliminations at any node can proceed as soon as those at the child nodes have completed and the resulting contributions from the children have been summed (assembled) with data at the parent node. This is the only synchronization that is required and means that operations at nodes that are not ancestors or dependants are completely independent. The parallelism resulting from this observation will be referred to as tree parallelism.

We note that the factorization at each node is done using full linear algebra and direct addressing so that factorization computations within a node can use the BLAS. All the indirect addressing is confined to the assembly process.

Amestoy and Duff have developed a parallel multifrontal code for the solution of symmetrically structured unsymmetric equations (Amestoy and Duff 1989, Amestoy 1991, Amestoy and Duff 1993, Amestoy et al. 1995). Recently, a library version of the experimental code called MUPS, has been included in Release 12 of the Harwell Subroutine Library (HSL, 1996). This HSL code is called MA41, and the results in this section are from runs with this code.

During the LU factorization, if only the tree parallelism is exploited, the speed-up is very disappointing. The actual speed-up depends on the problem but is typically only 2 to 3 irrespective of the number of processors. This poor performance is caused by the fact that the tree parallelism decreases as the computation proceeds towards the root of the tree. Moreover, Amestoy and Duff (1993) have observed that typically 75% of the work is performed in the top three levels of the assembly tree. It is thus necessary to obtain further parallelism within the large nodes near the root of the tree by using parallel versions of the BLAS in the factorizations within the nodes. Amestoy and Duff call this node parallelism. When both tree and node parallelism are combined the situation becomes much more encouraging.

In Table 3, we show typical performance of MA41 for a range of RISC-based computers. The RISC BLAS is used on the DEC and the MEIKO and the manufacturer-supplied one in the other cases. A medium size sparse matrix, BCSSTK15 from the Harwell-Boeing set (Duff, Grimes and Lewis, 1992), is used in this table. This is a matrix of order 3948 with 117816 nonzeros from a structural analysis application. A minimum degree ordering is used and the number of floating-point operations for the factorization is 443 million. The performance achieved by MA41 is usually more than 60% of that of DGEMM.

Although we have concentrated in this section on the use of higher level

Computer	Peak perf. Mflops	DGEMM Mflops	**MA41** Mflops
DEC 3000/400-AXP	133	49	34
HP 715/64	128	55	30
IBM RS6000/750	125	101	64
IBM SP2 (thin node)	266	213	122
MEIKO CS2-HA	100	43	31

Table 3. Performance summary of the multifrontal LU factorization MA41 on matrix BCSSTK15 on a range of RISC processors. The average performance of DGEMM is also shown (averaged over square matrices of order 32, 64, 96, and 128).

BLAS in a multifrontal method designed for symmetrically structured sparse matrices, the use of such computational kernels in sparse direct codes is now very widespread and is arguably the main contributor to the efficiency of modern codes for the direct solution of sparse equations. They have also been used in, for example, the multifrontal factorization of unsymmetric matrices (Davis and Duff 1993) and in supernodal codes for unsymmetric problems (Demmel, Eisenstat, Gilbert, Li and Liu 1995). A fuller discussion on the use of high level computational kernels in sparse direct methods can be found in Duff (1996).

6 Solution of Partially Separable Linear Systems Using Element-by-Element Preconditioners

6.1 Introduction

We study the solution of large unassembled partially separable systems by methods that aim to exploit their inherent structure. In particular, we consider the linear systems that arise from the minimization of the partially separable (Griewank and Toint, 1982) objective function

$$f(\boldsymbol{x}) = \sum_{i=1}^{p} f_i(\boldsymbol{x}^i), \tag{2}$$

where each set of *local* variables, $\boldsymbol{x}^i \in \Re^{n_i}$, is a subset of the *global* variables, $\boldsymbol{x} \in \Re^n$, and $n_i \ll n$.

In unconstrained optimization, we often require an (approximate) solution, $\boldsymbol{d}$, to the Newton equations

$$\nabla_{xx} f(\boldsymbol{x}) \boldsymbol{d} = -\nabla_x f(\boldsymbol{x}) \tag{3}$$

When f has the form (2), equation (3) can be expressed as

$$\left(\sum_{i=1}^{p} \nabla_{xx} f_i(\boldsymbol{x}^i) \right) \boldsymbol{d} = - \sum_{i=1}^{p} \nabla_x f_i(\boldsymbol{x}^i). \tag{4}$$

The Hessian matrix of each f_i is a low-rank, sparse matrix. The overall Hessian is thus frequently also sparse. Putting this in a more general context, we consider the solution of structured systems of linear equations of the form

$$Ax = b, \tag{5}$$

where

$$A = \sum_{i=1}^{p} A_i. \tag{6}$$

and A is large and normally positive definite. Similar linear systems arise when solving constrained optimization problems using augmented Lagrangian methods, and, of course, when using finite-element methods to solve elliptic partial differential equations.

Although direct methods may be appropriate for solving (5), we consider here the use of the conjugate gradient method combined with Element-by-Element preconditioners. Additionally, we show that amalgamating elements before constructing such a preconditioner can dramatically improve the speed and numerical behaviour of the method.

6.2   Element-by-Element Preconditioners

Element-By-Element (EBE) preconditioners were introduced by Hughes, Levit and Winget (1983) and Ortiz, Pinsky and Taylor (1983) and have been successfully applied in a number of applications in engineering and physics (see, for example, Hughes, Ferencz and Hallquits (1987), and Erhel, Traynard and Vidrascu (1991)). A detailed analysis of this technique is given by Wathen (1989). These preconditioners have some nice features. They can be computed element-wise and most of them do not require assembly. Furthermore, they permit efficient parallelization.

We describe here only the EBE preconditioner since, in our experience, this is one of the most promising. Further details can be found in Daydé et al. (1994b).

Assuming that A is positive definite, we may rewrite A as:

$$A = \sum_{i=1}^{p} M_i + \sum_{i=1}^{p} (A_i - M_i) = M + \sum_{i=1}^{p} (A_i - M_i), \tag{7}$$

where $M_i = diag(A_i)$ and $M = \sum_{i=1}^{p} M_i$. Now, let $M = L_M L_M^T$ be the Cholesky factorization of M (L_M is simply a diagonal matrix). Then,

$$A = L_M \left(I + \sum_{i=1}^{p} L_M^{-1}(A_i - M_i)L_M^{-T} \right) L_M^T = L_M \left(I + \sum_{i=1}^{p} E_i \right) L_M^T, \tag{8}$$

where $E_i = L_M^{-1}(A_i - M_i)L_M^{-T}$. Using the approximation $I + \sum_{i=1}^{p} E_i \approx \prod_{i=1}^{p}(I + E_i)$, we obtain:

$$A \approx L_M \prod_{i=1}^{p} (I + E_i)L_M^T. \tag{9}$$

A further approximation gives the EBE preconditioner

$$P_{EBE} = L_M \left(\prod_{i=1}^{p} L_i \right) \left(\prod_{i=1}^{p} D_i \right) \left(\prod_{i=p}^{1} L_i^T \right) L_M^T, \tag{10}$$

where the L_i and D_i factors come from the LDL^T factorization of the matrices $I + E_i$ (also known as the Winget decomposition).

Clearly, the efficiency of the EBE preconditioner depends on the the partitioning of the initial matrix and on the magnitude of the off-diagonal elements of the elementary matrices. As the decomposition of A is, in general, not unique, different decompositions may significantly affect the performance of the preconditioner.

6.3 Preprocessing of Unassembled Linear Systems

The effectiveness of the preconditioner depends crucially on the overlap between elements. Daydé et al. (1994b) show that amalgamating elements before constructing such a preconditioner can dramatically improve the speed and numerical behaviour of the method. The amalgamation process typically reduces the overlap between elements, and it is this that leads to improvements in performance.

Our preprocessing step consists of grouping the elements into sets, assembling the elements within each set into a *super-element*, and then applying an Element-by-Element technique to the super-elements instead of the original A_i. Although, as we have already said, we focus on the use of the EBE preconditioner, most of the conclusions are true for the other ones.

Amalgamation Algorithm. A variety of amalgamation techniques are considered in detail by Daydé et al. (1994b). Here, we only consider the most successful of these.

Let $\mathcal{G}_i$ denote an element, $\mathcal{V}_i$ denote the set of indices of variables used by the element $\mathcal{G}_i$, and $|\mathcal{V}_i|$ denote the cardinal of $\mathcal{V}_i$. Let $tim(i)$ refer to the time spent:

- in a matrix-vector product of order i for the diagonal (DIAG) preconditioner (strategy **amalg1** in Table 5); or
- in a matrix-vector product and in two triangular solves of order i for the EBE preconditioner (strategy **amalg2** in Table 5).

The amalgamation process we have used computes the *benefit*

$$b(\mathcal{G}_i, \mathcal{G}_j) = tim(|\mathcal{V}_i|) + tim(|\mathcal{V}_j|) - tim(|\mathcal{V}_i \cup \mathcal{V}_j|), \tag{11}$$

for all pairs of elements and amalgamates the pair with the largest benefit so long as it is larger than a *threshold* value.

We note that $tim(i)$ only depends on i and can be computed once and for all. These machine-dependent costs are stored in files and determined during the installation of the software. If we are only interested in reducing the time per iteration, the best value for the threshold would be zero, but negative values will result in further amalgamations and, hence, possibly better preconditioners.

6.4 Numerical Experiments

We describe, in Table 4, a set of test matrices that will be used in our experiments. n is the order of the matrices, p the number of elements, and κ the condition number. The matrices come either from the Harwell-Boeing collection, see Duff et al., 1992 (CEGB2802), or are problems in SIF format from the CUTE collection, see Bongartz, Conn, Gould and Toint, 1993 (CBRATU3D, NOBNDTOR, and NET3).

The pattern of CEGB2802 arises from a structural engineering problem, NOBNDTOR is a quadratic elastic torsion problem arising from an obstacle problem on a square, NET3 is a very ill-conditioned example from the optimization of a high-pressure gas network, and CBRATU3D is obtained by discretizing a complex 3D PDE problem in a cubic region.

Problem name	n	p	Min elt size	Max elt size	Mean elt size	Degree of overlap	κ
CBRATU3D	4394	4394	5	8	7.5	7.5	3.4×10^1
CEGB2802	2694	108	42	60	58.7	2.4	5.7×10^4
NET3	512	531	1	6	2.6	2.7	2.4×10^9
NOBNDTOR	480	562	1	5	4.2	4.9	1.8×10^2

Table 4. Summary of the characteristics of each test problem

One important characteristic is the degree of overlap. It is defined as the average number of elements sharing each variable and is an indicator as to how well Element-by-Element preconditioners will behave.

In practice, we use a Modified Cholesky factorization close to the one described in Section 4.3 to guarantee that the preconditioners are positive definite.

Detailed results of amalgamation for various preconditioners can be found in Daydé et al. (1994b, 1995, and 1996). These results indicate that Element-by-Element preconditioners are effective in terms of the numbers of iterations required and the clustering of eigenvalues of the preconditioned Hessian, particularly if the overlap between blocks is small. But, except for ill-conditioned problems, EBE is not significantly more efficient than diagonal preconditioning. One reason is because of the structure of the elements. When there is low overlap,

EBE appears much more efficient than diagonal preconditioning. Amalgamating elements may reduce the number of iterations by decreasing the degree of overlap in the new partition.

We show, in Table 5, the effect of amalgamating elements for diagonal and EBE preconditioning with a threshold equal to 0 for runs on an HP 715/64. The solution time, t_{sol}, includes the time for computing the preconditioner and the time for iterating to convergence. The time for computing the preconditioners (both diagonal and EBE) is negligible on all the problems (less than 0.03 seconds) except when using EBE on CBRATU3D and CGEB2002 where it is around 0.5 and 0.7 seconds respectively.

Problem	Amalg.	Amalg. time	p	Avg size elt	Diagonal			EBE		
					#its	t_{sol}	MF	#its	t_{sol}	MF
CBRATU3D	none		4394	7.5	53	3.5	9.5	20	5.1	4.6
	amalg1	4.9	3400	9.1	53	3.5	10.7	23	5.6	5.6
	amalg2	8.0	1950	14.0	53	3.8	13.0	24	5.8	7.6
CEGB2802	none		108	58.7	661	31.3	16.7	120	15.6	12.4
	amalg1	0.2	108	58.7	661	31.5	16.6	120	16.0	12.1
	amalg2	0.2	106	59.4	660	31.7	16.5	111	14.6	12.3
NET3	none		538	2.6	1559	5.2	4.9	723	8.7	1.9
	amalg1	0.1	143	5.6	1580	3.2	8.8	268	1.5	4.9
	amalg2	0.2	57	11.1	1668	3.2	12.2	216	1.0	8.7
NOBNDTOR	none		562	4.2	68	0.3	7.2	30	0.5	3.0
	amalg1	0.2	107	11.7	68	0.2	13.4	36	0.3	8.2
	amalg2	0.3	57	17.3	68	0.3	15.3	32	0.3	9.8

Table 5. Comparison of DIAG and EBE preconditioners without amalgamation and with amalgamation strategies **amalg1** and **amalg2** on HP 715/64. Amalg. time is the time for performing the symbolic amalgamation plus the numerical assembly of the super-elements. p is the number of elements and MF is the performance expressed in Mflops.

Large gains in execution time are obtained using amalgamation when the elements are initially small and overlap significantly. With amalgamation, EBE is more efficient than diagonal preconditioning as soon as the problem is sufficiently hard to solve. For diagonal preconditioning, the gains from amalgamation are due to the better execution rates whereas, for EBE, the gains are due both to a better execution rate and a smaller number of iterations.

The amalgamation procedure is currently rather costly, but it is hoped that good heuristics will decrease this preprocessing cost with roughly the same effect. As we may have to solve many systems with the same structure in the course of a nonlinear optimization calculation, a good preprocessing step may pay handsome dividends in the long run.

Use of Sparse Storage. When using such an algorithm, the super-elements are stored as full matrices and thus some zeros may be explicitly stored. The density of the super-elements depends on the structure of the initial elements and, because of the dependence of the amalgamation on $tim(.)$, also on the target computer. For example, on an ALLIANT FX/80 vector computer using a threshold equal to zero, we typically obtain elements with a density around 0.3, which is quite large for the elements to be treated as sparse. On RISC architectures, the density of the elements obtained can be larger, since the amalgamation process is stopped earlier (long vectors are not required for efficiency as on a vector processor). However, we have suggested that continued amalgamation is often beneficial to the quality of the preconditioner, and this results in a reduction in the density of the super-elements. Thus it may well be advantageous to use a sparse representation of the super-elements. We define the average density to be the ratio

$$d = \frac{\sum_{i=1}^{p} nz_i}{\sum_{i=1}^{p} s_i^2}, \tag{12}$$

where p is the number of elements, nz_i is the number of nonzero entries in element i, and s_i is the number of variables in the element i.

In Table 6, we show the construction time and the time spent in the solves of the EBE preconditioner using both full and sparse storage of the super-elements for the test problem NET3. As matrix-vector products and dot-products are performed in the same way in both cases — using the full initial elements, the construction time and the time spent in the solves are the more meaningful parameters to compare. When using sparse storage, the factorization of the Winget decomposition of each super-element is obtained using the sparse symmetric solver MA27 (Duff and Reid, 1983) from the Harwell Subroutine Library. We compare two different orderings for the elimination of the variables: the minimum degree ordering and the natural ordering of the elements. Except for very small and full elementary matrices, it appears that minimum degree is always better.

The sparse storage becomes more efficient than full storage for threshold values smaller than -0.00012, which corresponds to a density smaller than 0.26. This is especially true for the triangular systems for which there is less overhead using sparse storage than for the factorization.

If we consider a vector computer, such as the ALLIANT FX/80, amalgamation is useful with full super-element storage. But again, there is a significant overhead in using sparse elements since the number of iterations is initially small and not significantly reduced by amalgamation, and because of the large amount of fill-in during the factorizations. Therefore, in this case, we prefer to use long vectors rather than sparse elements.

When solving a well conditioned problem, sparse elements do not appear to be very useful. For an ill-conditioned problem, amalgamation often reduces the number of iterations. Sparse elements can then be effective provided a lot of amalgamation occurs. We believe that the main use of sparse super-elements is

				Dense storage		Sparse storage				
							Min. deg.		Nat. ord.	
Thresh.	Nb elt	Avg size	Avg dens.	# its	Cons. time	Time solve	# its	Cons. time	Time solve	Cons. time	Time solve
0.01	465	2.7	1.00	684	0.01	3.58	675	0.14	6.49	0.12	6.49
0.0	104	7.0	0.51	243	0.02	0.76	231	0.07	1.29	0.06	1.29
-0.00004	52	11.9	0.34	187	0.02	0.75	183	0.07	0.87	0.05	1.02
-0.00012	39	15.1	0.26	142	0.02	0.63	138	0.06	0.63	0.06	0.79
-0.0002	29	19.6	0.19	123	0.02	0.69	122	0.06	0.54	0.07	0.76
-0.001	19	28.7	0.11	87	0.04	0.71	84	0.06	0.36	0.08	0.64
-0.004	13	40.9	0.06	57	0.07	0.87	58	0.06	0.23	0.16	0.63
-0.02	9	57.6	0.02	17	0.22	0.65	16	0.06	0.08	0.77	0.41
-1000.0	8	64.0	0.01	1	0.66	0.14	1	0.07	0.02	2.35	0.06

Table 6. Results of amalgamation obtained for the problem NET3 using different thresholds on a SPARC-10 workstation using full and sparse kernels in the the factorizations and triangular solves.

in those large scale ill-conditioned problems for which direct factorization gives rise to too much fill-in.

Element-by-Element preconditioners may be extremely effective for sparse structured systems of linear equations that arise in partial differential equations and partially separable nonlinear optimization applications. Furthermore, they seem to offer great possibilities of vectorization/parallelization on multiprocessor architectures. They also allow for the use of efficient computational kernels such as the BLAS and blocked factorizations. We believe that further experimentation is necessary to assess the full potential of the methods. The amalgamation technique can also be applied to other Element-by-Element preconditioners or to block methods.

7 The Sparse BLAS

Dodson, Grimes and Lewis (1991) proposed a sparse extension of the BLAS some years ago, but this only considered the Level 1 BLAS for vector-vector operations, such as a sparse SAXPY. In keeping with the theme of this paper, we do not discuss these kernels here but concentrate instead on proposals for sparse extensions to higher level BLAS that offer the promise of providing efficient building blocks in a similar fashion to the full kernels discussed earlier.

It is important to stress at the outset that such kernels are not designed for use within sparse direct codes. Indeed it is firmly our belief that the most efficient way to design sparse direct codes is to remove all indirect addressing from the innermost loop and to use full BLAS for the actual elimination operations as we indicated in Section 5. Instead, we envisage the most common use of the sparse BLAS kernels in the iterative solution of sparse equations, and we illustrate such use in Section 7.2.

7.1 Definition of Sparse BLAS

We concentrate here on the **User** Level definition of the higher level sparse BLAS as defined by Duff, Marrone, Radicati and Vittoli (1995). The actual implementation for a particular data structure on a particular architecture would be performed using the lower level toolkit codes of Carney, Heroux and Li (1993).

The proposal of Duff et al. (1995) defines standard interfaces for the following functions:

(1) a routine for performing the product of a sparse and a dense matrix,
(2) a routine for solving a sparse upper or lower triangular system of linear equations for a matrix of right-hand sides,
(3) a routine to check the input data, to transform from one sparse format to another, and to scale a sparse matrix, and
(4) a routine to permute the columns of a sparse matrix and a routine to permute the rows of a full matrix.

We note that (1) and (2) define an extension of the Level 3 BLAS, but they include operations on vectors as a trivial subset. These may be coded separately at the machine dependent level.

The data preprocessing routine (3) is essential to this proposal. It is intended that this routine be called before the body of the computation. The interface is designed to accept many different data formats and produce many others. In particular, it can interrogate the machine it is running on and transform the data into a format that is particularly suited for that machine.

Many algorithms require the permutation of matrices. Additionally, some efficient implementations of sparse matrix-vector products, and of the solution of sparse triangular systems on vector or parallel processors, require the vectors to be reordered. If high efficiency is required, it is necessary to avoid explicit vector permutations in the inner loops and, to enable this, routines have been added (4) to permute sparse matrices and full matrices appropriately. The permutation routines can also be called outside the body of the computation in order to increase efficiency by avoiding permutations within the main loop of the algorithm. This facility is discussed more by Duff et al. (1995).

Although the routines in (3) and (4) are an integral and important part of the proposal, we will here confine ourselves to a further discussion of the routines in (1) and (2).

The main additional issue for the sparse case over the dense one lies in the data format used for the matrices. In the sparse case, there are many different formats which are chosen as natural for the application, for compactness, for clarity, or for efficiency. Indeed, the data structure used may change depending on which criterion is emphasized.

In a Fortran 77 environment, we choose to represent the sparse matrix using no less than six arrays in order to accommodate most commonly used data formats. The principal arrays are the real entries and two integer arrays which may, for example, hold the row and column indices of the respective entries if a coordinate storage scheme were being used. In the following, these arrays are designated by A, IA1, and IA2. A character string, FIDA, indicates the storage scheme being used (for example, FIDA = "COO" for coordinate format) and a character array, DESCRA, gives attributes of the matrix (for example, symmetry, triangularity). Finally, a further short integer array, INFOA, supplies further information concerning the matrix, for example the number of entries in the case of the coordinate scheme. In the Fortran 90 environment, also defined by Duff et al. (1995), all these arrays are included in a derived data type to which is also added left and right permutation arrays, PL and PR, respectively. This Fortran 90 derived data type is shown in Figure 10. The sparse matrix has order M by K.

```
MODULE TYPESP
    TYPE SPMAT
        INTEGER M,K
        CHARACTER*5 FIDA
        CHARACTER*1 DESCRA(10)
        INTEGER    INFOA(10)
        DOUBLE PRECISION,POINTER :: A(:)
        INTEGER,POINTER :: IA1(:),IA2(:),PL(:),PR(:)
    END TYPE SPMAT
END MODULE TYPESP
```

Fig. 10. Fortran 90 derived data type for sparse matrices

With this definition of a sparse matrix, the routines for the matrix-matrix products in (1), which are defined by

$$- C \leftarrow \alpha P_R A P_C B + \beta C$$
$$- C \leftarrow \alpha P_R A^T P_C B + \beta C$$

and for solving triangular systems of equations with multiple right-hand sides in (2), which are defined by

$$- C \leftarrow \alpha D P_R T^{-1} P_C B + \beta C$$
$$- C \leftarrow \alpha D P_R T^{-T} P_C B + \beta C$$
$$- C \leftarrow \alpha P_R T^{-1} P_C D B + \beta C$$
$$- C \leftarrow \alpha P_R T^{-T} P_C D B + \beta C$$

where

- A is a sparse matrix
- T is a triangular sparse matrix
- B and C are dense matrices
- D is a diagonal matrix

- P_R and P_C are permutation matrices
- α and β are scalars,

are as shown in Figures 11 and 12 respectively.

_CSMM (TRANS, M, N, K, ALPHA, PR, FIDA, DESCRA, A, IA1, IA2, INFOA, PC,
      B, LDB, BETA, C, LDC, WORK, LWORK, IERROR)

TRANS = 'N'	TRANS = 'T'
$C \leftarrow \alpha P_R\ A\ P_C\ B + \beta C$	$C \leftarrow \alpha P_R\ A^T\ P_C\ B + \beta C$

Fig. 11. _CSMM, sparse matrix times dense matrix kernel

_CSSM(TRANS, M, N, ALPHA, UNITD, D, PR, FIDT, DESCRT, T, IT1, IT2, INFOT,
     PC, B, LDB, BETA, C, LDC, WORK, LWORK, IERROR)

	TRANS = 'N'	TRANS = 'T'
UNITD = 'U'	$C \leftarrow \alpha\ P_R\ T^{-1}\ P_C\ B + \beta C$	$C \leftarrow \alpha\ P_R\ T^{-T}\ P_C\ B + \beta C$
UNITD = 'L'	$C \leftarrow \alpha P_R\ D\ T^{-1}\ P_C\ B + \beta C$	$C \leftarrow \alpha P_R\ D\ T^{-T}\ P_C\ B + \beta C$
UNITD = 'R'	$C \leftarrow \alpha P_R\ T^{-1}\ D\ P_C\ B + \beta C$	$C \leftarrow \alpha P_R\ T^{-T}\ D\ P_C\ B + \beta C$
UNITD = 'B'	$C \leftarrow \alpha P_R D^{\frac{1}{2}} T^{-1} D^{\frac{1}{2}} P_C B + \beta C$	$C \leftarrow \alpha P_R D^{\frac{1}{2}} T^{-T} D^{\frac{1}{2}} P_C B + \beta C$

Fig. 12. _CSSM, solution of sparse triangular systems of equations

7.2 Use in Iterative Methods

As we mentioned earlier, the primary reason for the design of the sparse high level BLAS is for use in the iterative solution of sparse linear equations. We note that our proposal will fit equally well whether the iterative software performs a call to a matrix-vector multiply routine or whether reverse communication is used since in either case a call can be made to a given sparse matrix-full matrix multiplication routine; the call is made by the routine in the former case and by the user in the latter.

We feel that the best interface for iterative solvers, particularly for flexibility and efficiency is to use reverse communication so that the typical use of our sparse BLAS within an iterative solver would be as indicated in the skeleton code in Figure 13.

```
C   Perform an iteration of the BiConjugate Gradient  method

        IFLAG = 0
        DO 10 ITER=1, MAXIT

C   Call to solve routine
        CALL SOLVER(IFLAG, ......

C   Successful termination
        IF (IFLAG.EQ.1) THEN
          GO TO 20
        END IF

C   Error return
        IF (IFLAG.LT.0) THEN
          GO TO 30
        END IF

        IF (IFLAG.EQ.2) THEN
C   Perform the matrix-vector product
          CALL DCSMM(.....
          GO TO 10
        END IF

        IF (IFLAG.EQ.2) THEN
C   Perform the preconditioning operation
          CALL DCSSM(.....
          GO TO 10
        END IF

  10  CONTINUE

C   Code to handle successful termination
  20  ....

C   Code executed if error returns
  30  ....
```

Fig. 13. Use of kernels in iterative solution of sparse equations

8   Conclusion

The studies described in this paper demonstrate how portable and efficient mathematical software can be designed on high performance computers by making heavy use of computational kernels. The main computational kernels that we consider are the Level 3 BLAS, and we show how they can be used, not only in the solution of full systems of linear equations but also in the direct solution of sparse equations. For the iterative solution of sparse equations, we show how advantage can be taken of full blocks within an Element-by-Element preconditioner and how an extension of the BLAS for sparse matrices can be used in the iterative solution code.

Acknowledgements

The authors are very grateful to Patrick Amestoy and Jérome Décamps for their support in getting some of the results displayed in the paper.

References

Amestoy, P. R. (1991), Factorization of large sparse matrices based on a multifrontal approach in a multiprocessor environment, Phd thesis, Institut National Polytechnique de Toulouse. Available as CERFACS report TH/PA/91/2.

Amestoy, P. R. and Duff, I. S. (1989), 'Vectorization of a multiprocessor multifrontal code', *Int. J. of Supercomputer Applics.* **3**, 41–59.

Amestoy, P. R. and Duff, I. S. (1993), 'Memory allocation issues in sparse multiprocessor multifrontal methods', *Int. J. of Supercomputer Applics.* **7**, 64–82.

Amestoy, P. R., Daydé, M. J., Duff, I. S. and Morère, P. (1995), 'Linear algebra calculations on a virtual shared memory computer', *Int Journal of High Speed Computing* **7**, 21–43.

Anderson, E., Bai, Z., Bischof, C., Demmel, J., Dongarra, J., DuCroz, J., Greenbaum, A., Hammarling, S., McKenney, A., Ostrouchov, S. and Sorensen, D. (1992), *LAPACK Users' Guide.*, SIAM.

Bodin, F. and Seznec, A. (1994), Cache organization influence on loop blocking, Technical Report 803, IRISA, Rennes, France.

Bongartz, I., Conn, A. R., Gould, N. I. M. and Toint, P. L. (1993), CUTE: Constrained and Unconstrained Testing Environment, Technical Report TR/PA/93/10, CERFACS, Toulouse, France.

Carney, S., Heroux, M. A. and Li, G. (1993), A proposal for a sparse BLAS toolkit, Technical Report TR/PA/92/90 (Revised), CERFACS, Toulouse, France.

Choi, J., Demmel, J., Dhillon, I., Dongarra, J., Ostrouchov, S., Petitet, A., Stanley, K., Walker, D. and Whaley, R. C. (1995a), ScaLAPACK: A portable linear algebra library for distributed memory computers - design issues and performance, Technical Report LAPACK Working Note 95, CS-95-283, University of Tennessee.

Choi, J., Dongarra, J., Ostrouchov, S., Petitet, A., Walker, D. and Whaley, R. C. (1995b), A proposal for a set of parallel basic linear algebra subprograms, Technical Report LAPACK Working Note 100, CS-95-283, University of Tennessee.

Conn, A. R., Gould, N. I. M. and Toint, P. L. (1992), LANCELOT: *a Fortran package for large-scale nonlinear optimization (Release A)*, number 17 *in* 'Springer Series in Computational Mathematics', Springer Verlag, Heidelberg, Berlin, New York.

Davis, T. A. and Duff, I. S. (1993), An unsymmetric-pattern multifrontal method for sparse LU factorization, Technical Report RAL 93-036, Rutherford Appleton Laboratory.

Daydé, M. J. (1996), A block version of the eskow-schnabel modified cholesky factorization, Technical Report RT/APO/95/8, ENSEEIHT-IRIT.

Daydé, M. J. and Duff, I. S. (1989), 'Level 3 BLAS in LU factorization on the CRAY-2, ETA-10P and IBM 3090-200/VF', *Int. J. of Supercomputer Applics.* **3**, 40–70.

Daydé, M. J. and Duff, I. S. (1991), 'Use of level 3 BLAS in LU factorization in a multiprocessing environment on three vector multiprocessors, the ALLIANT FX/80, the CRAY-2, and the IBM 3090/VF', *Int. J. of Supercomputer Applics.* **5**, 92–110.

Daydé, M. J. and Duff, I. S. (1996), A block implementation of level 3 BLAS for RISC processors, Technical Report RT/APO/96/1, ENSEEIHT-IRIT.

Daydé, M. J., Duff, I. S. and Petitet, A. (1994a), 'A parallel block implementation of Level 3 BLAS kernels for MIMD vector processors', *ACM Transactions on Mathematical Software* **20**, 178–193.

Daydé, M. J., L'Excellent, J. Y. and Gould, N. I. M. (1994b), On the use of element-by-element preconditioners to solve large scale partially separable optimization problems, Technical report, ENSEEIHT-IRIT, Toulouse, France. RT/APO/94/4, to appear in SIAM Journal on Scientific Computing.

Daydé, M. J., L'Excellent, J. Y. and Gould, N. I. M. (1995), Solution of structured systems of linear equations using element-by-element preconditioners, *in* 'Proceedings 2nd IMACS International Symposium on Iterative Methods in Linear Algebra', pp. 181–190. Also ENSEEIHT-IRIT Technical Report, RT/APO/95/1.

Daydé, M. J., L'Excellent, J. Y. and Gould, N. I. M. (1996), Preprocessing of sparse unassembled linear systems for efficient solution using element-by-element preconditioners, *in* L. Bougé, P. Fraigniaud, A. Mignotte and Y. Robert, eds, 'Proceedings of Euro-Par 96, Lyon', Vol. 2 of *Lecture Notes in Computer Science, Vol. 1124*, Springer Verlag, Heidelberg, Berlin, New York, pp. 34–43. Also ENSEEIHT-IRIT Technical Report RT/APO/96/2.

Demmel, J. W., Eisenstat, S. C., Gilbert, J. R., Li, X. S. and Liu, J. W. H. (1995), A supernodal approach to sparse partial pivoting, Technical Report UCB//CSD-95-883, Computer Science Division, U. C. Berkeley, Berkeley, California.

Dennis, J. and Schnabel, R. (1983), *Numerical Methods for Unconstrained Optimization and Nonlinear Equations*, Prentice Hall, Englewood Cliffs, N.J.

Dodson, D. S., Grimes, R. G. and Lewis, J. G. (1991), 'Sparse extensions to the Fortran Basic Linear Algebra Subprograms', *ACM Transactions on Mathematical Software* **17**, 253–263.

Dongarra, J. and Whaley, R. C. (1995), A users' guide to the blacs, Technical Report CS-95-281, University of Tennessee, Knoxville, Tennessee, USA.

Dongarra, J. J. (1992), Performance of various computers using standard linear algebra software, Technical Report CS-89-85, University of Tennessee, Knoxville, Tennessee, USA.

Dongarra, J. J. and Grosse, E. (1987), 'Distribution of mathematical software via electronic mail', *Comm. ACM* **30**, 403–407.

Dongarra, J. J., Du Croz, J., Duff, I. S. and Hammarling, S. (1990), 'Algorithm 679. a set of Level 3 Basic Linear Algebra Subprograms.', *ACM Transactions on Mathematical Software* **16**, 1–17.

Dongarra, J. J., Duff, I. S., Sorensen, D. C. and van der Vorst, H. A. (1991a), *Solving Linear Systems on Vector and Shared Memory Computers*, SIAM, Philadelphia.

Dongarra, J. J., Mayes, P. and Radicati di Brozolo, G. (1991b), Lapack working note 28 : The IBM RISC System/6000 and linear algebra operations, Technical Report CS-91-130, University of Tennessee.

Duff, I. S. (1996), Sparse numerical linear algebra: direct methods and preconditioning, Technical Report RAL 96-047, Rutherford Appleton Laboratory. Also CERFACS Report TR-PA-96-xxx.

Duff, I. S. and Reid, J. K. (1983), 'The multifrontal solution of indefinite sparse symmetric linear systems', *ACM Transactions on Mathematical Software* 9, 302–325.

Duff, I. S. and Reid, J. K. (1984), 'The multifrontal solution of unsymmetric sets of linear systems', *SIAM Journal on Scientific and Statistical Computing* 5, 633–641.

Duff, I. S., Grimes, R. G. and Lewis, J. G. (1992), Users' guide for the Harwell-Boeing sparse matrix collection (Release I), Technical Report RAL 92-086, Rutherford Appleton Laboratory.

Duff, I. S., Marrone, M., Radicati, G. and Vittoli, C. (1995), A set of Level 3 Basic Linear Algebra Subprograms for sparse matrices, Technical Report TR-RAL-95-049, RAL.

Erhel, J., Traynard, A. and Vidrascu, M. (1991), 'An element-by-element preconditioned conjugate gradient method implemented on a vector computer', *Parallel Computing* 17, 1051–1065.

Eskow, E. and Schnabel, R. B. (1991a), 'Algorithm 695: Software for a new modified cholesky factorization', *ACM Transactions on Mathematical Software* 17, 306–312.

Eskow, E. and Schnabel, R. B. (1991b), 'A new modified cholesky factorization', *SIAM Journal on Scientific and Statistical Computing* 11, 1136–1158.

Gallivan, K., Jalby, W. and Meier, U. (1987), 'The use of blas3 in linear algebra on a parallel processor with a hierarchical memory', *SIAM J. Sci. Stat. Comput.* 8, 1079–1084. Timely communications.

Gallivan, K., Jalby, W., Meier, U. and Sameh, A. (1988), 'Impact of hierarchical memory systems on linear algebra algorithm design', *Int Journal of Supercomputer Applications* 2(1), 12–48.

Gill, P. and Murray, W. (1974), 'Newton-type methods for unconstrained and linearly constrained optimization', *Mathematical Programming* 28, 311–350.

Gill, P., Murray, W. and Wright, M. (1981), *Practical Optimization*, Academic Press, London and New York.

Griewank, A. and Toint, P. L. (1982), On the unconstrained optimization of partially separable functions, *in* M. J. D. Powell, ed., 'Nonlinear Optimization', Academic Press, London and New York.

HSL (1996), *Harwell Subroutine Library. A Catalogue of Subroutines (Release 12)*, AEA Technology, Harwell Laboratory, Oxfordshire, England. For information concerning HSL contact: Dr Scott Roberts, AEA Technology, 552 Harwell, Didcot, Oxon OX11 0RA, England (tel: +44-1235-434714, fax: +44-1235-434136, email: Scott.Roberts@aeat.co.uk).

Hughes, T. J. R., Ferencz, R. M. and Hallquits, J. O. (1987), 'Large-scale vectorized implicit calculations in solid mechanics on a CRAY X-MP/48 utilizing EBE preconditioned conjugate gradients', *Computational Methods in Applied Mechanics and Engineering* 61, 215–248.

Hughes, T. J. R., Levit, I. and Winget, J. (1983), 'An element-by-element solution algorithm for problems of structural and solid mechanics', *Compututational Methods in Applied Mechanics and Engineering* 36, 241–254.

Kågström, B., Ling, P. and Loan, C. V. (1993), Portable high performance GEMM-based Level-3 BLAS, *in* 'Proceedings of the Sixth SIAM Conference on Parallel Processing for Scientific Computing', SIAM, pp. 339–346.

L'Excellent, J. Y. (1995), Utilisation de préconditionneurs élément-par-élément pour la résolution de problèmes d'optimisation de grande taille, PhD thesis, INPT-ENSEEIHT.

Ortiz, M., Pinsky, P. M. and Taylor, R. L. (1983), 'Unconditionally stable element-by-element algorithms for dynamic problems', *Compututational Methods in Applied Mechanics and Engineering* **36**, 223–239.

Schlick, T. (1993), 'Modified Cholesky factorizations for sparse preconditioners', *SIAM Journal on Scientific and Statistical Computing* **14**, 424–445.

Schnabel, R. B., Koontz, J. E. and Weiss, B. E. (1985), 'A modular system of algorithms for unconstrained minimization', *ACM Transactions on Mathematical Software* **11**, 419–440.

Wathen, A. J. (1989), 'An analysis of some element-by-element techniques', *Computational Methods in Applied Mechanics and Engineering* **74**, 271–287.

Parallel Implementation of a Symmetric Eigensolver Based on the Yau and Lu Method [*]

Stéphane Domas[1], Françoise Tisseur[2],

[1] Laboratoire de l'Informatique du Parallélisme, URA 1398 du CNRS and INRIA Rhône-Alpes, 46 Allée d d'Italie, 69364 Lyon Cedex 07, France.
`sdomas@lip.ens-lyon.fr`

[2] Equipe d'Analyse Numérique de St-Etienne,UMR 5585, 23, rue Paul Michelon, 42023 Saint-Etienne, France. `ftisseur@anumsun1.univ-st-etienne.fr`

Abstract. In this paper, we present preliminary results on a complete eigensolver based on the Yau and Lu method. We first give an overview of this invariant subspace decomposition method for dense symmetric matrices followed by numerical results and work in progress of a distributed-memory implementation. We expect that the algorithm's heavy reliance on matrix-matrix multiplication, coupled with FFT should yield a highly parallelizable algorithm. We present performance results for the dominant computation kernel on the Intel Paragon.

1 Introduction

As quantitative analysis becomes increasingly important in sciences and engineering, the need for faster methods to solve bigger and more realistic problem grows. Large order symmetric eigenvalues problems occur in a wide variety of applications, including the dynamic analysis of large-scale structures such as aircraft and spacecraft, the prediction of structural responses in solid and soil mechanics, the study of solar convection, the modal analysis of electronic circuits, and the statistical analysis of data.

There are many algorithms for solving the symmetric eigenvalue problem [13]. Much recent work has been devoted on parallel solvers, both on traditional methods [8, 9, 10] and in the development of new methods [3, 7]. The traditional method for computing the eigensystem of a real dense symmetric matrix A consists in three steps [11]. First, A is reduced to tridiagonal form. Second, the eigenvalues and eigenvectors of the tridiagonal matrix are computed. Third, the eigenvectors are back transformed via the reduction transformation.

In this paper, we investigate the parallelization of a new eigensolver for real dense symmetric matrices. Our algorithm is based on a recent method attributed to Yau and Lu [16], which reduces the symmetric eigenvalue problem to a number of matrix multiplications. Yau and Lu's method involves approximating invariant subspaces of a special matrix using an FFT. The computation of the special

---

[*] This work is partly supported by the European project KIT 108 and Eureka Euro-TOPS project.

matrix and the vectors for the FFT is rich in matrix-matrix multiplications Matrix multiplications can be implemented efficiently on most high-performance machines, and is often available as an optimized implementation in a level 3 BLAS library [2, 4].

The rest of this paper will give an overview of the Yau and Lu's method. We will present the algorithm and valid it with numerical results. Then, we will investigate the parallelization of this new eigensolver and present performance results for the dominant computation kernel on the Intel Paragon.

2 Yau and Lu method

For computing invariant subspaces of a symmetric $n \times n$ matrix A with eigenvalues $\lambda_1, \ldots, \lambda_n$ and eigenvectors $x_1, \ldots, x_n$, Yau and Lu use a polynomial acceleration method.

Consider the unitary matrix $B = e^{iA}$ whose eigenvalues all lie on the unit circle. Note that A and B have the same invariant subspaces.

Let $P_N(z) = \sum_{j=0}^{N-1} \beta_j z^j$ be a polynomial of degree $N - 1$ that has a peak at $z = 1$ and is close to zero on the unit circle away from a vicinity of $z = 1$. Such a polynomial exists and its coefficients $\beta_j, j = 0, N - 1$ can be obtained by a recursive formula (see [16]).

Starting from an initial vector expanded in terms of the eigenvectors as $v_0 = \sum_{i=1}^{n} \alpha_i x_i$, we can define the function $u : [0, 2\pi] \to \mathbb{R}^n$ by

$$u(\lambda) = P_N(e^{-i\lambda} B)v_0 = \sum_{j=1}^{n} \alpha_j P_N(e^{i(\lambda_j - \lambda)})x_j.$$

If λ is chosen close to a particular λ_k and the other eigenvalues of B are not close to λ then the coefficient of x_j will be small except when $j = k$. Thus, $u(\lambda)$ can be viewed as an approximation of the eigenvector of B associated with the eigenvalue $e^{i\lambda_k}$.

Setting $v_j = B^j v_0$, the function $u(\lambda)$ can be written as

$$u(\lambda) = P_N(e^{-i\lambda} B)v_0 = \sum_{j=0}^{N-1} \beta_j B^j v_0 e^{-ij\lambda} = \sum_{j=0}^{N-1} \beta_j v_j e^{-ij\lambda},$$

where $\beta_j v_j$ are the Fourier coefficients of u. Therefore, the FFT can be used to compute $u(\lambda)$ at many different values of λ simultaneously. Then, we need to select vectors $u(\lambda)$ that can be taken as eigenvectors, group them into p orthogonal clusters and add more vectors if necessary. These p clusters form an orthogonal basis $W = [W_1, \ldots, W_p]$ whose elements span invariant subspaces of A and hence, application of A to W decouples the spectrum :

$$W^T A W = \begin{pmatrix} A_1 & & 0 \\ & \ddots & \\ 0 & & A_p \end{pmatrix}.$$

So, the initial problem is reduced to a small symmetric matrix eigenvalue problem in each cluster. Note that the subproblems $A_1, \ldots, A_p$ can be solved totally independently. The algorithm is presented in next section.

3 Numerical algorithm

Consider a real symmetric matrix A. The following steps find the eigenvalues and eigenvectors of A to the desired precision.

1- **Scaling and Translation:** Compute upper and lower bounds of the spectrum of A and use these bounds to scale and translate the spectrum of A in $[0, 2\pi)$.

2- **Polynomial computation:** Let T_{N-1} be the Chebyshev polynomial of degree $N - 1$. The degree N is chosen such that

$$\frac{1}{T_{N-1}\left((3 - \cos\frac{\pi}{n})/(1 + \cos\frac{\pi}{n})\right)} \leq \kappa,$$

where κ is a measure of the desired accuracy of the computed invariant subspace. Compute the coefficients $\beta_0^{(N-1)}, \ldots, \beta_{N-1}^{(N-1)}$ of the polynomial P_N by the recursive formulas :

$$\beta_0^{(0)} = 1 \; , \; \beta_0^{(1)} = b \; , \; \beta_1^{(1)} = a$$
$$\beta_0^{(k+1)} = a\beta_1^{(k)} + 2b\beta_0^{(k)} - \beta_0^{(k-1)}$$
$$\beta_1^{(k+1)} = a\left(2\beta_0^{(k)} + \beta_2^{(k)}\right) + 2b\beta_1^{(k)} - \beta_1^{(k-1)}$$
$$\beta_j^{(k+1)} = a\left(\beta_{j-1}^{(k)} + \beta_{j+1}^{(k)}\right) + 2b\beta_j^{(k)} - \beta_j^{(k-1)} \text{ for } j > 1$$
$$\beta_j^{(k+1)} = 0 \text{ for } j > k + 1.$$

where

$$a = \frac{2}{1 + \cos(\pi/n)}, \quad b = \frac{1 - \cos(\pi/n)}{1 + \cos(\pi/n)}.$$

3- **Unitary matrix:** Compute matrices $\cos(\pi X)$ and $X^{-1}\sin(\pi X)$ where $X = \frac{A}{\pi} - I$ using the following Chebyshev expansion :

$$\cos(\pi x) \simeq c_0 + c_1 T_2(x) + c_2 T_4(x) + \cdots + c_5 T_{10}(x)$$
$$+ T_{10}\left(c_6 T_2(x) + c_7 T4(x) + \cdots + c_{10} T_{10}(x)\right)$$
$$\frac{\sin(\pi x)}{x} \simeq s_0 + s_1 T_2(x) + s_2 T_4(x) + \cdots + s_5 T_{10}(x)$$
$$+ T_{10}\left(s_6 T_2(x) + s_7 T4(x) + \cdots + s_{10} T_{10}(x)\right),$$

where $c_0, \ldots, c_{10}, s_0, \ldots, s_{10}$ are Chebyshev coefficients.

4- **Computation of vectors** $v_j = e^{ijA}v_0, j = 0, 2N-1$ The real and imaginary part of v_1 are obtained with the approximation of $\cos(\pi X)$ and $X^{-1}\sin(\pi X)$:

$$v_1 = \cos(\pi X)v_0 + X^{-1}\sin(\pi X)(Xv_0).$$

The remaining vectors can be computed following these $M = (\log_2 N - 1)$ steps:

step 1 : $\quad C_1 = \cos(A)$
$\qquad\qquad v_2 = 2C_1 v_1 - v_0$

step 2 : $\quad C_2 = 2C_1^2 - I$
$\qquad\qquad (v_3, v_4) = 2C_2(v_1, v_2) - \overline{(v_1, v_0)}$

step 3 : $\quad C_3 = 2C_2^2 - I$
$\qquad\qquad (v_5, v_6, v_7, v_8) = 2C_3(v_1, v_2, v_3, v_4) - \overline{(v_3, v_2, v_1, v_0)}$

$\qquad\qquad \vdots \qquad\qquad \vdots$

step M : $\quad C_p = 2C_{p-1}^2 - I$
$\qquad\qquad (v_{\frac{N}{2}+1}, \ldots, v_N) = 2C_p(v_1, \ldots, v_{\frac{N}{2}}) - \overline{(v_{\frac{N}{2}-1}, \ldots, v_0)}.$

5- **Evaluation of** $u(\lambda)$**:** Via the FFT, compute the vectors

$$u_k = u(\lambda)_{\lambda=\frac{k\pi}{N}} = Re \sum_{j=0}^{N-1} \beta_j v_j e^{-i\frac{jk\pi}{N}}, \qquad \text{for } k = 0, \ldots, 2N-1.$$

6- **Selection and refinement:** Select the most useful vectors from the $2N$ vectors $u_0, \ldots, v_{2N-1}$. Group them into a number of orthogonal clusters, add more vectors if necessary and reduce the initial problem to a small symmetric matrix eigenvalue problem in each cluster.

The most time consuming part of the algorithm is the computation of the $2N$ vectors v_j. We need $\log_2 N - 1$ multiplications of real symmetric matrices and $\log_2 N - 1$ more multiplications between a symmetric matrix and a rectangular one. This part needs $(\log_2(N) - 1)n^3 + 4Nn^2$ floating point operations. The computation of e^{iA} to the desired accuracy requires 6 multiplications of real symmetric matrices for $C\cos\pi X$ and one more for $S = X^{-1}\sin\pi X$, that is, $7n^3$ operations. The step of computing the u_k is still efficient because of the FFT algorithm and can be done in $nN\log_2(N)$ operations. When necessary, the work for the supplementary vectors involves $\frac{8}{3}n^3$ more operations since we need to construct an orthogonal matrix by a QR factorization. The reduction $W^T AW$ involves two matrix multiplications or $3n^3$ operations. Usually, the subproblems in each cluster only involve small matrices and the cost is negligible compared to the total work. So, the total number of operations is given by

$$(\frac{17}{3} + \log_2(N))n^3 + 4Nn^2 + O(n^2).$$

Since N is typically a small multiple of n, we see from this operation count that the sequential complexity of the Yau and Lu algorithm is considerably greater than the QR algorithm. However, note that steps 3,4 and 6 of the algorithm are all based on matrix-matrix multiplications and if we suppose that $N \simeq 8n$ then

$$\frac{\text{Total operations involving matrix multiplications}}{\text{Total operations}} \geq \frac{35 + 3\log_2(n)}{38 + 3\log_2(n)}.$$

So, for matrices of dimension between 500 and 1000, we find that matrix multiplications account for more than 90% of the total operation count. For larger dimensions of the matrix A, this percentage will of course increase. The efficiency of level 3 BLAS routines can justify the use of the extra multiplications. The Reuse-Ratio defined by the rapport between the number of flops and the size of memory reference bounds the performances. Its value is 2/3 for level 1 BLAS, 2 for level 2 BLAS and n/2 for level 3 BLAS. So, high level BLAS 3 improve performances.

4 Numerical results

All the test results presented in this section where performed on a SUNsparc 512, MP. The arithmetic was IEEE standard double precision with a machine precision of $\varepsilon = 2^{-53} \simeq 2.22044 \times 10^{-16}$ and over/underflow threshold $10^{\pm 307}$.

We have tested our algorithm on a large set of test matrices using the LAPACK [1] test generation routine DLATMS. This routine constructs symmetric matrices of the form

$$A = U^T D U$$

where U is a random orthogonal matrix and $D = \text{diag}(\lambda_1, \ldots, \lambda_n)$ a diagonal matrix. We can define the elements of D and then simulate more or less critical situations that is, well separated spectrum, clusters of eigenvalues, ...

We quantified accuracy in the computed eigenvalues by computing the relative error

$$\max_{1 \leq i \leq n} \frac{|\lambda_i - \hat{\lambda}_i|}{|\lambda_{max}|}$$

where λ_i denotes an exact eigenvalue and $\hat{\lambda}_i$ the corresponding computed eigenvalue (see Fig.1).

Accuracy in the residuals for a given matrix A is quantified by computing the maximum normalized 2-norm residual

$$\max_i \frac{\|A\hat{x}_i - \hat{\lambda}_i \hat{x}_i\|_2}{\|A\|_F}, \qquad \text{with} \qquad \|\hat{x}_i\|_2 = 1$$

where $\hat{x}_i$ is the computed eigenvector corresponding to the computed eigenvalue $\hat{\lambda}_i$ (see Fig.2).

We have computed (see Fig. 3) the departure from orthogonality given by

$$\max_{i,j} |(Q^T Q - I_n)_{ij}|$$

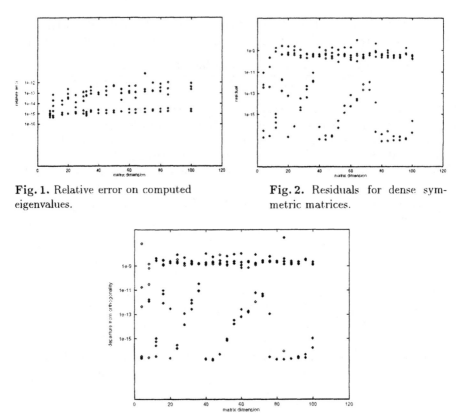

Fig. 1. Relative error on computed eigenvalues.

Fig. 2. Residuals for dense symmetric matrices.

Fig. 3. Departure from orthogonality for dense symmetric matrices.

where Q is the matrix of eigenvectors.

The accuracy of invariant subspaces is controlled in step 2 of the algorithm. As we have chosen N such that $\kappa \leq 10^{-9}$ we expect at most nine corrects significant digits for the eigenvectors. If we choose N such that $\kappa \leq 10^{-16}$, we increase the computational cost but obtain better accuracy (see Fig. 4 and Fig. 5).

5 Parallel implementation

There are two forms of parallelism in the Yau and Lu method. The first one corresponds to data parallelism with the heavy reliance on matrix-matrix multiplications. The second one is a kind of functional parallelism with the reduction of the initial problem to a number of small symmetric eigenvalue problems that can be solved totally independently on each processor. That is why we say that Yau and Lu method yields to a highly parallelizable algorithm.

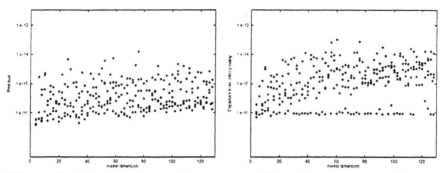

Fig. 4. Residuals on computed eigenvalues for dense symmetric random matrices.

Fig. 5. Departure from orthogonality for dense symmetric random matrices.

5.1 Parallel Yau and Lu algorithm

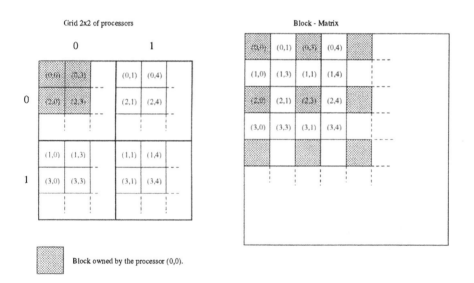

Fig. 6. Block cyclic distribution on a 2x2 processor grid.

For an estimation of the upper and lower bound of the spectrum of A we use a method developed by Rojo and Soto [15]. This method is based on matrix multiplications but does not increase the total number of operations since the computed matrix-matrix product is reused for the construction of C and S.

We prefer to focus our parallelization on the most cost effective part of the algorithm, that is the construction of the matrices C, S and the Fourier's coefficients. In order to ensure a good load balancing, performance and scalability of our code, we use a 2-dimensional block cyclic distribution on a $P \times Q$ processor grid (see Fig.6). This kind of distribution encompasses a large number (but not all) data distribution schemes.

```
/*Construction of T_1, T_2, T_4, T_8, T_10.                    */
T_2 ← 2A * A − I                          /* Parallel level 3 BLAS */
FOR i=4 TO 10 STEP 2
        T_i ← 2 T_{i-2} * T_2 − T_{i-4}    /* Parallel level 3 BLAS */
ENDFOR
/*Chebyshev expansion of C and S                                */
C ← c_0 I
S ← s_0 I
FOR i=1 TO 5
        C ← c_i T_{2i} + C
        S ← s_i T_{2i} + S
        CC ← c_{i+5} T_{2i} + CC
        SS ← s_{i+5} T_{2i} + SS
ENDFOR
C := −T_10 * CC + C                        /* Parallel level 3 BLAS */
S := −T_10 * SS + S                        /* Parallel level 3 BLAS */
```

Table 1. Algorithm for computing C and S.

In Tab.1, we present the parallel algorithm for the computation of the matrices $C = \cos \pi X$ and $S = X^{-1} \sin \pi X$ which defined the unitary matrix e^{iA}. It is only based on 7 calls of parallel level 3 BLAS.

The computation of Fourier's vectors is the most cost effective part in computation and communication times. At each step $k, k = 1, \ldots, \log_2(N/2)$, the following matrix-matrix multiplication is performed:

$$C_k = 2C_{k-1} * C_{k-1},$$
$$V_k = 2C_k * U_k - W_k,$$

where

$$C_0 - \cos \pi X, W_0 = v_0 \text{ and } U_0 = v_1$$
$$W_k = [\text{perm}(\overline{V_{k-1}}), W_{k-1}], \quad U_k = [U_{k-1}, V_{k-1}].$$

The three matrices V_k, U_k, W_k are rectangular ones and their sizes increase from iteration to iteration. The rectangular matrix W_k depends on a permutation of V_k's columns and communications between processor columns are necessary for its construction. For a good load balancing of the computation, we impose the matrices V_k, U_k, W_k to have a column block size partitioning equal to 2. The distribution we use for the computed Fourier's coefficients v_j avoid the necessary communications for the bit-reversal that is the first step of the FFT. We present the parallel algorithm in Tab.2.

For the parallel 1-D FFT [6], we use a communication computation overlap algorithm.

```
/* Initialization                                                      */
W ← v₀
U ← Cv₀ + iSv₀                          /* Parallel level 2 BLAS */
V ← 2C * U − W                          /* Parallel level 3 BLAS */

/* Main loop                                                           */
FOR i = 1 TO log₂(N/2) DO
    C ← 2C * C − I                      /* Parallel level 3 BLAS */
    W ← [perm(V̄), W]    /* update of W, send/recv between processor col.*/
    U ← [U, V]                                        /* update of W
    V ← 2C * U − W                      /* Parallel level 3 BLAS */
ENDFOR
```

Table 2. Algorithm for computing v_j for $j = 0, \ldots, N - 1$.

For the last part of the algorithm, each processor selects the most useful computed vectors u_k and forms a basis. When the number of selected vectors is less than n, we use a parallel QR factorization in order to complete the basis. after projection onto this basis, each processor solves its own small symmetric eigenvalue problem using a standard symmetric eigensolver (for example, DSYEV from LAPACK [1]).

5.2 Symmetric matrix-matrix product

In Section 3, we have shown that the computational cost of the algorithm is dominated by dense matrix-matrix multiplications. Thus, the performance of this algorithm will depend heavily on the matrix multiplication code. We need two different types of matrix-matrix products. The first one is the product between symmetric and rectangular matrices and the second one is product between two symmetric matrices which commute. So the result will be a symmetric matrix. We want to develop a double precision distributed matrix code for the symmetric matrix product that take into account the special properties of these matrices (which is not currently available in ScaLAPACK [5]).

The algorithm below is based on an idea presented by Snyder in [12]. It uses a block scattered distribution of the matrices. The whole matrices are distributed and not only the upper or lower triangular part. The symbols are the followings. M is the matrix size, distributed on a $P \times Q$ grid of processors. There are $N_b \times N_b$ blocks, and each processor has $P_b \times P_b$ blocks.

$N_b = \lceil \frac{M}{B_s} \rceil$
$P_b = \lceil \frac{N_b}{P} \rceil$
/* Computation of the diagonal blocks of C */
FOR $i = 0$ TO N_b DO
    $cur\_row = mod(i, P)$
    $cur\_col = mod(i, Q)$
    $b_r = \lceil \frac{i}{P} \rceil$
    $b_c = \lceil \frac{i}{Q} \rceil$
    IF ($my\_col = cur\_col$) THEN
        computes the diagonal blocks of C :
        DSCMM $\Rightarrow \alpha A_{.,b_c}^T \times B_{.,b_c} + \beta C_{b_r,b_c} \to C_{b_r,b_c}$
        global sum of C_{b_r,b_c} :
        DGSUM2D $\Rightarrow$ the result is left on proc. (cur_row, cur_col)
    ENDIF
ENDFOR
/* Computation of the blocks of the upper triangular part of C */
FOR $i = 0$ TO $N_b - 1$ DO
    $cur\_row = mod(i, P)$
    $cur\_col = mod(i, Q)$
    $b_r = \lceil \frac{i}{P} \rceil$
    $b_c = \lceil \frac{i}{Q} \rceil$
    IF ($my\_col = cur\_col$) THEN
        DGEBS2D $\Rightarrow$ broadcasts $A_{.,b_c}$ to all processors of the cur_row row
        computes the ith block row of the upper triang. part of C :
        DGEMM $\Rightarrow \alpha A_{.,b_c}^T \times B_{.,(b_c+1,...,P_b)} + \beta C_{b_r,(b_c+1,...,P_b)} \to C_{b_r,(b_c+1,...,P_b)}$
        global sum of $C_{b_r,(b_c+1,...,P_b)}$:
        DGSUM2D $\Rightarrow$ the result is left on proc. (cur_row, my_col)
    ELSE
        DGBR2D $\Rightarrow$ receives $A_{.,b_c}$
        computes the ith block row of the upper triang. part of C :
        DGEMM $\Rightarrow \alpha A_{.,b_c}^T \times B_{.,(b_c,...,P_b)} + \beta C_{b_r,(b_c,...,P_b)} \to C_{b_r,(b_c,...,P_b)}$
        global sum of $C_{b_r,(b_c,...,P_b)}$:
        DGSUM2D $\Rightarrow$ the result is left on proc. (cur_row, my_col)
    ENDIF
ENDFOR
/* transpose and copy the upper triangular part of C in the lower part.*/

Table 3. Symmetric matrix multiplication algorithm.

In the first phase, the diagonal blocks of the matrix C are computed. Only the upper triangular part of each block is computed. For this, a FORTRAN subroutine (SCMM) has been developed since no level 3 BLAS subroutine exists to achieve such a computation. On the Intel Paragon, this compiled subroutine is as efficient as the optimized level 3 BLAS subroutine.

Since the matrices are symmetric, each diagonal block is the product of two block columns that are distributed on the same column of processors. But after the multiplication, each processor has a part of the result. Then, a global sum on this column of processor gives the total result. For example, on a 2×3 grid, the C_{11} block is the product of column $A_{.1}$ by the column $B_{.1}$ and these two columns are distributed on processors 1 and 4. The result of the global sum is left on processor 4 which owns the C_{11} block.

The second phase is hardly different from the first. It computes the remaining blocks of the upper triangular part of C. Therefore, the A and B block columns to multiply are not always on the same processors. For example, on a 2×3 grid, C_{13} is the product of $A_{.1}$ which is distributed on processors 1 and 4, and $B_{.3}$ which is on 0 and 3. Consequently, the $A_{.i}$ block column has to be broadcast and multiplied by $B_{.,(i,...,N_b)}$ to compute $C_{i,(i,...,N_b)}$. As in the first part, each processor has a partial result after the multiplication and a global sum is needed.

The last step consits to transpose the strictly upper triangular part of C in order to obtain the full matrix.

The Fig. 7 shows a comparison between the PDGEMM routine that computes a full matrix multiplication and our routine. In solid line, this is the time in seconds taken by PDGEMM for different matrix sizes. In dashed line, this is the same for our routine. In dotted line, this is the division of the two times. We can see that our code is very efficient for large sizes. The ratio between a complete and a symmetric product can even reach 2. With smaller matrix sizes, the gain decreases. A preliminary theoretical analysis shows that a ratio of 2 is not possible for small matrix sizes. This is due to the block scattered distribution. Each processor has not exactly the half of the computation to achieve a symmetric product. But this ratio is above 1.5 most of the time whatever the matrix size.

5.3 Implementation on the Intel Paragon

All the tests have been done on an Intel Paragon with 30 nodes. Each node is composed of two i860, one for the computations and one for the communications. The nodes are connected by a bidirectional 2d-torus that allows a sustained bandwidth of 69 Mbytes and a latency of $60\mu s$.

Preliminary results on the Paragon have been obtained (see Fig. 8 and Fig. 9). Measures concern the main computational kernel of the code with the matrix-matrix product **pdgemm** of ScaLAPACK.

Because of the large amount of memory needed for the algorithm, the maximum problem size is 256 on one processor and 512 on 4. Even if we have not yet incorporated our symmetric matrix-matrix product in the code, we obtain an efficiency close to 1 and speed-up close to 4 for middle problem size (256) on 4 processors. For smaller problem sizes (50-100), the speed-up stays above 2.

Fig. 10 shows a comparison of the execution times between the ScaLAPACK routine PDSYEVX and the main computational kernel of our code. The routine PDSYEVX [8] is based on a bisection method followed by inverse iterations. We

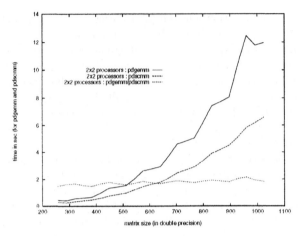

Fig. 7. Comparison between PDGEMM and symmetric matrix multiplication performances.

conclude that our code may be competitive with the bisection method for the computation of all the eigenvalues and all the eigenvectors.

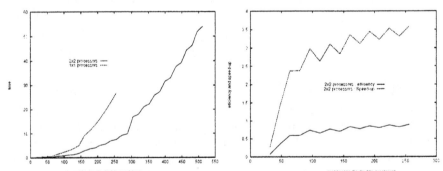

Fig. 8. Execution times for a 2x2 grid according to the size of the matrix on Paragon.

Fig. 9. Efficiency and speed up for a 2x2 grid according to the size of the matrix on Paragon.

6   Conclusion

We studied a new approach to compute the eigenvalues and eigenvectors of a real symmetric matrix. The algorithm parallelized in this paper is not efficient on a

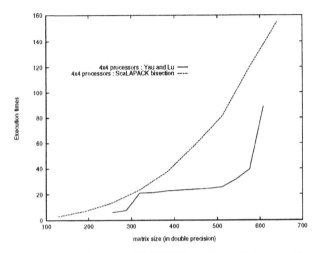

Fig. 10. Execution times for PDSYEVX and Yau & Lu on a Paragon.

sequential machine but can fully take advantage of parallel machines because of less data dependences and the use of matrix products as the most important computed kernel. We obtain good performances for our code concerning efficiency and execution time. Theoritical study [14] showed that our code should scale nicely on parallel machines with a very large number of processors. Of course, the experiments carried out with a grid of 2×2 processors are not conclusive in this respect. That why we want to test our code on large size problems. Future target machine are the SP2 and the T3D.

Acknowledgment I would like to thank the referees and F. Desprez for their helpful comments and suggestions.

References

1. E. Anderson, Z. Bai, C. H. Bischof, J. W. Demmel, J. J. Dongarra, J. J. Du Croz, A. Greenbaum, S. J. Hammarling, A. McKenney, S. Ostrouchov, and D. C. Sorensen. *LAPACK Users' Guide, Release 2.0.* Society for Industrial and Applied Mathematics, Philadelphia, PA, USA, second edition, 1995.
2. Christian Bischof, William George, Steven Huss-Lederman, Xiaobai Sun, Anna Tsao, and Thomas Turnbull. SYISDA users' guide, Version 2.0. Technical report, Mathematics and Computer Science Division, Argonne National Laboratory, Argonne, IL, USA, 1995.
3. Christian Bischof, Steven Huss-Lederman, Xiaobai Sun, Anna Tsao, and Thomas Turnbull. Parallel performance of a symmetric eigensolver based on the invariant subspace decomposition approach. In *proceedings of Scalable High Performance Computing Conference'94, Knoxville, Tennessee*, pages 32–39, May 1994. (also PRISM Working Note #15.

4. Jaeyoung Choi, James Demmel, I. Dhillon, Jack J. Dongarra, Susan Ostrouchov, Antoine P. Petitet, k. Stanley, David W. Walker, and R. Clint Whaley. ScaLA-PACK: A portable linear algebra library for distributed memory computers - design issues and performances. LAPACK Working Note 95, Oak Ridge National Laboratory, Oak Ridge, TN, USA, 1995.

5. Jaeyoung Choi, Jack J. Dongarra, Roldan Pozo, and David W. Walker. ScaLA-PACK: A scalable linear algebra library for distributed memory concurrent computers. Technical Report CS-92-181, Department of Computer Science, University of Tennessee, Knoxville, TN, USA, November 1992. LAPACK Working Note 55.

6. C.W. Cooley and J.W. Tuckey. An Algorithm for the Machine Calculation of Complex Fourier Series. *Math. Comput.*, 19:297–301, 1965.

7. J. J. M. Cuppen. A divide and conquer method for the symmetric tridiagonal eigenproblem. *Numer. Math.*, 36:177–195, 1981.

8. James W. Demmel and K. Stanley. The performance of finding eigenvalues and eigenvectors of dense symmetric matrices on distributed memory computers. Technical Report CS-94-254, Department of Computer Science, University of Tennessee, Knoxville, TN, USA, September 1994. LAPACK Working Note 86.

9. J. Dongarra and D. Sorensen. A fully parallel algorithm for the symmetric eigenvalue problem. *SIAM J. Sci. Stat. Comput.*, 8:139–154, 1987.

10. D. Giménez, R. van de Geijn, V. Hernández, and A. M. Vidal. Exploiting the symmetry on the jacobi method on a mesh of processors. In *4th EUROMICRO Worshop on Parallel and Distributed Processing, Braga, Portugal*, 1996.

11. Gene H. Golub and Charles F. Van Loan. *Matrix Computations*. Johns Hopkins University Press, Baltimore, MD, USA, second edition, 1989.

12. C. Lin and L.Snyder. A matrix product algorithm and its comparative performance on hypercubes. In *Scalable High Performance Computing Conference SHPCC92*, pages 190–193. IEEE Computer Society, 1992.

13. Beresford N. Parlett. *The Symmetric Eigenvalue Problem*. Prentice-Hall, Englewood Cliffs, NJ, USA, 1980.

14. Makan Pourzandi and Françoise Tisseur. Parallèlisation d'une nouvelle méthode de recherche de valeurs propres pour des matrices réelles symétriques. Report TR94-37, LIP, ENS Lyon, 1994.

15. Oscar Rojo and Ricardo L. Soto. A decreasing sequence of eigenvalue localization regions. *Linear Algebra and Appl.*, 196:71–84, 1994.

16. Shing-Tung Yau and Ya Yan Lu. Reducing the symmetric matrix eigenvalue problem to matrix multiplications. *SIAM J. Sci. Comput.*, 14(1):121–136, 1993.

Preconditioned Conjugate Gradient Methods for Semiconductor Device Simulation on a CRAY C90 Vector Processor

Stefan Thomas

Technische Universität Hamburg-Harburg, Technische Informatik 2,
Harburger Schloßstr. 20, 21071 Hamburg, Germany
s.thomas@tu-harburg.d400.de

Abstract. The finite difference discretization of the semiconductor equations yields symmetric, positive definite block-tridiagonal linear systems, which can be solved efficiently by the conjugate gradient method (CG). We have investigated several preconditioners with respect to vectorization to improve the simulation runtime. The performance of the different strategies has been evaluated on a CRAY C90 vector processor. We have found, that diagonal scaling can hardly be improved by additional incomplete Cholesky and polynomial preconditioners, because the reduction in the total number of iterations is usually compensated by the increased complexity of the preconditioned CG iteration. However, if the CG method is embedded in a nonlinear outer iteration, runtime savings have been obtained in some cases, because the preconditioned algorithms have produced a stable outer iteration with less stringent stopping criteria for the inner CG iterations.

1 Device Simulation

Semiconductor devices are commonly described by the drift-diffusion model, which consists of the following set of equations [12]. Poisson's equation

$$-\nabla \cdot \varepsilon \nabla \varphi + q(n - p - N) = 0 \tag{1}$$

with the electrostatic potential φ, mobile carrier concentrations n, p, net impurity concentration N, permittivity ε and elemental charge q relates the total space-charge to the divergence of the electric field. The stationary continuity equations for electrons and holes are given by

$$\frac{1}{q}\nabla \mathbf{J}_n - R = 0\,, \quad -\frac{1}{q}\nabla \mathbf{J}_p - R = 0 \tag{2}$$

with the current densities $\mathbf{J}_n$, $\mathbf{J}_p$ and the net recombination rate R. The current densities are given by

$$\mathbf{J}_n = q\mu_n(-n\nabla\varphi + kT/q\nabla n)\,, \quad \mathbf{J}_p = q\mu_p(-p\nabla\varphi - kT/q\nabla p)\,, \tag{3}$$

where μ_n, μ_p are the carrier mobilities, k is Boltzmann's constant and T is the carrier temperature. The discretization of the continuity equations (2) with the current densities (3) produces non-symmetric systems. Slotboom suggested the transformed state variables v, w for the carrier concentrations n, p [13]

$$v = \frac{n}{n_i} \exp\left(-\frac{q\varphi}{kT}\right) , \quad w = \frac{p}{n_i} \exp\left(\frac{q\varphi}{kT}\right) . \tag{4}$$

Poisson's equation and the current densities become

$$\nabla \cdot \varepsilon \nabla \varphi = q \left[\exp\left(\frac{q\varphi}{kT}\right) v - \exp\left(\frac{q\varphi}{kT}\right) w - N\right] , \tag{5}$$

$$\mathbf{J}_n = \mu_n n_i kT \exp\left(\frac{q\varphi}{kT}\right) \nabla v , \quad \mathbf{J}_p = \mu_p n_i kT \exp\left(-\frac{q\varphi}{kT}\right) \nabla w , \tag{6}$$

where n_i is the intrinsic carrier concentration. Substitution of the current densities (6) into the continuity equations (2) yields self-adjoint differential operators, which can be discretized to symmetric, positive definite systems. The exponential variations $\exp(q\varphi/kT)$ and $\exp(-q\varphi/kT)$ in the current-densities are taken into account by the Scharfetter-Gummel discretization scheme [11] and diagonal scaling (see below).

This nonlinear, coupled set of partial differential equations is solved numerically because of the complexity of two- and three-dimensional device geometries. The equations are linearized with a block nonlinear iteration, the so-called Gummel iteration [5]. Here, the Poisson and continuity equations are linearized and solved separately one after the other to update only one set of state variables (*inner iteration*), while the other variables are fixed. This procedure is repeated until convergence (*outer iteration*).

The current-voltage characteristics of a semiconductor device are derived incrementally by solving the drift-diffusion equations for different bias conditions at the contacts, which are modeled by Dirichlet boundaries. The simulation starts with the solution of the semiconductor equations in thermal equilibrium. Solutions of the previous bias steps are then extrapolated to an initial guess for the Gummel iteration of the next bias step.

2 Preconditioning of the Conjugate Gradient Method

Two-dimensional finite difference discretization of the elliptic operators of the semiconductor equations in $n = n_x \times n_y$ grid points yields large, sparse linear systems

$$Ax = b, \quad A \in \mathrm{I\!R}^{n \times n}, \mathrm{x}, \mathrm{b} \in \mathrm{I\!R}^n \tag{7}$$

with symmetric, positive definite coefficient matrix A, which has a block-tridiagonal structure with 5 off-diagonals.

The conjugate gradient algorithm is an effective method to solve these linear systems [3]. Here, the size of the discretized systems and the exponential coupling of the variables result in large condition numbers for even moderate mesh sizes, so that preconditioning becomes attractive

- to reduce the number of iterations and - hopefully - runtime,
- to improve the stability of the iteration process (*inner* and *outer* iteration).

In the following, we investigate diagonal preconditioners, factorization methods and polynomial preconditioning with respect to vectorization properties.

2.1 Diagonal Scaling

While the electrostatic potential φ varies in the order of several volts across the simulated domain, the thermal voltage k_BT/q is only 0.026V at room temperature. The exponential expressions $\exp(q\varphi/kT)$ and $\exp(-q\varphi/kT)$ in the current-densities can therefore vary over several orders of magnitude. Additionally, nonuniform mesh spacings increase the range of the entries of the coefficient matrices. Here, the simplest preconditioning is by *diagonal scaling*

$$D^{-1/2}AD^{-1/2}D^{1/2}x = D^{-1/2}b, \qquad (8)$$

with matrices $D^{-1/2} = \text{diag}(a_{11}^{-1/2},\ldots,a_{nn}^{-1/2})$, $D^{1/2} = \text{diag}(a_{11}^{1/2},\ldots,a_{nn}^{1/2})$, preserving the symmetry of the original matrix A and producing a constant main diagonal. The diagonally scaled conjugate gradient algorithm will be denoted by DSCG in the following.

2.2 Incomplete Cholesky Factorizations

Here, the conjugate gradient method is applied to the transformed system

$$M^{-1}Ax = M^{-1}b, \quad M \in \mathbb{R}^{n\times n}. \qquad (9)$$

In each iteration k the linear system

$$M\tilde{r}^{(k)} = r^{(k)} \qquad (10)$$

with the original and transformed residual vectors $r^{(k)}$ and $\tilde{r}^{(k)}$ has to be solved. One usually chooses M as the incomplete Cholesky factorization $M = LL^T \approx A$ of the coefficient matrix A, which yields the incomplete Cholesky conjugate gradient (=ICCG) algorithm. The linear systems (10) can then be solved by forward and backward elimination

$$M\tilde{r} = LL^T\tilde{r} = r \quad \Leftrightarrow \quad Ly = r, \quad L^T\tilde{r} = y. \qquad (11)$$

Because of the data dependencies and the small number of off-diagonal elements the straight forward implementation of (11) does not vectorize efficiently.

Greenbaum and Rodrigue showed that some part of the work in solving a system $Ly = r$ can be vectorized by taking advantage of the block-tridiagonal structure of the coefficient matrix arising from two-dimensional finite difference discretizations in $n = n_x \times n_y$ grid points [4]. The root-free incomplete Cholesky factorization $M = LDL^T$ yields a lower block-diagonal matrix $L = [C_i, B_i, 0]$ with bidiagonal matrices $B_i = [b_{ij}, 1, 0]$ and diagonal matrices C_i. The forward elimination $y = L^{-1}r$ can be computed by the first-order recurrence

$$y_1 = B_1^{-1}r_1$$

$$y_i = B_i^{-1}(r_i - C_iy_{i-1}), \quad i = 1,\ldots n_y,$$

in which the computation of $(r_i - C_iy_{i-1})$ for the i-th block is now vectorizable and the application of the bidiagonal B_i^{-1} still requires forward elimination.

Van der Vorst extended this approach and introduced a Neumann series approximation

$$B_i^{-1} \approx I + N_i + N_i^2 + N_i^3 , \quad N_i = I - B_i$$

for the remaining first-order recurrence, leading to a fully vectorizable algorithm [14]. Jordan has proved the superior performance of the vectorized variants on a CRAY 1 vector processor [6]. In the following, we will denote these partly and fully vectorized variants of the incomplete Cholesky factorization with ICCGv1 and ICCGv2, respectively.

2.3 Polynomial preconditioning

Here, the conjugate gradient method is applied to the transformed linear system

$$s(A)Ax = s(A)b \tag{12}$$

with a matrix polynomial $s(A)A = c_m A^m + c_{m-1} A^{m-1} + \ldots + c_1 A$ of small degree m. To yield an efficient preconditioner, the resulting matrix $s(A)A$ should have an eigenvalue distribution which is favorable to the conjugate gradient method. Therefore, one usually tries to find a polynomial $s(A)A$, which minimizes

$$\|1 - s(\lambda)\lambda\|$$

for the spectrum $\lambda \in \sigma(A)$ of the eigenvalues. The strategies to construct an optimal polynomial differ in the choice of the norm, an optional weight function and a-priori knowledge about the eigenvalue distribution of the spectrum $\sigma(A)$.

Least squares polynomials (LSQ) associated with Jacobi weights have been tabulated by Saad in [9]. These polynomials can be derived from an estimate of the largest eigenvalue λ_n, which is obtained easily by application of Gershgorin's theorem during the assembly of the coefficient matrix A.

Chebyshev polynomials (CHEB) depend on an accurate estimate α, β of the bounds λ_1, λ_n of the spectrum $\sigma(A)$. Unfortunately, the obvious approximation of the smallest eigenvalue λ_1 by $\alpha = 0$ is not applicable in principle. One can show that Chebyshev polynomials constructed from the exact extremal eigenvalues λ_1, λ_n yield the *best* preconditioner with respect to the condition number, which is minimized. However, several authors have reported that these polynomials are not optimal in view of the iteration count, because the distribution of the eigenvalues has a stronger impact on the convergence properties than the condition number alone. Saad pointed out, that the choice of the interval $[\alpha, \beta]$ slightly inside $[\lambda_1, \lambda_n]$ produces more efficient preconditioners in many cases. As the exact extremal eigenvalues are usually not available, the approximation with the Lanczos procedure seems especially suitable, as the estimates α, β converge incrementally from inside towards λ_1, λ_n.

Adaptive weighted polynomials (ADAPT), as have been proposed by Fischer and Freund [2], are based on the eigenvalue distribution $f(\lambda)$ of the spectrum. One can obtain a relative good approximation of the distribution by application of several Lanczos steps. The distribution-*density* is then used as a weight function $w(\lambda)$ for a weighted Chebyshev approximation problem, which can be solved by the Remez algorithm.

We have to point out, that all the favorable vectorization properties of the modified incomplete Cholesky and polynomial preconditioned conjugate gradient algorithms are due to the diagonal structure of the linear system. Ruggiero has compared the performance of incomplete Cholesky and polynomial preconditioners on a CRAY Y-MP vector processor for a number of numerical examples of finite element discretizations of flow problems and finite difference discretizations of the diffusion equation [8]. In most of the test examples the matrices have been stored in compressed form and the vectorization performance has been rather poor in general. In one test example the matrix has also been stored by diagonals and the runtime of the polynomial preconditioners has been improved by a factor of 5!

In our application we could take advantage of the symmetry of our 5-diagonal matrices, which have been stored in an two-dimensional array

```
real, dimension (n,3) :: A,
```

where `A(:,1)` holds the main diagonal and `A(1:n-1,2)`, `A(1:n-nx,3)` store the 1st and n_x-th upper off-diagonals.

3   Results

The different preconditioning techniques have been investigated on a CRAY C90 parallel vector processor. Each processor is clocked at 4ns. With parallel dual-vector pipelines for addition and multiplication each of the maximum 16 processors is capable of 1 GFLOP/s peak performance. The register length is 128 and the half-performance vector lengths for saxpy-operations and inner products are approximately $n_{saxpy}^{1/2} \approx 200$, $n_{sdot}^{1/2} \approx 300$ [7]. The device simulator has been implemented in **Fortran 90** (Cray Research Inc., Release 1.0). All calculations have been performed in 64-bit arithmetic. The timings in clock periods have been obtained with the intrinsic **irtc** function. The total runtime and MFLOP/s performance have been derived with the **hpm** utility.

3.1   Test Problem

We have evaluated the performance of the preconditioners for two finite difference discretizations of a MOS transistor in $n = 43 \times 47 = 2,021$ and $n = 85 \times 82 = 6,970$ grid points. We have assumed Dirichlet boundary conditions for the drain-, source- and bulk-contacts, homogeneous Neumann boundary conditions for insulated surfaces and a mixed boundary condition for the gate-contact. Figure 1 depicts the coarser rectangular mesh and electron concentration in thermal equilibrium without gate bias.

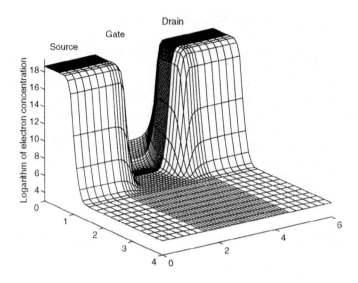

Fig. 1. Finite difference discretization of a MOS transistor in $n = 43 \times 47$ grid points. Electron concentration in thermal equilibrium without gate bias.

3.2 Preconditioning and Vectorization Performance

For the investigation of the preconditioning and vectorization performance of the different algorithms we have compared the number of iterations and runtime of the CG algorithm to solve the first Newton step of the nonlinear Poisson equation in thermal equilibrium with a gate bias of $U_G = 1.5V$. Diagonal scaling has reduced the condition number of the smaller and larger linear system from $\kappa = 15,800$ to $\kappa = 80$ and from $\kappa = 6,600$ to $\kappa = 85$, respectively. Because diagonal scaling results in little additional overhead, we applied the incomplete Cholesky and polynomial preconditioners to the scaled linear systems with constant main diagonal $a_{ii} = 4$. In this case, an upper bound for the largest eigenvalue λ_n is given by $\beta = 8$.

Preconditioning Polynomials. Figure 2.a shows the eigenvalue distribution $f(\lambda)$ and weight function $w(\lambda)$ of the smaller linear system, which have been approximated after 20 Lanczos steps. The corresponding weighted Chebyshev polynomial (ADAPT), least squares polynomial (LSQ) and Chebyshev polynomial (CHEB), which has been constructed from an estimate of the extremal eigenvalues after 20 Lanczos steps are depicted in Figure 2.b.

The Chebyshev polynomial (CHEB) equioscillates about 1 across the interval $[\lambda_1, \lambda_n]$. The estimated value α of the smallest eigenvalue determines the steepness and the extension of the polynomial. Here, the obvious approximation

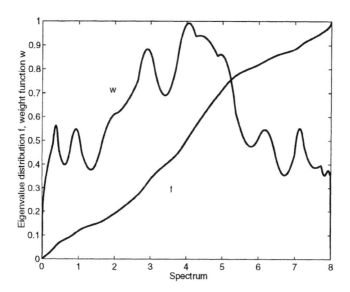

a) Approximated eigenvalue distribution f and derived weight function w.

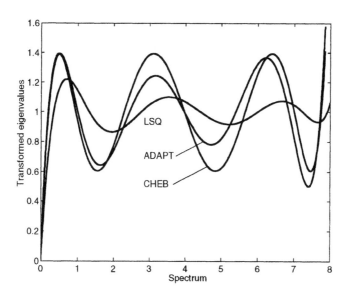

b) Transformation of the estimated spectrum by a least squares (LSQ), Chebyshev (CHEB) and weighted Chebyshev polynomial (ADAPT) of degree $m = 7$.

Fig. 2. First Newton step of Poisson's equation at thermal equilibrium.

$\alpha = 0$ would produce a polynomial, which oscillates between $[0, 2]$ and is inappropriate as a preconditioner, because it may map some eigenvalues to 0. The weighted Chebyshev polynomial (ADAPT) focuses on the concentration of the inner eigenvalues. The least squares polynomial (LSQ) is biased to suppress the larger eigenvalues. It is worth noting, that the eigenvalue distribution f corresponds to an almost uniform distribution of the eigenvalues and that the least squares polynomial achieves the best overall concentration across the interval.

Preconditioning Performance and Runtime. The number of iterations and total runtime of the different preconditioned CG algorithms are depicted in Figure 3. The convergence criterion has been formulated as

$$\text{iterate } k = 1, \dots \text{ until:} \quad \frac{||r^{(k)}||}{||r^{(0)}||} < \epsilon = 10^{-6} . \tag{13}$$

Preprocessings include: the incomplete Cholesky factorization (ICCG, ICCGv1, ICCGv2); the Lanczos procedure for estimating the extremal eigenvalues (CHEB); the approximation of the eigenvalue distribution and solution of the weighted Chebyshev problem with the Remez algorithm (ADAPT). Postprocessing of the polynomial preconditioned CG variants includes a single matrix-polynomial vector product for back transformation.

Diagonal scaling alone has led to a substantial reduction in the number of iterations and runtime as might have been expected from the improvement in the condition numbers. The incomplete Cholesky factorization and the polynomial preconditioners resulted in a further decrease in the number of iterations but have mainly failed to reduce the runtime because of

- preprocessing overheads (ICCG variants, CHEB and especially ADAPT),
- increased arithmetic complexity of a single iteration, and
- poor vector performance (especially ICCG, ICCGv1).

The fully vectorized preconditioner ICCGv2 utilizes the CRAY C90 vector processor more effectively than the standard ICCG and partly vectorized ICCGv1 algorithm. The approximation of B_i^{-1} has not affected the number of iterations.

For the finer mesh, Figure 3.b, the disastrous impact of preprocessing has decreased slightly. Here, the least squares polynomials have been able to outperform the incomplete Cholesky preconditioners in the number of iterations *and runtime*.

The extra work, which has been invested in the adaptive construction of a weighted Chebyshev polynomial, has not been rewarded in the number of iterations and is prohibitive for these rather small linear systems because of the preprocessing overhead.

162

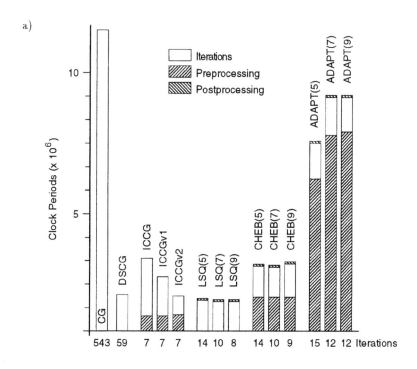

a)

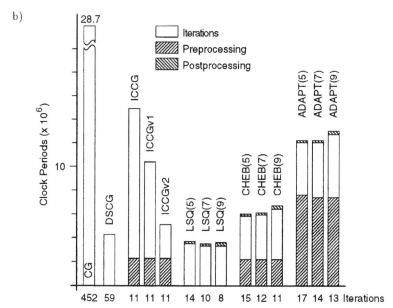

b)

Fig. 3. Runtime in clock periods for the solution of the first Newton step of Poisson's equation in thermal equilibrium, a) $n = 43 \times 47$, b) $n = 85 \times 82$.

Stability of the Adaptive Preconditioner. Figure 4.a shows the eigenvalue distribution and weight function of the linear system of the continuity equation for electrons in the first outer iteration for the first bias point of the current-voltage characteristic ($U_D = 0.02V$). Diagonal scaling reduced the condition number of the linear system from $\kappa > 100,000$ to $\kappa \approx 10,000$. The preconditioning polynomial, which has been constructed by the solution of the weighted Chebyshev approximation problem, is depicted in Figure 4.b.

The weighted Chebyshev polynomial extends wide into the negative range of numbers. The resulting preconditioned linear system is not positive definite any more and the CG iteration fails to converge in principal. For a more robust implementation we would suggest damping strategies for the derivation of the weight function $w(\lambda)$ from the eigenvalue distribution $f(\lambda)$.

The density of small eigenvalues close to 0 has helped the Lanczos procedure to converge towards the smallest eigenvalue. The estimate after 20 Lanczos steps has been used for the construction of the Chebyshev polynomial (CHEB), which now equioscillates rather wide 1 ± 0.8.

Here, the original CG algorithm without diagonal scaling fails to converge. With diagonal scaling the number of iterations is 378. The Chebyshev polynomial performs slightly better (60 iterations) than the least squares polynomial (69), which goes in line with the observation of Ashby et al, that Chebyshev polynomials perform superior if the spectrum of the eigenvalues is dense near the origin [1]. However, this advantage in the number of iterations is practically compensated in runtime by the preprocessing overhead of the Lanczos procedure.

3.3 Simulation of Current-Voltage Characteristics

As a consequence of the results of the preceding section, we have decided to focus our attention on the DSCG algorithm, ICCG variants and LSQ polynomial preconditioner. We have investigated the runtime and convergence properties for the simulation of the current-voltage characteristic of a MOS transistor with a gate bias of $U_G = 1.5V$. The drain voltage U_D has been incremented in 23 bias points from 0.0 to $3.0V$.

The convergence criterion for the nonlinear *outer* iteration has been stated as

$$\frac{\|\Delta\varphi^{(i)}\|}{\|\varphi^{(i)}\|} < \varepsilon_\varphi \;\wedge\; \frac{\|\Delta v^{(i)}\|}{\|v^{(i)}\|} < \varepsilon_{vw} \;\wedge\; \frac{\|\Delta w^{(i)}\|}{\|w^{(i)}\|} < \varepsilon_{vw} \;.$$

In this application, the choice of $\varepsilon_\varphi = 10^{-3}$ and $\varepsilon_{vw} = 10^{-5}$ has led to a stable Gummel iteration. For the *inner* iteration of the CG algorithm we have imposed the stopping criterion (13) with $\varepsilon_{\varphi,CG} = \varepsilon_{vw,CG} = 10^{-3}$ for the Poisson and continuity equations.

Table 1 denotes the runtime in seconds, MFLOP/s performance and total number of Gummel cycles for the simulation of the coarser mesh.

At first, the accuracy of the diagonally scaled CG algorithm alone was not sufficient for the convergence of the outer Gummel iteration, so we had to impose a stricter convergence criterion of $\varepsilon_{vw,GC} = 10^{-5}$ for the inner CG iteration.

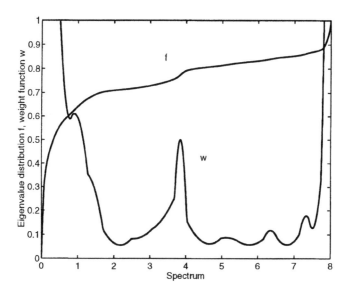

a) Approximated eigenvalue distribution f and derived weight function w.

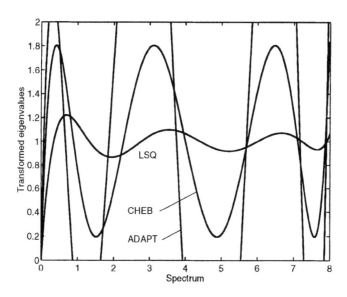

b) Transformation of the estimated spectrum by a least squares (LSQ), Chebyshev (CHEB) and weighted Chebyshev polynomial (ADAPT) of degree $m = 7$.

Fig. 4. First Gummel cycle of the continuity eq. of electrons at the first bias point.

Table 1. Runtime in seconds for the simulation of the current-voltage characteristics on the coarser mesh with 43×47 grid points.

	DSCG	ICCG		LSQ			Mixed
		v1	v2	$m = 5$	7	9	$m = 9$, v2
Time (s)	43.7	55.7	44.1	31.4	36.9	35.6	35.0	31.0
MFLOP/s	270	48	61	102	203	204	205	112
Cycles	397	397	397	397	398	397	397	397

Table 2. Runtime in seconds for the simulation of the current-voltage characteristics on the finer mesh with 85×82 grid points.

	DSCG	ICCG		LSQ			Mixed
		v1	v2	$m = 5$	7	9	$m = 9$, v2
Time (s)	123.6	268.7	195.7	109.4	114.9	110.0	109.9	107.2
MFLOP/s	333	53	73	158	268	271	267	167
Cycles	384	374	374	382	382	382	382	382

Here, the improvements in runtime have come from the increased stability of the preconditioned algorithms. While the ICCG and ICCGv1 algorithms have not been competitive in comparison with DSCG, the fully vectorized ICCGv2 variant with the Neumann series approximation has produced the best runtime performance and has converged as stable as the exact ICCG algorithms. The polynomial preconditioned CG algorithms have yielded much higher MFLOP/s performances but have not been able to beat the ICCGv2 algorithm. As the Poisson equation has shown a favorable eigenvalue distribution for the least squares polynomial preconditioner, we have also investigated a *mixed* approach, in which the Poisson equation has been solved by LSQ(m=9) and the continuity equations by ICCGv2. The runtime has been reduced further slightly.

The performance of the preconditioners for the finer mesh is denoted in Table 2. Here, the diagonally scaled CG algorithm has converged stable with the less stringent convergence criterion $\varepsilon_{vw,CG} = 10^{-3}$ using some additional Gummel cycles. Now, the approximation in the ICCGv2 algorithm has introduced additional Gummel cycles. Even though, ICCGv2 has performed superior compared to the ICCG and ICCGv1 variants. LSQ(9) has been almost as fast as ICCGv2 and, consequently, the mixed approach has been applied successfully again.

4 Conclusions

As the continuity equations are hard to solve, preconditioning becomes necessary for even comparatively small linear systems derived from meshes with only several 1000 grid points. Diagonal scaling alone constitutes a very efficient preconditioner in this application. Additional preconditioning with incomplete Cholesky factorizations or polynomial preconditioners can reduce the number of (inner, conjugate gradient) iterations considerably. However, this improvement is almost offset by the increased arithmetic complexity of the single iteration, as has been demonstrated in Figure 3. The benefits of preconditioning become more obvious if the conjugate gradient method is embedded into a nonlinear outer iteration, like Gummel's method for semiconductor device simulation. Here, a stable iteration can be achieved with a less stringent convergence criterion for the inner iteration, leading to a shorter total runtime.

To exploit the performance of a vector processor with an incomplete Cholesky preconditioner, the fully vectorized variant ICCGv2 has to be implemented. The approximation of B_i^{-1} by a Neumann series has had no greater impact on the stability of the Gummel iteration (382 iterations versus 374 of ICCG, ICCGv1) than the polynomial preconditioners (382 iterations).

For these small linear systems the preprocessing overhead for the estimation of the extremal eigenvalues and/or approximation of the eigenvalue distribution is prohibitive and does not guarantee an improvement in the number of iterations in general. The adaptive construction of the polynomials from an approximation of the eigenvalue distribution fails for the degenerated distribution of the continuity equations.

The best runtime has been achieved with a mixed approach, in which Poisson's equation has been solved by a least squares polynomial preconditioner and the continuity equations have been solved by the fully vectorized incomplete Cholesky preconditioner. We might expect a further improvement by application of Chebyshev polynomials for the continuity equations, if the polynomials are constructed from a fixed interval $[\alpha, \beta]$, which is estimated only *once* for each Gummel cycle or the total simulation, reducing preprocessing overheads.

References

1. Ashby, S. F., Manteuffel, T. A., Otto, J. S., *A Comparison of Adaptive Chebyshev and Least Squares Polynomial Preconditioning for Hermitian Positive Definite Linear Systems*, SIAM J. Sci. Stat. Comput., **13**, 1-29, 1992

2. Fischer, B., Freund, R. W., *On Adaptive Weighted Polynomial Preconditioning For Hermitian Positive Definite Matrices*, SIAM J. Sci. Comput., **15**, pp. 408-426, 1994

3. Golub, G., Ortega, J. M., *Scientific Computing - An Introduction With Parallel Computing*, Academic Press, Boston, Mass., 1993

4. Greenbaum, A., Rodrigue, G. H., *The Incomplete Cholesky Conjugate Gradient for the STAR (5-point) Operator*, Rep. UCID 17574, Lawrence Livermore National Laboratory, Livermore California, 1977.

5. Gummel, H. K., *A Self-Consistent Iterative Scheme for One-Dimensional Steady State Transistor Calculations*, IEEE Trans. Electron Devices, **11**, pp. 455-465, 1964.

6. Jordan, T. L., *Conjugate Gradient Preconditioners for Vector and Parallel Processors*, in: Birkhoff, G., Schoenstadt, A. (editors), *Elliptic Problem Solvers II*, Academic Press, Orlando, 1984.

7. Oed, W., *Cray Y-MP C90: System features and early benchmark results*, Parallel Computing, **18**, pp. 947-954, 1992.

8. Ruggiero, V., *Polynomial Preconditioning on Vector Computers*, Applied Mathematics and Computation, **59**, 131-150, 1993.

9. Saad, Y., *Practical use of polynomial preconditionings for the conjugate gradient method*, SIAM J. Sci. Stat. Comput., **6**, pp. 865-881, 1985.

10. Saad, Y., Krylov subspace methods on supercomputers, SIAM J. Sci. Stat. Comput., **10**, pp. 1200-1232, 1989.

11. Scharfetter, D. L., Gummel, H. K., *Large-Signal Analysis of a Silicon Read Diode Oscillator*, IEEE Trans. Electron Devices, **16**, pp. 64-77, 1969.

12. Selberherr, S., *Analysis and simulation of semiconductor devices*, Springer, Wien, 1984

13. Slotboom, J. W., *Iterative Scheme for 1- and 2-Dimensional D.C.-Transistor Simulation*, Electronics Letters, **5**, pp. 677-678, 1969.

14. Van Der Vorst, H. A., *A vectorizable version of some ICCG methods*, SIAM J. Sci. Stat. Comput., **3**, pp. 350-356, 1982.

A Parallel Implementation of the General Lanczos Method on the CRAY T3D

José Ignacio Aliaga[1]

Dpto. Informática, Universidad Jaume I, 12071-Castellón, Spain

e-mail : aliaga@inf.uji.es

Vicente Hernández[1][†]

Dpto. Stmas. Inform. y Comp., Univ. Politéc. de Valencia, 46071-Valencia, Spain

e-mail : vhernand@dsic.upv.es

José Luis Pérez[1][†]

Dpto. Stmas. Inform. y Comp., Univ. Politéc. de Valencia, 46071-Valencia, Spain

e-mail : jlperez@dsic.upv.es

Abstract

The general Lanczos method overcomes the breakdown and the block reduction problems which appear in the block nonsymmetric Lanczos method. In the general method we exploit the relation between Lanczos and the two-sided Gram-Schmidt processes. This paper presents a parallel implementation of the general Lanczos method on the massively parallel processor CRAY T3D, where good efficiencies and performances can be obtained.

1 Introduction.

The nonsymmetric Lanczos method was defined as a generalization of the symmetric case (Lanczos, 1950), where the method reduces the original matrix, A, to tridiagonal form by using a nonorthogonal transformation. In this case, two Lanczos vector sequences are obtained, respectively, related to the original matrix and its transpose. The method terminates when the inner product of two Lanczos vectors is zero. If none of the Lanczos vectors is null and the inner product is zero, a serious breakdown appears. Many people have studied ways to solve this problem, (Parlett, Taylor and Liu, 1985), (Boley and Golub, 1991), (Gutknecht, 1992), (Parlett, 1992), (Freund, Gutknecht and Nachtigal, 1993).

The block versions of the Lanczos methods permit us to improve the performance, and to find multiple eigenvalues of the matrix A. The projection matrices are defined as block matrices, such that the methods reduce the original matrix to block tridiagonal form. In the symmetric case (Golub and Underwood, 1977), the main problem is to reduce the block size when the new block of Lanczos vectors is singular. In the nonsymmetric case (Kim and Craig, 1988), it is necessary to study additional difficulties such as the different number of initial vectors in each Lanczos sequence and the serious breakdown problem, (Aliaga, 1995). Considering all these drawbacks, a general Lanczos algorithm is obtained (Aliaga, Boley, Freund and Hernández, 1996).

First, we describe how to deal with breakdown in the nonsymmetric Lanczos methods. Afterwards, we generalize the solution on the block nonsymmetric Lanczos method, obtaining the general Lanczos method. We also describe numerical aspects of the implementation of this method. Finally, we introduce its parallel implementation on the CRAY T3D, analyzing the performances obtained, and their relation with the performances of basic operations, namely sparse matrix-vector and inner products.

---

[1] Supported by the ESPRIT III Basic Research Programm of the EC under contract num. 9072. (Project GEPPCOM).

[†] Supported by the R+D Project GV-1076/93 of the Valencian Community.

2 The Nonsymmetric Lanczos Method.

2.1 Definition in Exact Arithmetic.

Given two initial vectors, $x \in \mathfrak{R}^n$ and $y \in \mathfrak{R}^n$, the nonsymmetric Lanczos method (Lanczos, 1950) allow us to reduce a matrix $A \in \mathfrak{R}^{n \times n}$ to the tridiagonal form T, by using a nonorthogonal similarity transformation $Q^{-1}AQ = T$,

$$T = \text{tridiag} \begin{pmatrix} & \gamma_1 & \gamma_2 & \cdots & \gamma_{n-1} & \\ \alpha_1 & \alpha_2 & \alpha_3 & \cdots & & \alpha_n \\ & \beta_1 & \beta_2 & \cdots & \beta_{n-1} & \end{pmatrix} \in \mathfrak{R}^{n \times n} .$$

To explain how the method works, two tridiagonalization processes associated, respectively, with A and A^t are defined,

$$\begin{cases} AQ = QH & , \quad Q = [q_1, \cdots, q_n] \quad , \quad x = q_1 \beta_0 \\ A^t P = PG & , \quad P = [p_1, \cdots, p_n] \quad , \quad y = p_1 \gamma_0 \end{cases}$$

where H and G are tridiagonal matrices, and β_0 and γ_0 are scale factors for the initial vectors, x and y. The product of the matrices Q and P is a nonsingular diagonal matrix, $P^t Q = \Omega = \text{diag}(\omega_1, \cdots, \omega_n)$, which allows us to obtain the relation between the matrices H and G,

$$\left. \begin{array}{c} P^t AQ = P^t QH = \Omega H \\ Q^t A^t P = Q^t PG = \Omega^t G \end{array} \right\} \quad \Rightarrow \quad \Omega H = G^t \Omega .$$

It is possible to scale the matrices Q and P such that $P^t Q = I$, and then $T = H = G^t$. In the rest of the paper, it is supposed that $P^t Q = I$.

Matrices Q and P define two Lanczos vector sequences, whose vectors are computed by using the following three-term recurrences, obtained from the tridiagonalization processes,

$$\begin{cases} Aq_j = \gamma_{j-1}q_{j-1} + \alpha_j q_j + \beta_j q_{j+1} & , \quad j = 1, 2, \cdots, n \quad , \quad \gamma_0 q_0 = \beta_n q_{n+1} = 0 \\ A^t p_j = \beta_{j-1}p_{j-1} + \alpha_j p_j + \gamma_j p_{j+1} & , \quad j = 1, 2, \cdots, n \quad , \quad \beta_0 p_0 = \gamma_n p_{n+1} = 0 . \end{cases}$$

The nonsymmetric Lanczos iteration is directly obtained from these recurrences, where the values β_j and γ_j define the scale factors for the matrices Q and P. In this case, the product of these values is equal to ω_j,

$$\alpha_j = p_j^t A q_j$$

$$u_{j+1} = (A - \alpha_j I)q_j - \gamma_{j-1}q_{j-1} \quad , \quad v_{j+1} = (A^t - \alpha_j I)p_j - \beta_{j-1}p_{j-1}$$

$$\omega_j = \beta_j \gamma_j = v_{j+1}^t u_{j+1} \quad , \quad q_{j+1} = u_{j+1}/\beta_j \quad , \quad p_{j+1} = v_{j+1}/\gamma_j .$$

The iteration terminates when the inner product of the new Lanczos vectors is null,

$$\omega_j = \beta_j \gamma_j = v_{j+1}^t u_{j+1} = 0 .$$

If the iteration terminates and none of the new Lanczos vectors is null,

$$\omega_j = 0 \quad , \quad u_{j+1} \neq 0 \quad , \quad v_{j+1} \neq 0$$

then a serious breakdown appears. This is an undesirable situation and several alternatives, based on look-ahead techniques, have been defined (Parlett, Taylor and Liu, 1985), (Boley and Golub, 1991), (Gutknecht, 1992), (Parlett, 1992), (Freund, Gutknecht and Nachtigal, 1993).

2.2 The Look-Ahead Techniques.

A way to solve the serious breakdown problem in the nonsymmetric Lanczos method is to exploit the relationship between Lanczos and two-sided Gram-Schmidt processes (Parlett, 1992). Given two sequences of vectors, represented by matrices,

$$U = [u_1, u_2, \cdots, u_r] \in \Re^{nxr}, \; V = [v_1, v_2, \cdots, v_s] \in \Re^{nxs},$$

the two-sided Gram-Schmidt method factorizes them as follows,

$$\begin{cases} U = QR & , \; Q = [q_1, q_2, \cdots, q_r] & , \; R \text{ is unit upper triangular} \\ V = PS & , \; P = [p_1, p_2, \cdots, p_s] & , \; S \text{ is unit upper triangular} \\ P^t Q = \Omega & , \; \Omega = \mathrm{diag}\left(\omega_1, \cdots, \omega_{\min\{r, s\}}\right). \end{cases}$$

These factorizations are equivalent to computing the LDR decomposition of $V^t U$, where L is a unit lower triangular matrix, D is a diagonal matrix and R is a unit upper triangular matrix,

$$V^t U = (PS)^t QR = S^t \left(P^t Q\right) R = S^t \Omega R \;\; \Rightarrow \;\; LDR\left(V^t U\right) = S^t \Omega R.$$

This factorization fails if a zero pivot appears in the LDR decomposition of $V^t U$, that is, one of the diagonal elements of Ω is null. The way to solve this problem is to use the block LDR decomposition, $LDR\left(V^t U\right) = S^t \Omega R$, where Ω is a block diagonal matrix.

It is possible to extend this strategy to the two-sided Gram-Schmidt method, such that the matrices U and V are factorized as follows,

$$\begin{cases} U = QR & , \; Q = [Q_1, Q_2, \cdots, Q_b] & , \; R \text{ is unit upper triangular} \\ V = PS & , \; P = [P_1, P_2, \cdots, P_b] & , \; S \text{ is unit upper triangular} \\ P^t Q = \Omega & , \; \Omega = \mathrm{diag}(\Omega_1, \cdots, \Omega_b) & , \; \Omega_j = P_j^t Q_j \end{cases}$$

where the size of the first (b-1) clusters of Gram-Schmidt vectors, Q_j and P_j, is the same and equal to the minimum number of vectors such that Ω_j is nonsingular. The last clusters, Q_b and P_b, may be empty or uncompleted. In the second case, their sizes may be different.

In order to see the relation of this method to the nonsymmetric Lanczos method, we define the columns of the matrices U and V as follows,

$$\begin{cases} u_1 = x = q_1 \beta_0 & , \; u_{j+1} = A q_j & , \; j = 1, 2, \cdots, n \\ v_1 = y = p_1 \gamma_0 & , \; v_{j+1} = A^t p_j & , \; j = 1, 2, \cdots, n. \end{cases}$$

Then,

$$\begin{cases} U = [x, AQ] = [x, QH] = Q[e_1 \beta_0, H] = QR \\ V = [y, A^t P] = [y, PG] = P[e_1 \gamma_0, G] = PS \end{cases}$$

where $e_1 \in \Re^n$ is the first unitary vector. Thus, if the two-sided Gram-Schmidt method is applied to the matrices U and V, whose columns are, respectively, defined from the expansion of the matrices Q and P produced by A and A^t, then the look-ahead nonsymmetric Lanczos method is obtained:

$$\begin{array}{l} \text{Look - Ahead Nonsymmetric} \\ \text{Lanczos Method} \end{array} \equiv \left\{ \begin{array}{l} \text{Compute U and V as expansion of Q and P ,} \\ \text{Apply two - sided Gram - Schmidt to U and V .} \end{array} \right.$$

For this reason, the breakdown problems are similar in the two-sided Gram-Schmidt process and in the nonsymmetric Lanczos method.

3 The General Lanczos Method.

3.1 Basic Algorithm.

The first block nonsymmetric Lanczos method was defined in (Kim and Craig, 1988), as a generalization of the scalar algorithm, with the assumption that the number of initial vectors of each sequence, X and Y, is the same. These authors extend to the nonsymmetric case the idea of the block symmetric Lanczos method introduced in (Golub and Underwood, 1977). Thus, their method reduces the matrix A to a block tridiagonal form, by using a nonorthogonal symilarity transformation Q,

$$Q^{-1}AQ = T = \text{tridiag} \begin{pmatrix} & C_1 & C_2 & \cdots & C_{m-1} & \\ A_1 & A_2 & A_3 & \cdots & & A_m \\ & B_1 & B_2 & \cdots & B_{m-1} & \end{pmatrix} \in \Re^{nxn} .$$

The blocks B_i and C_i are choosen as upper and lower triangular matrices, such that T is a band matrix, where its lower bandwith is equal to the number of columns of X and its upper bandwith is equal to the number of columns of Y.

The algorithm is also based on two tridiagonalization processes, which are defined by the following equations,

$$\left\{ \begin{array}{l} AQ = QH , \quad Q = [X_1, \cdots, X_m] , \quad X = X_1 B_0 \\ A^t P = PG , \quad P = [Y_1, \cdots, Y_m] , \quad Y = Y_1 C_0 . \end{array} \right.$$

Here B_0 and C_0 are upper triangular matrices obtained when X and Y are biorthogonalized, H and G are block tridiagonal matrices, and

$$P^t Q = \Omega = \text{diag}(\Omega_1, \cdots, \Omega_m) , \quad \Omega H = G^t \Omega .$$

where the Ω_j blocks are usually chosen as diagonal matrices. If these diagonal matrices are scaled, such that they are transformed into identity matrices, then,

$$\Omega_i = I \implies P^t Q = I , \quad T = H = G^t .$$

However, this algorithm does not solve the breakdown and the reduction of the block sizes.

3.2 The Generalization of the Look-Ahead Techniques.

The general Lanczos method extends the look-ahead nonsymmetric Lanczos method, and its definition is also related to the two-sided Gram-Schmidt method. Thus, the algoritm may be described as follows,

- Consider two sets of initial vectors, $X \in \Re^{nxh_x}$ and $Y \in \Re^{nxh_y}$.

- Compute the matrices U and V as follows:

$$
\begin{cases}
U = \left[U_1, \cdots, U_{m_x+1}\right] & : \quad U_1 = X \;, \quad U_{j+1} = AX_j \;, \quad j = 1, \cdots, m_x \\
V = \left[V_1, \cdots, V_{m_y+1}\right] & : \quad V_1 = Y \;, \quad V_{j+1} = A^t Y_j \;, \quad j = 1, \cdots, m_y.
\end{cases}
$$

where the blocks X_j and Y_j are obtained when the biorthogonalization process is applied to blocks U_j and V_j,

$$
Q = \left[X_1, X_2, \cdots, X_{m_x}\right] \;, \quad P = \left[Y_1, Y_2, \cdots, Y_{m_y}\right].
$$

- Apply the two-sided Gram-Schmidt method on the matrices U and V, obtaining the following factorization,

$$
\begin{cases}
U = QR \;, & Q = \left[Q_1, Q_2, \cdots, Q_b\right] \;, & \text{R is unit upper triangular} \\
V = PS \;, & P = \left[P_1, P_2, \cdots, P_b\right] \;, & \text{S is unit upper triangular} \\
P^t Q = \Omega \;, & \Omega = \mathrm{diag}(\Omega_1, \cdots, \Omega_b) \;, & \Omega_j = P_j^t Q_j
\end{cases}
$$

Thus, two different processes appear, working as a data flow algorithm:

- The two-sided Gram-Schmidt method is applied to the matrices U and V, which are initialized by matrices X and Y.
- When the vectors of one of the matrices U and V are fully biorthogonalized, we generate new vectors with the associated matrix, by using the previously defined expansion process.
- The method finalizes when the matrices U and V are fully biorthogonalized and it is not possible to make any expansion process.

From the previous definition of the block Lanczos method, we can show that,

$$
\begin{cases}
U = [X, AQ] = [X, QH] = Q[I_x B_0, H] = QR \\
V = [Y, A^t P] = [Y, PG] = P[I_y \Gamma_0, G] = PS
\end{cases}
$$

where I_x and I_y are identity matrices whose dimensions are related to the size of X and Y. That is,

$$
\begin{cases}
X \in \Re^{n \times h_x} & \Rightarrow \quad I_x = \left[e_1, e_2, \cdots, e_{h_x}\right] \\
Y \in \Re^{n \times h_y} & \Rightarrow \quad I_y = \left[e_1, e_2, \cdots, e_{h_y}\right]
\end{cases}
$$

where $e_j \in \Re^n$ denotes the j-th unitary vector.

3.3 Structure of Matrices H and G with Look-Ahead Techniques.

The utilization of the look-ahead techniques breaks the tridiagonal structure of the matrices H and G. To study the final structure of those matrices, the relationship of the Lanczos vectors is exploited,

- The "brother" of a column of a Lanczos matrix is the column with the same index in the other Lanczos matrix,

$$
q_j \xleftarrow{\text{brother}} p_j \;.
$$

- The "father" of a column of a Lanczos matrix is the column of the same matrix, such the former was obtained as an expansion of the second,

$$\begin{cases} u_j = Aq_k & \xrightarrow{\text{biorthogon.}} q_i \Rightarrow u_j \xrightarrow{\text{father}} q_k \quad, \quad q_i \xrightarrow{\text{father}} q_k \\ v_j = A^t p_k & \xrightarrow{\text{biorthogon.}} p_i \Rightarrow v_j \xrightarrow{\text{father}} p_k \quad, \quad p_i \xrightarrow{\text{father}} p_k \end{cases}$$

First, the structure of H and G in the nonsymmetric Lanczos method is studied, and then we generalize the study in the block nonsymmetric Lanczos method. Finally, we analyze how the look-ahead modifies their structure.

In the nonsymmetric Lanczos method, H and G are tridiagonal matrices, such that the Lanczos method work as follows:

- The Lanczos recurrences biorthogonalize the new Lanczos vectors, with regard to the two last computed Lanczos vectors of the other recurrence,

$$\begin{cases} u_{j+1} & \text{with regard to } \{p_{j-1}\, p_j\}, \\ v_{j+1} & \text{with regard to } \{q_{j-1}\, q_j\}. \end{cases}$$

- That is, the new Lanczos vectors are biorthogonalized with respect to the "father" and "grandfather" vectors of the other new Lanczos vector.

The generalization of this study on the block nonsymmetric Lanczos method is similar, where H and G are band matrices,

- The lower and upper band of H is equal to the number of initial vectors of X and Y, and the lower and upper band of G is equal to the number of initial vectors of Y and X,

$$H \text{ is a band matrix } (h_x, h_y) \quad, \quad G \text{ is a band matrix } (h_y, h_x).$$

- Thus, the new Lanczos vector is biorthogonalized with regard to a set of the last computed vectors of the other recurrence. The sizes of these sets are equal to the bandwiths,

$$\begin{cases} u_{j+1} & \text{with regard to } \{p_{j+1-h_x-h_y} \cdots p_{j+1-h_x} \cdots p_j\}, \\ v_{j+1} & \text{with regard to } \{q_{j+1-h_y-h_x} \cdots q_{j+1-h_y} \cdots q_j\}. \end{cases}$$

- To do that, first the "father" of the new Lanczos vector is obtained, and then the "father" of the "brother" of the previously obtained vector,

$$\begin{cases} u_{j+1} \xrightarrow{\text{father}} q_{j+1-h_x} \xrightarrow{\text{brother}} p_{j+1-h_x} \xrightarrow{\text{father}} p_{j+1-h_x-h_y}, \\ v_{j+1} \xrightarrow{\text{father}} p_{j+1-h_y} \xrightarrow{\text{brother}} q_{j+1-h_y} \xrightarrow{\text{father}} q_{j+1-h_y-h_x}. \end{cases}$$

We can see that the former study is a particular case of the second one, where the sizes of the initial block of vectors are equal to one.

The application of the look-ahead techniques breaks the structure of matrices H and G. To explain the final structure, we apply the same idea on the clusters:

- First, the "father" of the new Lanczos vector is obtained,

$$\begin{cases} u_{j+1} \xrightarrow{\text{father}} Q_\mu = \{\cdots q_{j+1-h_x} \cdots\}, \\ v_{j+1} \xrightarrow{\text{father}} P_\phi = \{\cdots p_{j+1-h_y} \cdots\}. \end{cases}$$

- After, the "brother" determines the corresponding cluster in the other matrix,

$$\begin{cases} Q_\mu = \left\{\cdots q_{j+1-h_x} \cdots\right\} \xrightarrow{\text{brother}} P_\mu = \left\{\cdots p_{j+1-h_x} \cdots\right\}, \\ P_\phi = \left\{\cdots p_{j+1-h_y} \cdots\right\} \xrightarrow{\text{brother}} Q_\phi = \left\{\cdots q_{j+1-h_y} \cdots\right\}. \end{cases}$$

- Finally, the cluster containing the "father" of the first vector of the previously determined cluster is obtained,

$$\begin{cases} P_\mu = \left\{p_\sigma \cdots\right\} \xrightarrow{\text{father}} P_\nu = \left\{\cdots p_{\sigma-h_y} \cdots\right\}, \\ Q_\phi = \left\{q_\varsigma \cdots\right\}. \xrightarrow{\text{father}} Q_\zeta = \left\{\cdots q_{\varsigma-h_x} \cdots\right\}. \end{cases}$$

- If the j-th column of the Lanczos matrices is in the b_j-th cluster, and this cluster is completed, then the new Lanczos are biorthogonalized with respect to the following clusters,

$$\begin{cases} u_{j+1} \text{ with respect to } \left\{P_\nu \cdots P_\mu \cdots P_{b_j-1} \, P_{b_j}\right\}, \\ v_{j+1} \text{ with respect to } \left\{Q_\zeta \cdots Q_\phi \cdots Q_{b_j-1} \, Q_{b_j}\right\}. \end{cases}$$

If the b_j-th cluster is uncompleted, we can not biorthogonalize the new Lanczos vector with regard this cluster, because they are included on it. In this cases, the new Lanczos vectors may be orthogonalized with respect this cluster, in order to improve the convergence of the method,

$$\begin{cases} u_{j+1} \text{ with respect to } \left\{P_\nu \cdots P_\mu \cdots P_{b_j-1} \, Q_{b_j}\right\}, \\ v_{j+1} \text{ with respect to } \left\{Q_\zeta \cdots Q_\phi \cdots Q_{b_j-1} \, P_{b_j}\right\}. \end{cases}$$

In (Aliaga, 1995), there is a mathematical development of this generalization.

3.4 Implementation Aspects.

The previous section describes the algorithm in exact arithmetic, where no roundoff error appears. In finite arithmetic, we must consider additional aspects related to the Lanczos methods, the breakdown problems and the block structure.

All variants of Lanczos methods present problems associated with loss of orthogonality, or biorthogonality, of the Lanczos vectors when they are implemented. There are many references where several techniques are defined to solve these problems for the symmetric case (Cullum and Willoughby, 1985), (Parlett and Scott, 1979), (Simon, 1984). For the nonsymmetric case, we have not found related references because the extension of the symmetric technique is not trivial. Thus, the most usual option is to apply the full rebiorthogonalization, where the new Lanczos vectors are rebiorthogonalized with respect to all previous completed clusters. To reduce this cost we propose to limit the number of clusters, η, which are used in this process,

$$\begin{aligned} u_j &= u_j - Q_k \left[\Omega_k^{-1}\left(P_k^t u_j\right)\right] \\ v_j &= v_j - P_k \left[\Omega_k^{-t}\left(Q_k^t v_j\right)\right] \end{aligned}, \quad k = \max\left\{1, b_j - \eta\right\}, \cdots, b_j - 1.$$

This alternative does not assure the fully biorthogonality of the clusters but the biorhogonality of the clusters is improved.

The main problem of the look-ahead algorithms is to locate when the serious breakdown appears. The first option is to study the singularity of the last cluster,

$$\sigma_{min}(\Omega_b) \leq tol \quad \Rightarrow \quad \text{serious breakdown appears}$$

where tol is a predefined value related with the roundoff unit, ε. The algorithms are very sensitive to the choice of tolerances, such that different tolerances can generate different partitions. Moreover, we must consider the problem of generating almost linearly dependent vectors. Freund et al, (Freund, Gutknecht and Nachtigal, 1993), propose a criterion to solve both problems.

3.5 The Deflation Techniques.

When a null Lanczos vector appears in a recurrence, we must throw it away of the associated Lanczos matrix, because we can obtain any information from it. Thus, the block size of the corresponding recurrence is reduced by one. In this case, we do not know how to obtain the "father" of each vector directly, and we have to trace the block size in each step for each recurrence. This can be carried out by using some auxiliar vectors of indexes, as is shown in (Aliaga, Boley, Hernández, Freund, 1996).

We have additional problems if near-zero Lanczos vectors appear. If we throw away these vectors, we loose some information. However if we maintain them, round-off errors can be too big. In (Aliaga, Boley, Hernández, Freund, 1996), it is proposed to throw away these vectors and biorthogonalize the new Lanczos vectors with respect to the cluster of the "father" of the near-zero vectors, to reduce the round-off errors.

4 CRAY T3D Overview.

The parallel algorithms developed in this work are evaluated on the massively parallel processor Cray T3D which has a peak performance of 38 gigaflops and whose main features are described below.

The Cray T3D hardware contains a group of processing elements, each one composed of a local memory and communication links with other processing units. There are Cray T3D systems with up to 32, 64, 128, 256, 512, 1025 or 2048 PE's available for users. The computer administrator usually divides the PE's into partitions, defining their size and the properties of the interconnection network. Each node consists of one microprocessor DECchip 21064 (Alpha).

The interconnection network has a T3D topology with 16-bit bidirectional links, a high bandwidth (150 MHz) and a dimensional routing algorithm across the three dimensional structure.

The Cray T3D system supports two programming modes to exploit its potential parallelism: virtual shared memory and message passing models (Cray, 1994a). The second is more efficient than the first one, but requires deeper knowledge of the architecture. In this work, we use the message passing model and particularly the PVM environment due to its portability. The Cray T3D PVM version (Cray, 1994b) has been implemented according to the machine features and is compatible with the 3.3 PVM version.

5. Parallel Algorithms.

5.1 Data Distribution.

Before studying parallel algorithms, it is important to consider the different data distributions that will be used. An appropriate data distribution will determine the load balancing of parallel algorithms and the communicating operations involved.

Data distribution in the general Lanczos algorithm must be studied for each type of operation involved. The data involved are the original matrix A, the different groups of vectors that appear in the method Q, P, U and V, the block diagonal matrix Ω and the upper triangular matrices R and S. The columns of matrices Q, P, U and V, are involved in saxpy operations, inner products and the matrix-vector product, whereas the matrix A is only involved in the matrix-vector product.

- Matrix A distribution: there are many possibilities for distributing this matrix because it is only used in the matrix-vector product.
- Groups of vectors distribution: there are fewer posibilities for distributing the groups of vectors Q, P, U and V, because they are involved in many stages of the algorithm.
- Structured matrices: Ω must appear on each processor, while the user is free to choose the distribution of R and S. In any case, a data structure has been designed for managing and storing efficiently these structures in memory.

Data distribution is divided into two steps: *data partitioning* and *mapping* to nodes.

Groups of vectors partitioning is very simple due to its unidimensional structure. It is only necessary to decompose the vectors in n_b partitions of a size equal to t_b, see figure 5.1. It is convenient that partition of groups of vectors of all matrices be identical for reducing the communicating cost of parallel algorithms in saxpy and inner product operations.

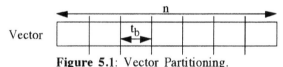

Figure 5.1: Vector Partitioning.

There exist different modes of *matrix partitioning*, but the two basic modes are the following:

- *Unidimensional* (row or column oriented): The matrix is divided into row (or column) blocks of a fixed size equal to t_b. The number of blocks is computed by means of the expression $n_b = n/t_b$, where n is the matrix dimension, see figure 5.2.a.
- *Bidimensional*: The matrix is divided into square blocks of size $t_b \times t_b$. The number of blocks is computed by means of the expression $n_b = n^2/t_b^2$. Bidimensional partitioning of the matrix improves processor locality in communication operations, although it is more complex, see figure 5.2.b.

In this work, parallel algorithms will make use of unidimensional partitioning of the original matrix due to the difficulty of doing bidimensional partitioning with sparse matrix stored in CSR format (Duff, Grimes and Lewis, 1989).

After studying the different types of data partitioning of the problem, it is interesting to study the different methods for *mapping* these data on the processors. The choice of an appropriate mapping strategy will allow a good load balancing of parallel algorithms. There are two types of data mapping among different processors:

- *Direct mapping*: One of the partitions is assigned to only one node .
- *Nondirect mapping*: A set of the partitions is assigned to each processor.

In the present work, we use a direct mapping of data partitioning among the nodes. This strategy is optimal if the matrix A is dense because the parallelization of

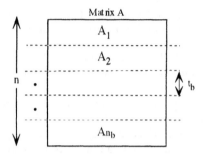
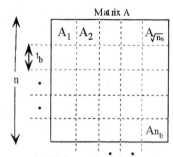

a) Unidimensional partitioning (row oriented)   b) Bidimensional partitioning
Figure 5.2: Different modes of matrix partitioning.

the matrix-vector product will be well balanced. If the matrix A is sparse, this strategy does not assure a good load balancing, and then, a preprocessing of matrix A must be applied before the data partitioning and mapping. This preprocessing generates a permutation matrix, Π, which is applied on rows and columns of A and on rows of X and Y. It is easy to see that the application of the method on the permuted data produces the same R and S, while the rows of matrices Q and P are only permuted. Thus, the numerical properties of the method are not changed (Aliaga, 1995).

5.2 Basic Operations.

Parallelization of the outer loop of the general Lanczos algorithms presents data dependency problems. Therefore, for these type of methods, the parallelization is only introduced inside the main loop of each basic operation. Three types of basic operations are performed by the method:

- Matrix-block of vectors product: This operation can be seen as several matrix-vector products.
- Inner product of vectors and saxpy operations.
- Solve a small linear system. (not parallelized).

The matrix-block of vectors product operation corresponds to the main computational cost and thus represents the main computational kernel of the method.

5.3 Parallelization of the Matrix-Vector Product.

Matrix-vector product $Ax = y$ is the main computational kernel of the Lanczos algorithms, due to its cost, $O(n^2)$, as opposed to the $O(n)$ cost of the vector operations. Then, the efficiency of the parallel Lanczos algorithms is mainly determined by the efficiency of the parallelization of this operation.

A unidimensional partitioning row-oriented of matrix A and a direct mapping of such a partition among the different processors will be used, see figure 5.3, which is equivalent to a unidimensional partitioning column-oriented of matrix A^t. So, it will be necessary to implement two matrix-vector product parallel algorithms, one for each of the two data distributions.

In the case of the *row-oriented algorithm,* we need to gather the distributed vector x by using a multibroadcast communication operation as shown in figure 5.4, and then do the partial products. As a result, the parallel unidimensional row-oriented matrix-vector product algorithm can be divided in two steps:

1 Comunication: multibroadcast operation of x.
2 Computation: do matrix-vector product $y[i] = A[i,:]x$, $1 \leq i \leq p$.

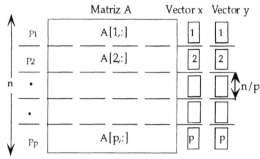

Figure 5.3: Unidimensional row oriented data distribution of matrix A, which results in a unidimensional column oriented distribution of matrix A^t, made up of blocks of columns $A[1,:]^t,...,A[p,:]^t$, where p is the number of processors.

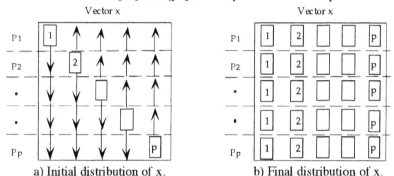

a) Initial distribution of x.   b) Final distribution of x.

Figure 5.4: Multibroadcast operation. In the end, each processor has an entire copy of the operand vector x.

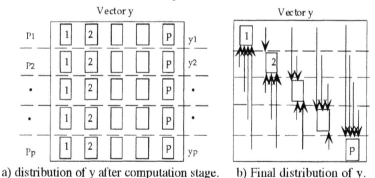

a) distribution of y after computation stage.   b) Final distribution of y.

Figure 5.5: Multigathering Operation. At the end of the Multigathering, each processor has its final block of the resulting vector y.

For the *column-oriented algorithm*, a multigathering communication operation must be implemented to distribute and accumulate correctly the solution vector y after computing the partial products, see figure 5.5. So the parallel unidimensional column-oriented matrix-vector product algorithm can be divided in two steps:

1 Computation: do matrix-vector product $y_i = A[i,:]^t x[i]$, $1 \leq i \leq p$, such that $(y=y_1+y_2+...+y_p)$.

2 Comunication: multigathering operation of vectors y_i to distribute and accumulate correctly vector y.

So the efficient implementation of the matrix-vector product parallel algorithms depends on an efficient implementation of the multibroadcast and multigathering comunicating operations. These operations are effiently implemented and evaluated for the Cray T3D system in (Pérez, Aliaga, and Hernández, 96).

The performances of the matrix-block-of-vector product is improved when it is computed as an only product. In this case, the communication operations works with groups of vectors and then the communication overhead is reduced. Parallel algorithms have better efficiency when the matrix is dense than when the matrix is sparse with a low number of nonzero elements, because, in the dense case, the cost of communications is relatively less than the global cost of the algorithm.

5.4 Parallelization of Vector Operations.

Parallelization of *saxpy operations* is very simple, if the data distribution of vectors shown in section 5.1 is used. These operations do not require any type of communication operations and the load balancing is perfect.

On the other hand, the parallelization of the *inner product operations* is rather complicated, because it requires a synchronisation of the different processors to accumulate the local inner product. This synchronisation can be made in two ways: by using a multibroadcast or a gather operation with an additional broadcast. Both approaches are expensive due to the reduced size of communicating messages involved, which produces a low transference rate and so, a decrease of parallel algorithm efficiency. A way of reducing the communication cost of the inner products is to gather more than one inner product in such a way that communication operations are made simultaneously. This increases the size of data messages and so also increases their transference rate. This idea can be used to join sequences of inner products that appear in the algorithm.

The structure of matrices H and G determines the number of saxpy operations and inner products in the algorithm. Additional vector operations will be necessary if rebiorthogonalization processes are used. The efficiency of the algorithm is reduced when the number of inner product increases.

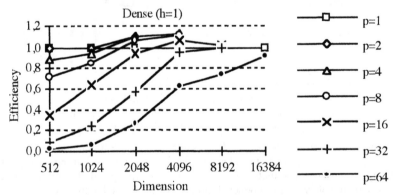

Figure 6.1: Efficiency of the dense matrix-vector product parallel algorithm with a unidimensional row oriented data distribution of the matrix.

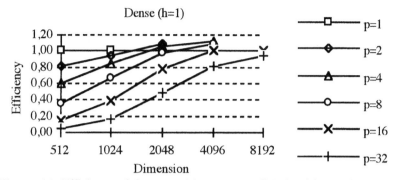

Figure 6.2: Efficiency of the general Lanczos parallel algorithm with up to 32 processors and using a dense matrix with, h=1.

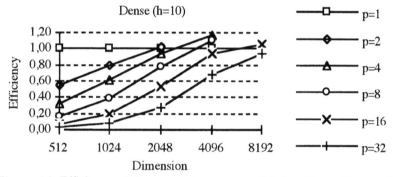

Figure 6.3: Efficiency of the general Lanczos parallel algorithm with up to 32 processors and using a dense matrix, h=10.

6 Results and Conclusions.

In this section we show the efficiencies of the basic operations studied previously associated with the general Lanczos method and the efficencies of the method itself. In the following figures p denotes the number of processors used and h denotes the size of the initial set of Lanczos vectors, X and Y.

6.1 Matrix-Vector Product.

In figure 6.1 are shown the efficiencies of the matrix-vector product parallel algorithm for the dense case. Note that they are high for any number of processors when the matrix dimension exceeds 2048. The performance in megaflops of the dense matrix-vector product parallel algorithm reaches 2.5 Gigaflops for the number of processors p=64, and the dimension of the matrix n=16384.

6.2 General Lanczos Algorithm.

A similar study has been made on the parallel general Lanczos algorithm for the dense case, figures 6.2 and 6.3, and for the sparse case with a 5% of nonzero elements, figure 6.4, both cases for randomly generated matrices. The efficiencies of the dense case when h=1 are similar but less than that of the matrix-vector product parallel algorithm, see figure 6.1. This is because in the general Lanczos parallel algorithm there is an overhead due to inner products and nonparallel operations. For

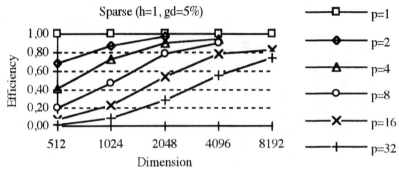

Figure 6.4: Efficiency of the general Lanczos parallel algorithm with up to 32 processors and using a sparse matrix with 5% of nonzero elements, h=1.

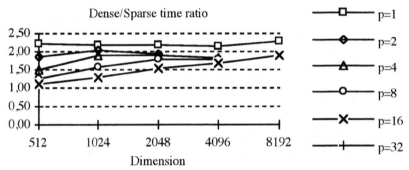

Figure 6.5: Sparse speed-up of the general Lanczos parallel algorithm with up to 32 processors and using a sparse matrix with 5% of nonzero elements, h=1.

the general Lanczos algorithm, see figure 6.2, good efficiencies are obtained when the dimension of the matrix is greater than n=2048. In some cases there are values of efficiency greater than one due to variations in the computational cost of the matrix-vector parallel algorithm (see also figure 6.1). In parallel algorithms, each processor works with only a partition of the full matrix, and then the overhead associated to memory access is reduced. So, computational cost of the parallel algorithm is reduced and efficiency may be higher than expected.

If the initial set of Lanczos vectors has a size of h=10, see figure 6.3, the efficiency is reduced for small matrices because of the increase of vector inner products in the algorithm, but for large matrices this overhead is compensated by a more efficient matrix-block-of-vectors parallel implementations.

For the sparse case, see figure 6.4, the efficiencies are more reduced than in the dense case because of the reduction of the computational cost of the algorithm. In any case, good efficiences are obtained for matrix dimensions greater than n=2048. In figure 6.5 we show the speed up for the sparse case.

6.3 Conclusions.

The efficiencies obtained by matrix-vector product parallel algorithms are high but the performances of inner product of vector parallel algorithms is lower. This reduction in the efficiency of one basic operation is not greatly reflected in the general Lanczos algorithm because of its low computational cost, and so, the performances of the parallel implementation of the general Lanczos method are very similar to those obtained for the matrix-vector product parallel algorithms. For sparse matrices, an efficient storage format has been used, CSR (Duff, Grimes and Lewis, 1989), in order to improve the computational cost of the algorithm.

7 Acknowledgements.

The authors express their gratitude to CRAY Research Inc., for the use of the CRAY T3D, sited in the Ecole Polytechnique Federale of Lausanne (Switzerland).

8 References.

J.I. Aliaga (1995), *Algoritmos paralelos basados en el método de Lanczos. Aplicaciones en problemas de control*, Tesis Doctoral, Departamento de Sistemas Informáticos y Computación, Universidad Politécnica de Valencia.

J.I. Aliaga, D.L. Boley, R.W. Freund and V. Hernández (1996), *A Lanczos-type method for multiple starting vectors*, TR-96-059, Dept. Comp. Science, Univ. Minnesota, (also submitted to Mathematics of Computation).

D.L. Boley and G.H. Golub (1991), *The nonsymmetric Lanczos algorithm and controllability*, Systems & Controls Letters, Vol. 16, pp. 97-105, North-Holland.

CRAY Research (1994a), *Cray MPP Fortran programmers manual.*

CRAY Research (1994b), *PVM and HeNCE Programmer's Manual.*

J.K. Cullum and R.A. Willoughby (1985), *Lanczos algorithms for large symmetric eigenvalue computations*, Vol. I Theory and Vol. II Programs, Birkhauser.

I.S. Duff, R. G. Grimes, J.H. Lewis (1989), *Sparse matrix test problems*, ACM Transactions on Mathematical Software, Vol. 15, pp. 1-14.

R.W. Freund, M.H. Gutknecht and N.M. Nachtigal (1993), *An implementation of the look-ahead Lanczos algorithm for non-Hermitian matrices*, SIAM Journal of Scientific Computation, Vol. 14, n° 14, pp. 137.158.

G.H. Golub and R. Underwood (1977), *The block Lanczos method for computing eigenvalues*, in Mathematical Soft. III, J.H. Rice Ed., Acad. Press, pp. 364-377.

M.H. Gutknecht (1992), *A completed theory of the unsymmetric Lanczos process and related algorithms, part I*, SIAM Journal on Matrix Analysis and Applications, Vol. 13, num. 2, pp. 594-639.

H.M. Kim and R.R. Craig Jr (1988), *Structural dynamics analysis using an unsymmetric block Lanczos algorithm*, International Journal for Numerical Method in Engineering, num. 26, pp. 2305-2318.

C. Lanczos (1950), *An iteration method for the solution of the eigenvalue problem of linear differential and integral operators*, Journal of Research of the National Bureau of Standards, Vol. 45, pp. 255-282.

B.N. Parlett (1992), *Reduction to tridiagonal form and minimal realizations*, SIAM Journal on Matrix Analysis and Applications, Vol. 13, num. 2, pp. 567-593.

B.N. Parlett, D.R. Taylor and Z.A. Liu (1985), *A look-ahead Lanczos algorithm for unsymmetric matrices*, Mathem. of Comput., Vol. 44, num. 169, pp. 105-124.

B.N. Parlett and D.S. Scott (1979), *The Lanczos algorithm with selective orthogonalization*, Mathem. of Computation, Vol. 33, num. 149, pp. 217-238.

J.L. Pérez, V. Hernández and J. Aliaga (1996), *Nucleos Computacionales de Métodos Iterativos de Resolución de Sistemas Lineales y Cálculo de Valores y Vectores Propios. Análisis e Implementación sobre el Multicomputador Masivamente Paralelo Cray T3D*, DSIC-II-28/96, DSIC, Univ. Polit. Valencia.

H.D.Simon (1984), *The Lanczos algorithm with partial reorthogonalization*, Mathematics of Computation, Vol. 42, num. 165, pp.115-142.

Parallel and Distributed Computations in a Parameter Inverse Problem

Henryk Telega, Institute of Computer Science,
Jagiellonian University ul. Nawojki 11, 30-072 Krakow, Poland

ABSTRACT: This paper presents an outline of a parallel algorithm for the identification of parameters in the Swartzendruber formula. This formula appears in the mathematical model of the prelinear filtration of ground water. The parameter inverse problem is formulated as an appropriate control problem. The idea of the proposed algorithm (called MSP) is to review large areas of the domain through proper, parallel investigation of many parts of it. The results of numerical tests are included. Some improvements to the algorithm are also proposed.

1. Introduction

The difficulties that arise when we try to solve inverse problems numerically come from such things as nonlinearity, bad conditioning, the often large scale of the task, and the existence of many local minima (when the problem is formulated as an optimal control problem). One of the strategies that can be helpful in overcoming these objections and to solve such complex problems in a shorter time (and more easily, if possible) is to take advantage of parallel computations (see for instance [11]).

In this paper we concentrate on the development and implementation of appropriate parallel and distributed nonlinear optimization methods for a test inverse problem.

2. The Test Inverse Problem

The problem that has been the basis for numerical computations consists of the identification of the parameters of the Swartzendruber formula, with later modification made by Schaefer [8, 12]. This formula determines a relationship between the filtration velocity and the hydraulic slope for the so-called prelinear filtration of ground water. Such a filtration process can be described by the piezometric height distribution h, that is the solution to a mixed Dirichlet/Neumann initial-boundary value problem for a nonlinear PDE (Partial Differential Equation) of a parabolic type. The Swartzendruber formula included in this PDE depends on a parameter vector $\tilde{p}$, which is a function of the space and the time variables.

When we know the parameters of the formula, we are able to solve important engineering problems like the stability analysis of dams or the design of boggy terrain drainages. Laboratory tests for the evaluation of these parameters are difficult and in practice they are possible only during the construction of the structures being considered. The quantity that can always be measured in a relatively easy way is the piezometric height distribution h. Knowing h and the initial-boundary conditions, we can find the parameter vector $\tilde{p}$ by methods of inverse analysis.

The identification of parameters can be formulated as follows (see [2,3] for details):

Find a $\tilde{p}^* \in P$ so that:

$$J\left(h\left(\tilde{p}^*, f, g\right)\right) \le J\left(h\left(\tilde{p}, f, g\right)\right) , \quad \forall \tilde{p} \in P, \tag{1}$$

where P denotes the admissible set of parameters, for the given pair (f, g), h is the solution of the following state PDE:

$$\begin{cases} \dfrac{\partial h}{\partial t} + A(t) h = f \text{ in } Q, \\ h(0) = g \text{ in } \Omega, \end{cases} \tag{2}$$

where $Q = \Omega \times (0, T)$, Ω is an open bounded subset of $\mathfrak{R}^3$ (the filtration area) with the regular boundary $\partial \Omega = \Gamma_1 \cup \Gamma_2$, $\Gamma_1 \cap \Gamma_2 = \varnothing$ and $\int_{\Gamma_2} d\sigma > 0$. Let us introduce the functional spaces $V = \left\{ \upsilon \in H^1(\Omega), \upsilon = 0 \text{ on } \Gamma_1 \right\}$, $H = L^2(\Omega)$, $Y = L^2(0, T; V)$, $Y' = L^2(0, T; V')$ and $W = \left\{ \upsilon \in Y, \upsilon' \in Y' \right\}$, where the partial derivative υ' with respect to t is understood in the distributional sense (see [2,3] for details). The pair $(f, g) \in Y' \times H$. $J: W \to \mathfrak{R}$ is the cost functional. It represents (in some sense) the distance between computed and observed data. $A(t): Y \to Y'$, $t \in [0, T]$ is the family of operators of the form

$$A(t) h = \sum_i^3 \frac{\partial}{\partial x_i}\left(v_i\left(p, x, t, \text{grad } h\right)\right), \tag{3}$$

where v_i denotes a component of the filtration velocity vector.
The constitutive Swartzendruber formula for the prelinear filtration [8,12] is given by:

$$v_i\left(p(t, x), Dh\right) = \varphi\left(\tilde{p}(t, x), |Dh|_L\right) \sum_{j=1}^3 l_{ij}(t, x) \frac{\partial h}{\partial x_j}, \quad i = 1, 2, 3 \tag{4}$$

where

$$|\eta|_L = \sum_{i,j=1}^3 l_{ij} \eta_i \eta_j, \text{ for } \eta \in \mathfrak{R}^3 .$$

We denote by p a part of the vector $\tilde{p}$ such that $\tilde{p}(t, x) = \left(\{l_{ij}\}, p\right)(t, x)$ and $p = (M, s_0, \theta)$. The physical meaning of the components is as follows:

l_{ij} - the anisotropy matrix,

M - maximal hydraulic conductivity of a soil,

s_0 - an analogue of the threshold gradient,

θ - index of nonlinearity (see [8]),

t, x -stand for time and position vectors, respectively.

The function φ is given by

$$\varphi(\tilde{p},s)=\begin{cases} M\left(1-\dfrac{s_0}{s}\left(1-\exp\left(-\dfrac{\theta s}{s_0}\right)\right)\right), & \text{if } E<s \\[4mm] \left[\dfrac{M}{E^2}\left(s_0-(s_0+\theta E)\exp\left(-\dfrac{\theta E}{s_0}\right)\right)\right]s+M\left(1-\dfrac{2s_0}{E}+\left(\dfrac{2s_0}{E}+\theta\right)\exp\left(-\dfrac{\theta E}{s_0}\right)\right), \\[4mm] \text{if } 0\le s\le E. \end{cases}$$

(5)

where E is an accuracy constant, s stands for the piezometric height gradient Dh. The constant E is not a physical parameter. It was introduced in order to assure the function $\varphi(s)$ to be strongly increasing for all $s\ge 0$ (see [8]). In most cases the value of E is close to zero and in practice it can be evaluated by methods given in [8].

For the one dimensional flow we can consider only $p=(M,s_0,\theta)$.

It has been proved (cf. [9]), that under suitable hypotheses, problem (2) has a unique solution and this solution can be effectively computed by mixed finite difference/finite elements schemes. Moreover, the inverse problem (1) admits a solution (see [2,3]).

One of the numerical methods for solving the problem (1) in an iterative way is to approximate it by a sequence of finite dimensional problems:

Find a $\tilde{p}_{n\tau}^{\,*}\in P_{n\tau}$ so that:

$$J_{n\tau}\left(h_{n\tau}\left(\tilde{p}_{n\tau}^{*},f,g\right)\right)\le J_{n\tau}\left(h_{n\tau}\left(\tilde{p}_{n\tau},f,g\right)\right),\quad \forall \tilde{p}_{n\tau}\in P_{n\tau}$$

$(1_{n\tau})$

and

$$R_\tau h_{n\tau}=f_r$$

$(2_{n\tau})$

where $(2_{n\tau})$ is the discrete version of the state equation (2), $h_{n\tau}$ is the piezometric height distribution approximated by finite elements in the space variable and by finite differences in the time variable. $P_{n\tau}$ is an approximation of P, $R=\{R_\tau\}$ is a discrete scheme operator (see for instance formula (17) in [9]), $n\tau$ represent discrete space and time indexes, respectively. Consult [9] for more details of the approximation method and the solver to $(2_{n\tau})$. Schaefer and Sędziwy [9] proved also that solutions to the approximated direct problems $(2_{n\tau})$ converge to the exact solution of (2). Moreover, the approximated inverse problem $(1_{n\tau})$ admits a solution (see [3]). Additional results concerning the stability of (1) with respect to uncertain data, and also concerning the convergence for (1_n) (another discrete version of (1) in which only the Galerkin approximation with respect to the space variable is provided) can be found in [2].

In the problem $(1_{n\tau})$, we assume that $p_{n\tau}$ is constant on some cluster of elements and time intervals. Physical observations imply that $P_{n\tau}$ constitutes a hypercube in $\Re^k$ for a specific natural k, namely each component $p^i_{n\tau}$ of the vector $p_{n\tau}$ satisfies the constraints:

$$0 < p^{iL}_{n\tau} \le p^i_{n\tau} \le p^{iU}_{n\tau} \quad i = 1, \ldots, k,$$

where $p^{iL}_{n\tau}$ and $p^{iU}_{n\tau}$ stand for the known lower and upper constant bounds, respectively.

In the sequel, for the sake of convenience, the index $n\tau$ will be omitted. We will concern ourselves with a one dimensional flow, for which l_{ij} is the unity matrix, and the parameter vector to be found will be $p = \left\{ \left(M^i, s^i_0, \theta^i \right), i = 1, .. k \right\}$. In the simplest real case (for a one-dimensional flow of water and taking into consideration one time interval) k is equal to 1, but when we want to recognize the tested area more precisely, k can be equal to tens or even hundreds (see [2, 13]).

3. Implementation of Standard Nonlinear Programming Methods

The numerical computations were based on the data collected during a field experiment carried out at the experimental range of the Warsaw Agricultural University [17]. A one-dimensional vertical flow of water under the central part of a testing embankment was considered. The filtration area was divided into three zones.

We assumed that the parameter vector $p = (M, s_0, \theta)$ was constant in each of the three zones and in each of the time intervals of the experiment. Therefore, the inverse problem under consideration consisted in estimating $9N$ parameters $\left(M^{ij}, s^{ij}_0, \theta^{ij} \right)$, $i = 1,2,3$, $j = 1, \ldots, N$, where N denotes the number of the time intervals of the experiment and i corresponds to the zone number. During the tests N was changed from 1 to 5.

The attempt to apply standard nonlinear programming methods (mainly variable metric method MIGRAD from MINUIT (CERN Program Library Entry D506)) [19] to the problem shows that the algorithm stops at many points of the solution domain (with or without convergence of MIGRAD). Some of these points can correspond to local minima, some may arise from bad conditioning. The results strongly depend on the starting points (see [2,13]). The tests showed that when we do not know a reasonable approximation of p, then it could take several hundreds or thousands of consecutive evaluations (starting from various points of P) to achieve the global minimum. Moreover, the number of evaluations grows significantly when N or the number of zones in the filtration area increases. The single evaluation of p performed on SUN SPARC-4 stations lasted from several minutes up to one hour, so an acceleration of computations and the effective use of computational resources is essential.

4. Strategies for Improving the Optimization Algorithms

We propose three general strategies to improve the optimization algorithms for the inverse problem considered:

- the first is to improve the local minimum search,
- the second one is to search large parts of the domain P,

- the third is to accelerate computations using parallel and distributed techniques.

These strategies refer to different "directions" of potential improvements. The aim of the first strategy is to make a search of local minima more stable i.e. less dependent on starting points. The hierarchic optimization proposed in [14] gives an improvement in this "direction", but makes the single evaluation of p even more expensive.

However, in this article we would like to focus on the second and the third strategies. The proposed algorithms are described in the two following sections.

5. Multiple Starting Point (MSP) Parallel Optimization

The idea is to review large areas of the domain P by parallel investigation. This can be achieved by running in parallel many scalar nonlinear programming processes starting from various points. It is the simplest strategy to deal with many local minima. Moreover, by this method, in some way, we can deal with bad conditioning — the arbitrary optimization process from the set of parallel ones, that has stopped without convergence, can be restarted with modified starting points. Instead of only one restarted process, we can generate several parallel processes that would scan some area of P, giving us a chance to obtain convergence.

All starting points form a grid (regular or irregular) in P. A proper choice and on-line modifications of the computational grid have influence on the efficiency of the search. The regions of P that are not "satisfactory" should be omitted, and the grid should be denser in these parts for which good results are more probable. In the numerical tests, the condensation of the grid was accomplished by generating new starting points randomly near the best of the former ones. The initial grid of starting points can be generated in various ways, for instance one can generate a number of random points uniformly in the domain, or one can use more sophisticated methods with a selection of subsets of random points that have the lowest functional value among all sample points within a certain neighborhood (see [11],[1],[6]).

Three versions of the *Multiple Starting Point* (MSP) algorithm have been tested—the basic parallel version (called MSP_B), the scalar version (MSP_S) and the modified parallel version (MSP_M). The scalar version has been prepared in order to estimate efficiency and speed-up of the other two others.

The implementation of MSP_B and MSP_M is based on the PVM library. The computations were distributed in a small computer network of six SUN Sparc 4 workstations and one SUN Sparc 5 connected by coaxial Ethernet 10 Mb/s.

5.1 MSP_B — Description of the Algorithm and Numerical Tests

The master (supervisor) process generates a grid of starting points (this process could be also performed in parallel) , spawns parallel nonlinear programming processes (slaves), and then waits for the results.

The algorithm of the master process is as follows:

(1) Generate the grid of starting points and remember it in the table GRID.
(2) DO nproc TIMES
- *Spawn a local optimization process (slave).*

- *Send to the slave coordinates of a starting point (from GRID).*

OD

(3) REPEAT

- *Wait for a result (the value of the cost functional J and coordinates of the vector p) from any slave.*
- *Compare J with the previous best one. Sort obtained results.*
- *IF J < J_{THR} THEN*
  - *Remember J and p.*
  - *Generate a new starting point randomly in a hypercubic neighborhood of p.*

  ENDIF

- *IF there are new points in GRID THEN*
  - *Spawn a new local optimization process (slave)*
  - *Send to the slave coordinates of a starting point (from GRID).*

  ENDIF

UNTIL global STOP CRITERION is satisfied.

(4) Kill all slaves that are still processed, stop computations and print out final results.

For tests the threshold J_{THR} has been fixed arbitrarily (after some experiments), but it can be also updated during computations. The global STOP CRITERION is as follows: the value of J is less than a global threshold J_{G_THR}, or the maximum time of computations was exceeded, or there are no more new points in GRID. We were able to estimate J_{G_THR} that would give satisfactory results because we knew the value of p from laboratory tests. We did not concerned here with more complicated stop criteria based on statistic methods like estimating the probability of finding the global minimum with accuracy to a certain set with a given measure or similar ([18], see also [6], [7]) (however, we intend to consider such criteria in the further research).

Each slave performs local optimization starting from a point of the grid. A new slave is spawned each time there are new points because the algorithm is prepared for further versions that will include load balancing and will have the possibility to spawn more processes on faster machines.

The MIGRAD [19] method was implemented in the slaves. Each local optimization consists in several (up to even several thousand) solver calls, so the time of computations in slaves is much greater than the time of communication between slaves and the supervisor. This is an example of *coarse-grain* parallelism. The sequential part of the task, i.e.:

- modifying the grid of starting points,
- sorting the results (this can be done in constant time when we assume that, for instance, only a maximum of 100 or 10 of the best points are remembered),
- checking of the global stop criterion, and
- begin/end, input/output operations,

together with the growth of the problem size (which means more starting points), becomes less and less in comparison with the parallel part of the task. One should then expect, that the *efficiency* (achieved speed-up as a fraction of the theoretical best speed-up) of parallel and distributed computations for this problem will warrant use of the proposed methods.

Numerical tests have been performed for grids of 10 and 100 points. The grids contained points close to experimentally determined "good starting values" to speed up the tests and to ensure global convergence. These "good" starting points were located in the final part of the table GRID in order to ensure that all points were used. As we are not concerned here with the convergence of the whole optimization task, but only speed-up and efficiency of parallel versions of MSP, all computations stop after all points of the grid have been processed (exactly 10 or 100 slaves have returned results).

The number of parallel processes was up to seven, one process (master or slave) per one processor (SUN Sparc 4 workstation). During all tests, computers were practically free from other tasks.

The representative results for initial grids of 10 points are presented in Table 1.

The notations are as follows:

TS - time of computations of the scalar version MSP_S,
TB - time of computations of the MSP_B,
nproc - number of slaves
TS/nproc - time of computations for the ideal parallelization

The ideal speed-up $s_i = nproc$, achieved speed-up $s = \dfrac{TS}{TB}$, efficiency $e = \dfrac{s}{s_i}$.

Test no.	TS [min]	nproc	TB [min]	Achieved speed-up (s)	TS/nproc [min]	Efficiency (e)
		2	153.75	1.652	127.00	0.826
1	254.00	5	71.48	3.553	50.80	0.711
		6	63.03	4.030	42.33	0.672
		2	132.5	1.536	101.78	0.768
2	203.55	5	68.85	2.956	40.71	0.591
		6	60.72	3.352	33.93	0.559

Table 1. 10 initial starting points

The difference between efficiency for different data (Tests no. 1 and 2 in Table 1) but the same number of slaves comes from the fact that the last processes are performed not entirely in parallel. For instance in the case of *nproc*=2, a part of the last process is most often performed alone. This is because the times of computations in slaves differ from each other significantly. For the relatively small number of points (here 10) this gives appreciable influence to the overall performance.

In accordance with Amdahl's law, the efficiency is less than 1 and it decreases with virtual machine size. However, we can expect that it will increase with the size of the problem (see remarks to the Amdahl's law in [4], see also [5]).

For grids of 100 points the representative results are as follows:

Test no.	TS [min]	nproc	TB [min]	Achieved speed-up (s)	TS/nproc [min]	Efficiency (e)
1		2	1251.87	1.842	1153.43	0.921
longest	2306,12	5	535.43	4.307	461.22	0.861
time		6	470.15	4.905	384.35	0.818
2		2	951.30	1.740	827.43	0.870
2	1654.85	5	395.82	4.181	330.97	0.836
		6	351.48	4.708	275.81	0.785
3		2	537.83	1.904	511.98	0.952
shortest	1023.97	5	248.58	4.119	204.79	0.824
time		6	216.28	4.734	170.66	0.789

Table 2. 100 initial starting points

The efficiency is generally greater than in the case of 10 starting points. As the required number of starting points can be much greater than 100 for the inverse problem considered, the efficiency of the network computations should be even better.

We noticed that the value of the cost functional was relatively quickly decreasing during first 1000–3000 solver calls in most slave processes that gave good results. Most slaves, that gave large values of J (for our data it was on the order of several tens of thousands and more) after 2000 solver calls, stopped with an unacceptable result of minimization. This gave rise to the idea of an additional acceleration: it can be obtained by a mechanism of stopping early all slow or "unpromising" processes. The algorithm based on this idea is called MSP_M.

5.2 MSP_M — Description of the Algorithm and Numerical Tests

The proposed algorithm of the master process is as follows:

(1) Generate the grid of starting points and remember it in the table GRID.
(2) DO nproc TIMES
- *Spawn a local optimization process (slave).*
- *Send to the slave coordinates of a starting point (from GRID).*

OD
(3) REPEAT
- *Wait for partial results from any slave (this includes the current value of the cost functional, current vector p, the number of iterations (IT) and the state of the local optimization process (ST)).*
- *Update thresholds J_{THR_1}, J_{THR_2} and J_{THR_3}.*
- *IF ST=CONVERGED THEN*
  - *Compare J with the previous best one. Sort obtained results.*
  - *IF $J < J_{THR\_1}$ THEN*
    - *Remember J and p.*
    - *Generate new starting point randomly in a hypercubic neighborhood of p.*

ENDIF

- *SPAWN_NEW_SLAVE()*
- *ELSE IF (J <= J_{THR_1} AND IT < ITER3) THEN Send message to the slave: Continue computations, send back final results.*
- *ELSE IF (IT <= ITER1) THEN*
  - *DECIDE_IF_THE_SLAVE_IS_PROMISING(J_{THR_3})*
- *ELSE IF (IT <= ITER2) THEN /* change threshold: $J_{THR\_2} < J_{THR\_3}$ */*
  - *DECIDE_IF_THE_SLAVE_IS_PROMISING(J_{THR_2})*
- *ELSE IF (IT < ITER3) THEN*
    - *Kill the slave*
    - *SPAWN_NEW_SLAVE():*
- *ELSE (IT >= ITER3)*
  - *IF J <= J_{THR_1} THEN*
      - *Remember J and p.*
      - *Generate new starting point randomly in a hypercubic neighborhood of p.*
    ENDIF
  - *Kill the slave*
  - *SPAWN_NEW_SLAVE():*
  ENDIF
UNTIL global STOP CRITERION is satisfied.
(4) *Kill all slaves that are processed yet, stop computations and print out final results.*

The two procedures:
DECIDE_IF_THE_SLAVE_IS_PROMISING(J_X)

- *CASE*
  - *$J_{THR\_1} < J < J_X$: Send message to the slave: Continue computations, send back new partial results.*
  - *$J >= J_X$:*
    - *Kill the slave*
    - *SPAWN_NEW_SLAVE():*
 ENDCASE
ENDPROCEDURE

SPAWN_NEW_SLAVE():
- *IF there are new points in GRID THEN*
    - *Spawn a new local optimization process (slave)*
    - *Send to the slave coordinates of a starting point (from GRID).*
  ENDIF
ENDPROCEDURE

The global stop criterion is the same as in MSP_B. Values of ITER1, ITER2 and ITER3 are set to 1000, 3000 and 8000 solver calls respectively.

The supervisor kills processes that are "unpromising". A slave process is "unpromising" when the current value of J is greater than a threshold J_{THR_3}, while the number of solver calls (IT) is less than or equal to ITER1, or the value of J is greater than a threshold J_{THR_2} while ITER1 < IT <= ITER2, or $J > J_{THR\_1}$ while IT > ITER2 ($J_{THR\_1} < J_{THR\_2} < J_{THR\_3}$).

All local optimization processes, which are extremely slow (IT >= ITER3) but indicate a small value of the cost functional ($J < J_{THR\_1}$) are also killed, but the results are remembered and one new starting point is generated in the close neighborhood of each such process.

Slaves, for which $J < J_{THR\_1}$ and IT < ITER3 are treated as potentially best. These slaves are allowed to continue computations to the end without sending partial results.

All other slave processes continue computations and should send new partial results to the supervisor.

The threshold values J_{THR_1}, J_{THR_2} and J_{THR_3} are updated during computations.

The example method of updating is:

$$\text{new } J_{THR\_i} = \max (\text{old } J_{THR\_i}, \ k_i * \text{current best } J), \ i=1,2,3$$

The beginning values of thresholds are set arbitrarily, k_i are 20, 400 and 1600 respectively.

The slave processes perform local minimization using the MIGRAD method as it was in MSP_B. Every certain number of iterations (in tests 1000) or after convergence is obtained, a slave sends partial results to the master, unless the message from the supervisor has been received: "Continue computations and send back final results". In this case a slave does not send partial results any more.

Numerical tests have been performed for the same initial grids as in the case of MSP_B. Number of processes and other conditions of tests were the same as in the previous section. For the tests, values of J_{THR_i} $i=1,2,3$ were fixed on previously estimated (during other tests) good levels (100, 2000, 8000, respectively).

The results for initial grids of 10 points are presented in Table 3.

Test no.	TS [min]	nproc	TB [min]	Achieved speed-up* (s*)	TS/nproc [min]	Efficiency (e*)
1	254.00	2	78.63	3.230	127.00	1.615
		5	41.13	6.175	50.80	1.235
		6	36.52	6.956	42.33	1.159
2	203.55	2	46.45	4.382	101.78	2.191
		5	25.68	7.925	40.71	1.585
		6	25.68	7.925	33.93	1.321

Table 3. 10 initial starting points

TM - time of computations of the MSP_M,

TS/nproc - time of computations for the ideal parallelization *without killing "unsatisfactory" processes.*

The ideal speed-up (without killing the mechanism) $s_i = nproc$,

achieved speed-up $s^* = \dfrac{TS}{TM}$, note that it can be greater than s_i (!). Efficiency

$e^* = \dfrac{s^*}{s_i}$.

* - Special kind of speed-up, calculated as the ratio $\dfrac{TS}{TM}$, where TS denotes the time of the scalar version MSP_S *without killing unsatisfactory processes,* so efficiency (denoted here e^*) can be greater than 1. A scalar version of MSP with the "killing" mechanism was not tested. Note, that the method of updating the thresholds $J_{THR\_i}, i = 1,2,3$ on-line (when new partial results are obtained) would give different results in a scalar implementation. The killing mechanism takes advantage of parallel computations.

In Test 2, the computation time is the same for 5 and 6 parallel processes. It follows from the fact that this was the time of computations in one of the slaves. No further improvement can be obtained by increasing the number of processors.

For grids of 100 points the representative results are as follows:

Test no.	TS [min]	nproc	TB [min]	Achieved speed-up* (s^*)	TS/nproc [min]	Efficiency* (e^*)
1	2306.12	5	485.42	4.751	461.22	0.950
		6	406.23	5.677	384.35	0.946
2	1654.85	5	239.52	6.909	330.97	1.382
		6	208.88	7.922	275.81	1.320
3	1023.97	5	169.47	6.042	204.79	1.208
		6	147.03	6.964	170.66	1.161

Table 4. 100 initial starting points

The results show that the introduced speed-up s^* can be greater than *nproc.* In some cases it was even more than a factor of 2. This is caused by the fact that the acceleration follows not only from parallelization of computations, but also by the mechanism of early killing of "unsatisfactory" processes. The factor of this additional acceleration strongly depends on the initial grid of starting points (how many points are from these regions of P that give "unsatisfactory" slaves), and also depends on the algorithm that determines which and when a local optimization process should be killed. In the case of the algorithm presented above, this acceleration depends also on parameters ITER1, ITER2, ITER3 and $J_{THR\_i}, i = 1,2,3$ thresholds. The values of these parameters and thresholds have been set after some experiments in order to obtain maximum acceleration, but not so as to lose local minima with small values of the cost functional.

The grids of 100 points have been chosen in such a way that the first one (Test no. 1, Tables 2 and 4) contains many "good" starting points, such that the value of the cost functional quickly decreases during first 1000 solver calls and then the local optimization process is slow (often more than 5000 solver calls to obtain convergence). This is the reason why the time of computations is the biggest in this case. Also the acceleration obtained is relatively small. The efficiency e^* is less than 1, however it is greater than for MSP_B. The third grid contains many points, that are not "good", and such that local optimizers (slaves) often stop computations with less than 1000 solver calls. The time of computations in Test no. 3 is therefore the smallest, but also the additional acceleration is less than in Test no. 2. The second grid has contained no "special kind" of points and was most random. The additional acceleration of MSP_M is the biggest here.

MSP_M allows us to, when we know the appropriate narrow bounds for the components of p (which means we can set at most several hundred starting points within the searched domain P), then we can solve the considered one-dimensional inverse problem in a reasonable time (below 10-20 hours as in the tests) using the computational power of several SUN workstations. Further acceleration could be possibly obtained by using other local methods of nonlinear programming in the slaves (the MSP is practically transparent with respect to local methods). The acceptable value of the cost functional can be obtained after several dozen or several hundred trials from different points.

Proper bounds for the components of p can be sometimes estimated when we know the beginning value of p from the laboratory/field experiments carried on during the construction of the structure considered. However, this is often not accessible.

Moreover, the next problem arises when we consider two or three dimensional flow: the number of zones in the filtration area can be equal to several dozen and the dimension of p will be in the tens or hundreds. When the dimension of the optimization task (the dimension of p) grows and we cannot define proper bounds for the components of p, the searching domain becomes too large even for distributed computations. The number of initial starting points in a regular grid increases exponentially with the dimension of the task. Searching through millions of points using the MSP_B and also MSP_M can be ineffective.

The proper setting of the initial starting point grid seems to be one of the most promising methods to improve the process of identification in such cases. The method of condensation of the grid, applied in MSP_B and MSP_M, is too simple to give significant improvement. Therefore a further modification of the MSP is proposed.

6. Proposition of a Parallelized Törn Global Optimization Method for the Generation of Starting Points for the MSP

The generation of the proper initial grid of starting points for the Multiple Starting Point parallel optimization algorithm is a very important problem. We propose to apply the parallel version of the Törn global optimization method [15,16] in cases where the domain of the searches is so large that the number of points in the regular grid can be equal to thousands or more.

The idea of the Törn method is based on the observation that when we start local optimization processes from many points (called by Törn the *global points*), we often obtain the same local minimum. Instead of wasting time for exact computations from each starting point, we can compute only several iterations of a very simple optimization method. The points obtained are called the intermediate points. A set of intermediate points that "belong" to one local minimum is called a *cluster*. For further exact computations, we can choose one or two points from each cluster.

The Törn method can accelerate searching through large areas. The parallelization and distribution of computations can give the additional speed up and can make this method more powerful. We propose to parallelize two threads of the Törn method:

- the minimization processes that are started from global points can be executed in parallel,
- the delimitation of clusters can be quicker when we use the proper parallel algorithm.

This is currently under investigation.

7. Conclusions

- The MSP_B algorithm accelerates searching through the domain (carried on in order to find the global minimum). The distribution of computations gives the advantage of a more effective use of computer resources. The problem is *coarse-grained* with the time of communication between processes being much less than the time of computations within those processes (see [4]). It warrants efficient use of parallel techniques.

- The resulting acceleration of the MSP_M algorithm is more than *nproc* times in comparison to scalar methods, where *nproc* stands for the number of parallel slave processes. The additional acceleration follows from the mechanism of early killing "unsatisfactory processes".

- The additional acceleration (more than *nproc* times) of MSP_M depends on the initial starting grid and the algorithm of killing "unsatisfactory" processes. The "killing" algorithm, proposed in this paper, seems to be quite efficient, though the proper setting of thresholds for time of computations in slaves and for the value of the cost functional require some experience and tests. However, we do not pretend that this algorithm is optimal and further modifications are not excluded.

- For the test problem of the identification of parameters in the Swartzendruber formula (in the case of one-dimensional flow), if we set the proper initial grid of the starting points, then MSP_M provides us with the possibility to solve the task within an acceptable time (at most several dozen hours) using the computational power of only several SUN workstations.

- We have implemented the MIGRAD method in slaves. Of course one can use quicker (maybe simpler) methods, but it does not change considerations concerning applied parallel techniques. The MSP is practically transparent with respect to local methods

- We suggest that the proposed method can accelerate the numerical solving of other problems with many local minima, both in inverse analysis and optimization. An important class of such problems—which are of great significance—are molecular configuration problems. They have hundreds or thousands of parameters and typically many local minima with function values very close to the global one. Moreover, local minima often have small basins of attraction [11].

- Further improvements of the algorithms described concerning the generation of the grid of starting points are now being investigated. We expect that the application of the parallel version of the Törn global optimization method will give good results in cases of more dimensional vector p (two or three dimensional flow).

- We are currently also working on hybrid genetic/parallel Törn algorithms for searches through large domains [10].

References

[1] R.H. Byrd, C.L. Dert, A.H.G. Rinnoy Kan and R.B. Schnabel. Concurrent stochastic methods for global optimization. *Math. Programming* 46 (1990) 1-30.

[2] Z. Denkowski, S. Migórski, R. Schaefer and H. Telega. Theoretical and practical aspects of inverse problems for nonlinear filtration process. In H.D. Bui et al. (eds), *Inverse Problems in Engineering Mechanics*, A.A. Balkema, Rotterdam, (1994) 403-409.

[3] Z. Denkowski, R. Schaefer and H. Telega. On identification problems for prelinear filtration of ground water. In K. Morgan et al. (eds), *Finite Elements in Fluids*, Barcelona, Pineridge Press, (1993) 878-886.

[4] H. S. Morse *Practical parallel computing*. AP Professional, Cambridge USA 1994.

[5] Pranay Chaudhuri *Parallel Algorithms Design and Analysis*. Prentice Hall of Australia Pty Ltd 1992.

[6] A.H.G. Rinnoy Kan and G.T. Timmer. Stochastic methods for global optimization, *Amer. J. Math. Management Sci.* (1984) 7-40.

[7] A.H.G. Rinnoy Kan and G.T. Timmer. Global optimization. *Handbooks in Operations Research and Management Science, Vol. I: Optimization*, G.L. Nemhauser, A.H.J. Rinnoy Kan and M.J. Todd, eds. North-Holland, (1989) 631-662.

[8] R. Schaefer. *Numerical models of the prelinear filtration.* Jagiellonian University Press, Cracow, 1991 (in polish).

[9] R. Schaefer and S. Sędziwy. Filtration in cohesive soils: modeling and solving. In K. Morgan et al. (eds), *Finite Elements in Fluids*, Barcelona, Pineridge Press, (1993) 887-891,.

[10] R. Schaefer and H. Telega. A hybrid genetic approach to the hydraulic conductivity identification in earthen dams. In J. Arabas (ed) Warsaw University of Technology, *Algorytmy ewolucyjne (Evolution algorithms), Proc. of I Polish Conference on Evolution Algorithms*, Murzasihle, 1996.

[11]   R. B. Schnabel. A view of the limitations, opportunities, and challenges in parallel nonlinear optimization. *Parallel Computing*, 21 (1995) 875-905.

[12]   D. Swartzendruber. Modification of Darcy's law for the flow of water in soils. *Soil. Sci.* 93, (1961) 23-29.

[13]   H. Telega. Nonlinear programming approach to the inverse problem of water percolation through cohesive soils. In K. Morgan et al. (eds), *Finite Elements in Fluids*, Barcelona, Pineridge Press, (1993) 892-899,.

[14]   H. Telega. Distributed and Parallel computations for solving identification problems. In M. Kasiak et al. (eds), *Proc. of XII Polish Conference on Computer Methods in mechanics*, Warsaw-Zegrze (1995).

[15]   A.A. Törn. Cluster analysis using seed points and density-determined hyperspheres as an aid to global optimization. *IEEE Transactions on Systems, Man, and Cybernetics*, p.610, (1977).

[16]   R. Wit. *Metody programowania nieliniowego (Methods of Nonlinear Programming)*. WNT, Warszawa, (1986) (in Polish).

[17]   W. Wolski, A. Szymański, A. Mirecki, Z. Lechowicz, R. Larson, J. Hartlen, K. Garbulewski, U. Bergdahl. Two stage - constructed embankment on organic soils, *Swedish Geotechnical Institute Reports*, no 32, Linköping, (1988).

[18]   R. Zieliński, P. Neumann. *Stochastische Verfahren zur Suche nach dem Minimum einer Funktion*. Akademie-Verlag, Berlin (1983) (Polish translation *Stochastyczne metody poszukiwania minimum funkcji*, WNT Warszawa 1986).

[19]   MINUIT, Function Minimization and Error Analysis. Reference Manual, CERN Program Library Long Writeup D506 Geneva (1992).

Automated Optimal Design
Using CFD and High Performance Computing

Doyle D. Knight

Department of Mechanical and Aerospace Engineering
Rutgers University - The State University of New Jersey
New Brunswick, NJ 08903

Abstract. The advent of high performance computing has led to an increased role of CFD in automated optimal design, particularly in aeronautics. This paper presents a brief description of the optimization problem, strategies for its solution, algorithms for optimization, references to recent applications, impact of high performance computing and parallelization, and two examples of application to the design of high speed inlets.

1 Introduction

Design is an inherent part of engineering. In practice, the design process incorporates analytical, computational and experimental results to achieve an optimal product[1] subject to a set of constraints. The optimal design may be achieved *manually* where the decisions are made by the designer, or *automatically* where the decisions are made by a computer. In the latter case (and sometimes, in the former), the best design is chosen on the basis of optimizing[2] an *objective function* $f(\mathbf{x})$ which is a scalar function of the design variables $\mathbf{x}$ subject to a set of *constraints*. One example is the determination of the shape of a scramjet inlet to achieve maximum total pressure recovery [1].

Computational Fluid Dynamics (CFD) has principally been employed in the *manual* design role to date. Typically, the CFD specialist is provided the geometry and boundary conditions, and performs a numerical simulation of the flowfield. The computed flowfield is then analyzed (*e.g.*, using contour plots) and the geometry and/or boundary conditions modified (oftentimes, heuristically) to improve the design.

Four factors have led to a significant expansion of CFD in *automated* design. First, the continued rapid improvements in computer performance (typically denoted as the era of High Performance Computing (HPC)) enable routine numerical simulations of increasing sophistication and complexity. Second, improvements in the accuracy, efficiency and robustness of CFD algorithms (see, for

---

[1] The definition of "optimal" is oftentimes one of the most critical and difficult aspects of the design process.

[2] We may assume optimizing is equivalent to minimizing the objective function (see below). For the purposes herein, we assume that the minimization is not known analytically, and therefore an iterative method is required.

example, Hirsch [2]) likewise contribute to the capability for simulation of more complex flows. Third, the development of efficient and robust optimizers enable automated search of design spaces [3]. Fourth, the development of sophisticated shell languages (*e.g.*, Perl [4]) enable effective control of pathological events in the automated design process using CFD (*e.g.*, handling the "crash" of a CFD code due to divide by zero, square root of a negative number, segmentation error, etc.).

The expanded role of CFD in automated design is not new. There is ample precedent for this role for CFD, particularly in design of airfoils (see, for example, the review by Lores and Hinson [5]).

The objectives of this paper are to provide a brief review of the optimization problem, to present a short survey of recent papers in automated design using CFD, to describe strategies and algorithms for optimization, to discuss the opportunities for using high performance computing (in particular, parallelism) and to present some recent examples of automated design.

2 Optimization

2.1 Definition

The general *nonlinear optimization problem* (also known as the *nonlinear programming problem*) may be defined[3] as [3, 6, 7]

$$\text{minimize } f(\mathbf{x}) \tag{1}$$

where $f(\mathbf{x})$ is the scalar *objective function*[4] and $\mathbf{x}$ is the *vector of design variables* with elements $\{x_i\}_{i=1,...,n}$. The optimization is typically subject to a limit on the allowable values of $\mathbf{x}$

$$\mathbf{a} \leq \mathbf{x} \leq \mathbf{b} \tag{2}$$

and m additional *linear* and/or *nonlinear constraints*

$$\begin{aligned}
c_i(\mathbf{x}) &= 0, \quad i = 1, 2, \ldots, m' \\
c_i(\mathbf{x}) &\leq 0, \quad i = m' + 1, \ldots, m
\end{aligned} \tag{3}$$

Special cases include the *linear programming problem* where f and c_i are linear, and the *quadratic programming problem* where f is quadratic and $c_i(\mathbf{x})$ are linear.

There are no general methods for guaranteeing that the *global* minimum of an *arbitrary* $f(\mathbf{x})$ will be found [6], although indeed the global minimum is sometimes found in practice. Typically, methods focus on determining a *local* minimum with additional heuristics to attempt to prevent the optimization algorithm from becoming "stuck" in a local minimum which is not the global minimum.

---

[3] For a (nonunique) taxonomy of optimization problems, see Moré and Wright [3].
[4] There is no loss of generality in assuming the function $f(\mathbf{x})$ is to be minimized. To maximize a function $\tilde{f}(\mathbf{x})$, set $f = -\tilde{f}$.

Formally, a point $\mathbf{x}^*$ is a (strong) local minimum [6] if there is a region surrounding $\mathbf{x}^*$ wherein the objective function is defined and $f(\mathbf{x}) > f(\mathbf{x}^*)$ for $\mathbf{x} \neq \mathbf{x}^*$.

Provided $f(\mathbf{x})$ is twice continuously differentiable[5], necessary and sufficient conditions for the existence of a solution to (1) to (3) may be obtained. A complete description is beyond the scope of this paper and the reader is referred to Gill *et al* [6]. For example, in the one-dimensional case with no constraints the sufficient conditions for a minimum at x^* are

$$g = 0 \text{ and } H > 0 \text{ at } x = x^* \tag{4}$$

where $g = df/dx$ and $H = d^2f/dx^2$. For the multi-dimensional case with no constraints[6]

$$|g_i| = 0 \text{ and } H \text{ is positive definite at } x = x^* \tag{5}$$

where $g_i = \partial f/\partial x_i$, $|g_i|$ is the norm of the vector g_i, and H is the *Hessian matrix*

$$H = \begin{pmatrix} \frac{\partial^2 f}{\partial x_1^2} & \cdots & \frac{\partial^2 f}{\partial x_1 \partial x_n} \\ \vdots & & \vdots \\ \frac{\partial^2 f}{\partial x_1 \partial x_n} & \cdots & \frac{\partial^2 f}{\partial x_n \partial x_n} \end{pmatrix} \tag{6}$$

2.2 Strategies and Algorithms for Automated Optimization

Numerous optimization strategies have been developed, in many cases intended for specific optimization problems. An example is shown in Fig. 1 similar to Gelsey [8]. The optimizer consists of a Design Associate (DA) comprised of one or more optimizers as described below, and a Modeling and Simulation Associate (MSA) comprised of a geometry module, grid generator, flow solver and post processor. The DA and MSA are executed repeatedly in sequence until some predefined convergence criterion are met.

2.3 Algorithms for Optimization

The efficacy of an optimization algorithm depends strongly on the nature of the design space. For engineering problems, the design space oftentimes manifests various types of pathologies [9]. First, the objective function f may possess multiple local optima [10], arising from physical or numerical reasons. Examples of the latter include noise introduced in the objective function by grid refinement between successive flow simulations, and incomplete convergence of the flow simulator. Second, the objective function f and/or its gradient g_i may exhibit nonsmoothness for physical reasons. For example, the thrust of a turbojet engine

---

[5] This is not always true. For example, the one-dimensional trade study shown in Fig. 10 implies a design space with "gaps" where the objective function is unevaluable.

[6] A matrix A is positive definite if all of the eigenvalues of A are positive.

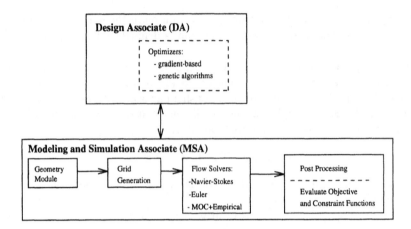

Fig. 1. Optimization strategy

changes discontinuously when the afterburner is ignited. Third, the objective function f may not be evaluable at certain points. This may be due to constraints in the flow simulator such as a limited range of applicability for empirical data tables.

A brief description of some different classes of optimizers is useful. In the interests of brevity, these methods are described for the unconstrained optimization problem. The reader is referred to Moré and Wright [3] for an overview of optimization algorithms and software packages, and to Gill [6] for a discussion of the constrained optimization problem.

Gradient optimizers are based on the assumption that the objective function f can be approximated in the vicinity of a point $\tilde{x}$ by a quadratic form

$$f \approx \tilde{f} + \tilde{g}_i(x_i - \tilde{x}_i) + \frac{1}{2}(x_i - \tilde{x}_i)\tilde{H}_{ij}(x_j - \tilde{x}_j) \tag{7}$$

where $\tilde{f}$, $\tilde{g}_i$ and $\tilde{H}$ imply evaluation at $\tilde{x}$ and the Einstein summation convention is implied. In the relatively simple *Method of Steepest Descent* [11], the quadratic term in (7) is ignored, and a line minimization is performed along the direction of $-g_i$. Once the minimum is found, the gradient g_i is computed at the new point and the procedure is repeated. This method is inefficient, however, for design spaces which are characterized by long, narrow "valleys".

The *Conjugate Gradient Methods* [11] perform a sequence of line minimizations along specific directions in the design space which are mutually orthogonal in the context of the objective function. Consider a line minimization of f along a direction $\mathbf{u} = \{u_i\}_{i=1,\dots,n}$. At the minimum $\tilde{x}$,

$$u_i \tilde{g}_i = 0 \tag{8}$$

by definition. Consider a second line minimization of f along a direction $\mathbf{v}$. From (7), the change in g_i along the direction $\mathbf{v}$ is $\tilde{H}_{ij}v_j$. Thus, the condition that the second line minimization also remain a minimization along the first direction $\mathbf{u}$ is

$$u_i \tilde{H}_{ij} v_j = 0 \tag{9}$$

When this condition is satisfied, $\mathbf{u}$ and $\mathbf{v}$ are denoted *conjugate pairs*. Conjugate Gradient Methods (CGM) generate a sequence of directions $\mathbf{u}, \mathbf{v}, \mathbf{w}, \ldots$ which are mutually conjugate. If f is exactly quadratic, then CGM yield an n-step sequence to the minimum.

In contrast to the Method of Steepest Descent and the Conjugate Gradient Methods, the *Sequential Quadratic Programming Methods* employ knowledge of the Hessian H which may be computed directly when economical or approximated from the sequence of g_i generated during the line search (the *quasi*-Hessian [3]). Given the gradient and Hessian, the location x_i^* of the minimum value of f may be found from (7) as

$$\tilde{H}_{ij}(x_j^* - \tilde{x}_j) = -\tilde{g}_i \tag{10}$$

For the general case where f is not precisely quadratic, a line minimization is typically performed in the direction $(x_i^* - \tilde{x}_i)$, and the process is repeated.

Control Theory utilizes the concept of an adjoint system of equations to determine the gradient g_i (see, for example, Jameson [12, 13], Baysal and Eleshaky [14] and Ta'asan *et al* [15]). The computation of the gradient using the adjoint requires approximately the same computer resources as the flow simulator, and is thus more efficient than a finite difference computation of the gradient which requires $n + 1$ evaluations (for one-sided differences). However, special considerations are required in the vicinity of flow discontinuities such as shocks [16].

Stochastic Optimizers incorporate a measure of randomness in the optimization process in order to avoid convergence to a local (rather than a global) minimum which is a common pitfall of gradient optimizers. *Simulated Annealing* [17] mimics the process of crystalization of liquids (or annealing of metals) by stochastically minimizing a function E analogous to the energy of a thermodynamic system. Consider a current state in the design space $\tilde{\mathbf{x}}$ and its associated "energy" $\tilde{E}$. A candidate next state $\mathbf{x}^*$ is selected by randomly perturbing typically one of the components $\tilde{x}_j, 1 \leq j \leq n$ of $\tilde{\mathbf{x}}$, and its energy E^* is evaluated[7]. If $E^* < \tilde{E}$ then $\tilde{\mathbf{x}} = \mathbf{x}^*$, *i.e.*, the next state is $\mathbf{x}^*$. If $E^* > \tilde{E}$ then the probability of selecting $\mathbf{x}^*$ as the next design state is

$$p = \exp(-(E^* - \tilde{E})/kT) \tag{11}$$

where k is the "Boltzman constant" (by analogy to statistical mechanics) and T is the "temperature" which is successively reduced during the optimization[8]. Thus, during the sequence of design states, the algorithm permits the selection

---

[7] Typically, each component of $\mathbf{x}$ is perturbed in sequence.

[8] Of course, only the value of the product kT is important. The probabilistic feature is implemented typically by calling a random number generator for a value r between 0 and 1. Then the state $\mathbf{x}^*$ is selected if $r < p$.

of a design state with $E > \tilde{E}$, but the probability of selecting such a state decreases with increasing $E - \tilde{E}$. This feature enables (but does not guarantee) the optimizer to "jump out" of a local minimum.

Genetic Algorithms (GA) mimic the process of biological evolution through introduction of stochastic mutations in a set of designs denoted the *population* [18]. At each iteration, the "least fit" member(s) of the population (*i.e.*, those designs with the highest value of f) are removed, and new members are generated by a recombination of some (or all) of the remaining members. Rasheed and Gelsey [19] present one specific implementation. In this approach, an initial population P of designs is generated by randomly selecting points $x_i, i = 1, \ldots, P$ satisfying (2). The two best designs (*i.e.*, with the lowest values of f) are joined by a straight line in the design space. A random point x' is chosen on the line connecting the two best designs. A mutation is performed by randomly selecting a point x_{P+1} within a specified distance of x'. This new point is added to the population. A member of the population is then removed according to a heuristic criterion, *e.g.*, among the k members with the highest f, remove the member closest[9] to x_{P+1}, thus maintaining a constant number of designs in the population. The process is repeated until convergence.

Both Simulated Annealing and Genetic Algorithms typically require evaluation of large numbers of designs. Consequently, they may not be suitable for optimization problems with cpu-intensive flow simulations (*e.g.*, Reynolds-averaged Navier-Stokes [RANS]).

2.4 Parameterization

In a given design problem, there are many different representations for the vector of design variables x, and the particular choice affects the cost and generality of the optimization method. Consider, for example, the optimal design of a two dimensional airfoil section. A typical objective function is the minimization (in the least squares sense) of the deviation of the surface pressure from a specified target pressure distribution. Venkataraman *et al* [20] define the airfoil geometry using four Bézier parametric curves [21] which result in a total of nineteen design variables representing the coordinates of the vertices of the Bézier polygons. Alternately, Obayashi and Tsukahara [10] store the airfoil coordinates for n airfoils in the form $y^1, \ldots, y^n$, and define the airfoil as

$$y = a_1 y^1 + \ldots a_n y^n \tag{12}$$

where $x = (a_1, \ldots, a_n)$ is the vector of design variables. In this example, $n = 4$.

2.5 Selection of Objective Function

The optimal mathematical formulation of the objective function is oftentimes a crucial element in automated design, as illustrated by Fejtek *et al* [22] in the optimal design of a turboprop nacelle. The desired objective function is the nacelle

---

[9] This method tends to prevent clustering of the population.

drag; however, the determination of the nacelle drag from the flow simulator (the panel code VSAERO and a streamline-based boundary layer method) is unreliable. Thus, Feyjek et al tested two different objective functions: the minimum value of pressure coefficient, and the boundary layer separation location along the cutwater profile[10].

3 Examples of Recent Applications

A few examples of recent applications of CFD to automated optimal design in aeronautics serve to demonstrate the developing capability in this field[11]. Siclari et al [24] used Simulated Annealing to optimize the placement of internal actuators in a transonic shockless airfoil to minimize drag at off-design conditions. Pandya and Baysal [25] developed and applied a gradient-based methodology to the optimization of a Mach 2.4 cranked wing. Merchant and Drela [26] optimized a suction supercritical airfoil, and observed that an appropriate distribution of boundary layer suction yields higher drag divergence Mach numbers for a given thickness ratio and design lift coefficient. Eyi et al [27] used a constrained optimization method to maximize the lift at fixed drag for multi-element airfoils. Jameson [28] used control theory to optimize the design of a swept transonic wing. Naik et al employed Navier-Stokes simulations for design of nacelles [29].

4 High Performance Computing

It is obvious that continued improvements in microprocessor performance (i.e., the doubling of processor speeds every eighteen months [30]) and ubiquity of parallel computing technology [31] provide enhanced opportunities for automated optimal design using CFD.

In many automated design applications using CFD, the flow solver (and, sometimes, the mesh generation) represents the principal demand for cpu resources, and thus parallelization of the flow solver improves performance of the optimizer. For example, Fejtek et al [22] employed a network of workstations for defining the surface of a turboprop nacelle, and Jameson and Alonso [32] used an IBM SP2 for both the flow simulation and solution of the adjoint equations for optimization of a business jet.

Straightforward opportunities for parallelism can be identified in the optimization algorithms. For gradient optimizers, computation of all of the elements g_i of the gradient using finite differences can obviously be performed in parallel. The adjoint formulation for determination of g_i can also be performed simultaneously with the flow solution. Regarding the line search, it is possible to consider

---

[10] The cutwater is the indented region between the propeller spinner and the inlet to the turboprop.

[11] There are similar examples of recent applications of CFD in multi-disciplinary optimization. Also, it is emphasized that aerodynamic optimization is not a new subject. Optimization of transonic airfoils, for example, has a long history (see [23]).

several increments along the search direction simultaneously, and select the most favorable, thereby achieving improved performance. For stochastic optimizers, both coarse- and fine-grained parallelism has been implemented in Genetic Algorithms, for example. The coarse-grained approach subdivides the population into distinct groups (*demes*) which are assigned to individual processing nodes where a standard GA search performed [33]. Information is transmitted between nodes by transferring one or more members of the population according to heuristic algorithms. Fine-grained approaches to GA have also been proposed [34].

5   Examples

Two examples of application of automated design in computational fluid mechanics are presented. In the first example, a single type of flow code (RANS) is employed, while in the second example, two types of flow codes (a simple physical model and a RANS code) are utilized.

5.1   Hypersonic Inlet

Description. In the late 1960s, a family of hypersonic inlets were designed by Seebaugh and his colleagues at Republic Aviation and NASA Ames Research Center [35, 36] for a Mach 10 to 12 aircraft (Fig. 2). In this design, the vehicle forebody provided the initial flow compression, decelerating the flowfield to approximately Mach 6 at the entrance to the series of rectangular inlet modules which are attached to the vehicle undersurface. The inlet modules provided the final compression of the flow to the required conditions at the entrance of the scramjet combustor.

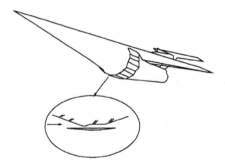

Fig. 2. Hypersonic aircraft

Two of the inlet designs are shown in Fig. 3 affixed to a wedge forebody. The freestream conditions at the entrance to the inlet are Mach number $M_\infty = 5.8$, total pressure $p_{t_0} = 2.69 \times 10^6$ Pa, and total temperature $T_{t_0} = 770°$ K. The incoming centerbody boundary layer thickness $\delta_0 = 1.1$ cm.

The inlets were designed by Seebaugh *et al* using the method of characteristics, coupled with a boundary layer correction and an empirical control volume model of the shock-boundary layer interaction. The P2 inlet was designed to provide a static pressure rise[12] of a factor of two[13] by means of a single shock wave generated by the deflection of the cowl. The design objective was to cancel the cowl shock at the point of impingement on the centerbody by appropriate contouring of the centerbody surface. Since the internal cowl surface was essentially flat, the achievement of this design objective would imply a constant static pressure at the throat – a condition generally considered desirable at the entrance to the combustor [36]. The P8 inlet was designed to provide a static pressure rise of a factor of eight by means of both a cowl-generated shock (of the same strength as the P2 inlet) and a distributed isentropic compression generated by curvature of the inner cowl surface. Again, the design objective was to cancel both the cowl shock and the distributed cowl-generated compression by appropriate contouring of the centerbody surface in the vicinity of the impingement of the cowl shock and cowl-generated compression.

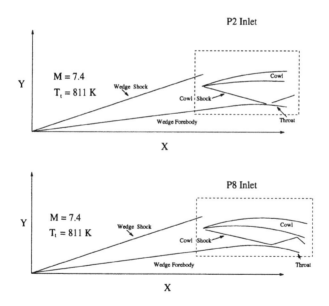

Fig. 3. P2 and P8 inlets

Experiments [35, 36] indicated that the design objective of cancelling the cowl shock and (in the case of the P8 inlet) the cowl-generated compression at the centerbody were not met. Reflected shock waves were observed (Fig. 3) in the experiment.

---

[12] Between the inlet entrance and the inlet throat.

[13] Hence the designation P2.

Design Methodology. In 1994, the Propulsion Design Group at Rutgers University began the development of an *automated* design methodology to (re)design the P2 and P8 inlets, respectively, using the same design objective – the cancellation of the cowl shock and (for the P8) cowl-generated compression by appropriate contouring of the inlet centerbody (Gelsey *et al* [37] and Shukla *et al* [38]). The design methodology, shown in Fig. 1 and described below, was based on previous work (see, for example, [8, 39, 40]).

The *Geometry Module* defined the centerbody geometry by a parameterized contour consisting of three sections shown in Fig. 4. The left section is a straight line rising at angle θ which terminates at point (x_l, y_l). The right section is a straight line turned through angle $\Delta\theta$ relative to the left line, starting at a point offset by $(\Delta x, \Delta y)$ from the end of the left section. The middle section is a smooth curve whose shape is uniquely determined by the requirement that it connect the left and right sections, have an angle of θ_1 at (x_l, y_l) and match the slope at $(x_l + \Delta x, y_l + \Delta y)$. This smooth curve is parametrically generated as follows:

$$\begin{pmatrix} x \\ y \end{pmatrix} = \begin{pmatrix} x_c \cos p + x_s \sin p + x_0 \\ y_c \cos p + y_s \sin p + y_0 \end{pmatrix}, \ 0 \le p \le \frac{\pi}{2} \tag{13}$$

and the six coefficients $(x_c, x_s, x_0, y_c, y_s, y_0)$ are uniquely[14] determined by the following six requirements:

p	$x(p)$	$y(p)$	angle
0	x_l	y_l	θ_1
$\pi/2$	$x_l + \Delta x$	$y_l + \Delta y$	$\theta + \Delta\theta$

The cowl contours for the P2 and P8 inlets were unchanged in order to achieve the same static pressure rise as the original designs.

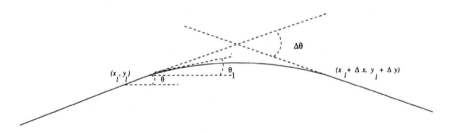

Fig. 4. Geometry model

Grid Generation was achieved by an algebraic method using cubic spline interpolation of the inner cowl surface and the parameterization of the centerbody described above.

Since interaction of the cowl shock with the centerbody boundary layer is a crucial element of the design, the *Flow Solver* employed a Reynolds-averaged

---

[14] Except in the case $\Delta\theta = 0$.

Navier-Stokes (RANS) code. NPARC [41] was used for the P2 and GASP© [42] for the P8 inlet. The Chien $k-\epsilon$ turbulence model [43] was employed in both cases. Details of the grid resolution and refinements studies are presented in [37, 38].

Validation of the RANS simulations for the original design is a prerequisite for redesign. Several studies (e.g., Knight [44], Ng et al [45], and Kapoor et al [46]) have demonstrated that RANS simulations can accurately predict the flowfield within the original P2 and P8 geometries including the reflected shock system observed in the experiments. Gelsey et al [37] and Shukla et al [38] also confirmed that the RANS computations accurately reproduced the flowfield within the original inlet configuration.

The *Post Processing* module evaluated the objective function and constraints. The mathematical form of the objective function which will eliminate the reflected wave structure is not obvious. Three different objective functions were used. The first is the static pressure distortion at the throat:

$$\sigma_p = \frac{1}{\bar{p}} \left[\frac{1}{H} \int_0^H (p - \bar{p})^2 dy' \right]^{1/2} \tag{14}$$

where $\bar{p}$ is the mean static pressure at the throat

$$\bar{p} = \frac{1}{H} \int_0^H p \, dy \tag{15}$$

where H is the height of the throat, and y is measured (vertically) from the lower surface. This objective function was used for both the P2 and P8 inlet. The second objective function is the area average static pressure distortion $\sigma_{\bar{p}}$:

$$\sigma_{\bar{p}} = \frac{1}{\bar{p}} \left[\frac{\sum (p(x, y) - \bar{p})^2}{n - 1} \right]^{\frac{1}{2}} \tag{16}$$

where $\sum$ indicates a sum over n grid cells downstream of the intersection of the cowl shock with the centerbody, and

$$\bar{p} = \frac{1}{n} \sum p(x, y) \tag{17}$$

The third objective function is a measure of the reflected shock strength:

$$\sigma_{sh} = \frac{1}{n_x} \sum_{n_x} \left(\frac{\Delta p}{\bar{p}_x} \right) \tag{18}$$

where

$$\bar{p}_x = \frac{1}{n_y} \sum_{n_y} p(x, y) \tag{19}$$

for all n_y cells at a given x location, where Δp is the static pressure jump across the shock, $p(x, y)$ is the static pressure distribution and n_x cells are used in the

x direction. The second and third objective functions were used for the P8 inlet only.

The constraint employed by the Post Processing module was the mean static pressure recovery:

$$r_p = \frac{1}{H p_\infty} \int_0^H p \, dy' \tag{20}$$

where p_∞ is the static pressure immediately upstream of the inlet entrance. The constraint required $r_p = 2.0 \pm 0.1$ for the P2 inlet and $r_p = 8.0 \pm 0.5$ for the P8 inlet.

Two additional measures were monitored during the design to insure overall quality. These are the total pressure distortion at the throat:

$$\sigma_{p_t} = \frac{1}{\bar{p}_t} \left[\frac{1}{H} \int_0^H (p_t - \bar{p}_t)^2 dy' \right]^{1/2} \tag{21}$$

where $\bar{p}_t$ is the mean total pressure at the throat

$$\bar{p}_t = \frac{1}{H} \int_0^H p_t \, dy' \tag{22}$$

and the mean relative total pressure at the throat

$$r_{p_t} = \frac{1}{H p_{t_\infty}} \int_0^H p_t \, dy' \tag{23}$$

where p_{t_∞} is the upstream freestream total pressure measured in the freestream flow at the entrance to the inlet.

Table 1. Design Variables for P2 and P8 inlets

Geometry P2	P8			
f	σ_p	σ_p	$\sigma_{\bar{p}}$	σ_{sh}
x_l	•	•	•	•
Δx	•	•	•	•
Δy	•	•	•	•
θ_1		•		
$\Delta\theta$	•		•	

The *design variables* for the P2 and P8 cases are indicated in Table 1. The P2 optimization employed four variables x_l, Δx, Δy and $\Delta\theta$. The P8 optimizations used from three to five variables depending on the optimization function f.

The *Design Associate (DA)* employed a gradient optimizer CFSQP [47] which minimizes an objective function subject to general smooth constraints. The package uses the sequential quadratic programming (SQP) method to solve a nonlinear programming problem by solving a sequence of quadratic programming problems. First, a quadratic programming problem is fitted to actual nonlinear problem by computing the Hessian[15] of the objective function with respect to design vector, and the gradient of each constraint function with respect to design vector. Second, the quadratic programming problem is solved using the package QLD which is an implementation of Powell's method. Third, a minimization is performed along the line defined by the current point and the minimum of the quadratic programming problem. CFSQP terminates when either 1) the current point is approximately the minimum, or 2) the improvement in the objective function during the line minimization is less than a certain tolerance. Further details are presented in [37, 38].

Results for P2 Inlet. The P2 inlet was successfully redesigned using the automated design methodology. Fig. 5 displays the static pressure contours for the original P2 geometry. The reflected shock is clearly evident in the computed flowfield, in agreement with experiment. Fig. 6 shows the static pressure contours for the optimized P2 geometry. There is no reflected shock. The original centerbody geometry is indicated by the dashed line. The objective function for the original and optimized inlets is shown in Table 2, and the optimal values of the design variables in Table 3.

The optimization using required 60 Navier-Stokes simulations (approximately 5 days on a Hewlett-Packard 735/125 workstation). However, it was noted that $\Delta x = 0.471$m (more than six times the distance from $x_l = 1.1189$m to the throat $x = 1.1938$ m), and $\Delta y = -0.0013$ m. Thus, the centerbody was effectively the wedge forebody connected to a horizontal line. Another optimization was performed using x_l as the single design variable ($\Delta x = 0, \Delta y = 0$ and $\Delta \theta$ chosen so that the centerbody is approximately horizontal for $x > x_l$). For the simpler geometry, only 10 Navier-Stokes simulations (0.8 day on Hewlett-Packard 735/125) were required to achieve a 74% reduction in σ_p which is virtually the same result.

---

[15] In practice, an approximation to the inverse of the Hessian (denoted the quasi-inverse Hessian) is used for efficiency. The quasi-inverse Hessian is updated on each iteration using the gradient of the objective function with respect to the design vector. CFSQP employs the Broyden-Fletcher-Goldfarb-Shanno update formula.

Table 2. P2 inlet optimization

Measure	Original	Optimal	Change
σ_p	0.0699	0.0134	-80.1%
r_p	2.0053	1.9238	-4.1%
σ_{p_t}	0.3942	0.4014	1.8%
r_{p_t}	0.4957	0.4926	0.6%

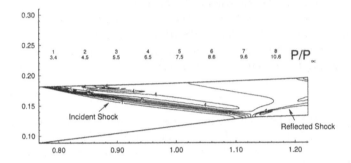

Fig. 5. Static pressure contours p/p_∞ for original P2 inlet

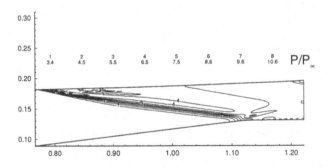

Fig. 6. Static pressure contours p/p_∞ for optimized P2 inlet

Table 3. Optimal P2 and P8 inlets

Parameter	P2	P8
x_l m	1.1189	1.13097
Δx m	0.4705	0.150
Δy m	-0.0013	-0.01687
θ_1 radians	0.11738	0.11738
$\Delta\theta$ radians	0.1249	0.291

Results for P8. The P8 inlet was also successfully designed using the auto-mated design methodology. The reflected shock strength objective function σ_{sh} was successful in virtually eliminating the reflected waves[16]. Fig. 7 displays the static pressure contours for the original geometry. The reflected shock is evident. Fig. 8 shows the static pressure contours for the optimized P8 geometry. There is virtually no reflected shock. The original P8 geometry is shown by the dashed line. The objective function for the original and optimized inlets is shown in Table 4 and the optimal values in Table 3.

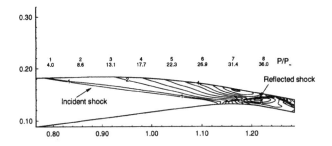

Fig. 7. Static pressure contours p/p_∞ for original P8 inlet

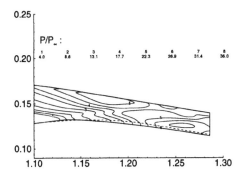

Fig. 8. Static pressure contours p/p_∞ for optimized P8 inlet

---

[16] The static pressure distortion σ_p and the area-weighted static pressure distortion $\sigma_{\bar{p}}$ were not successful [38].

The optimization for the P8 inlet required approximately 100 Navier-Stokes simulations for convergence with any measure of merit. Significant improvement in the objective function was achieved with 40 to 50 Navier-Stokes evaluations which corresponded to about 7 to 9 days on a DEC Alpha 250-4/266 workstation.

Table 4. P8 inlet optimization

Measure	Original	Optimal	Change
σ_{sh}	0.2688	0.0270	-89.9%
r_p	8.481	7.541	-11.1%
σ_{p_t}	0.4605	0.465	1.0%
r_{p_t}	0.4505	0.461	2.3%

5.2 Supersonic Inlet

Description. In 1995 the Propulsion Design Group (Rutgers University) and United Technologies Research Center initiated a collaborative research effort to develop an automated design methodology for 2-D/axisymmetric supersonic missile inlets [48]. The inlet, shown in Fig. 9, is a mixed-compression type[17] intended for a ramjet or combined rocket/ramjet missile. The function of the inlet is to provide the combustor with a airflow characterized by a specified mass flow rate, subsonic Mach number and total pressure.

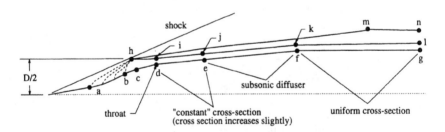

Fig. 9. Geometry model

The design objective chosen was to maximize the total pressure recovery r_{p_t}, defined as the ratio of the average total pressure at the inlet exit p_t to the upstream freestream total pressure p_{t_∞}, for a fixed external cowl shape and flight conditions of Mach 4 and 18.3 km altitude. This represents a simplification of

---

[17] A mixed compression inlet incorporates flow compression both upstream and downstream of the inlet entrance (defined as the cowl leading edge station).

the class of more general design problems (*e.g.*, maximize the range, payload or terminal velocity), but nonetheless is a practical design objective.

Design Methodology. The design methodology is based on the Design Associate and Simulation Associate models shown in Fig. 1. New issues arise in the context of supersonic missile inlet design and are addressed below.

A *Geometry Model* of an axisymmetric supersonic missile inlet was developed (Fig. 9) consisting of eight design variables (Table 5) and six fixed parameters (Table 6). The choice of the fixed parameters was based on typical supersonic missile designs. The external compression surface is a cone of angle θ_1 connected to an isentropic compression ramp $(a - b)$ which further turns the flow to a final angle θ_2. The surface is designed to provide a coalescence of the conical shock and isentropic compression waves at cowl lip. The region of internal compression $(c - d)$ is followed by the shock isolator[18] $(d - e)$, the subsonic diffuser $(e - f)$ and the constant area transfer duct $(f - g)$. In this study, the inlet was required to self-start at Mach 2.6, thereby determining the throat area for a given θ_1 and θ_2 [49]. The external cowl is formed by an ellipse which is considered to have low cowl wave drag [49].

Table 5. Parameters to Optimize

No.	Parameter	Definition
1	θ_1	initial cone angle
2	θ_2	final cone angle
3	x_d	axial location of throat
4	r_d	radial location of throat
5	x_e	axial location of end of "constant" cross section
6	θ_3	internal cowl lip angle
7	H_{ej}	height at end of constant cross section
8	H_{fk}	height at beginning of constant internal cross section

The *Flow Solver* employed a multi-level strategy in recognition of the high cost of Reynolds-averaged Navier-Stokes (RANS) simulations for this design. The total pressure recovery of a given geometry depends on the location of the terminal shock (in region $d - e$), which in turn is determined by the downstream static pressure p_g. The value of p_g is not known *a priori*, but rather is determined iteratively. Convergence of the value for p_g is achieved when a 1% increase in p_g

---

[18] The function of the shock isolator is to stabilize the terminal shock train downstream of the inlet throat $d - i$. In practice, the cross sectional area of the shock isolator increases slightly with downstream distance.

Table 6. Fixed Parameters

No.	Parameter	Definition
1	D	cowl diameter
2	r_f	centerbody radius of constant cross section region
3	x_g	length of inlet for computation
4	x_l	length of inlet for computation $(= x_g)$
5	x_n	length of inlet for computation $(= x_g)$
6	r_m	external diameter

would cause the terminal shock to be expelled out of the inlet. Thus, a sequence of several (typically, six) RANS simulations are needed for each geometry. This observation, coupled with the increased number of design parameters compared to the P2/P8 case (8 *vs.* typically 4) implies a new strategy is required.

The multi-level strategy consists of a simple inlet analysis code in the optimization loop (NAWC Inlet Design and Analysis Code [NIDA] [49]) and a Reynolds-averaged Navier-Stokes code (GASP$^{©}$ with the Chien k–ϵ low Reynolds number model) outside the optimization loop. The NIDA code was employed to search the design space, and the GASP code to verify the predictions of the NIDA code, filter the results and visualize the flowfield (thereby providing a physical description for performance improvements). NIDA, developed at United Technologies Research Center (UTRC), utilizes the method of characteristics for the supersonic flow upstream of the throat, and empirical correlations downstream of the throat for the terminal shock wave/turbulent boundary layer interaction and subsonic diffuser. It requires only a few cpu seconds per configuration on the DEC ALPHA 2100 4/275 work station.

Grid Generation was performed for the GASP simulations using an algebraic method. No grid generation was required for the NIDA simulations. The *Post Processing* module evaluated the objective function r_{p_t}.

The *Design Associate* employed three different optimizers: a genetic algorithm developed by Rasheed and Gelsey [19], the gradient-based CFSQP [47] with multiple starting points, and random probes. A genetic algorithm is a heuristic optimization method which attempts to mimic the process of natural selection [18]. The method used herein begins by taking a random population of points in the design space. The two "best" designs among this population (*i.e.*, with highest and second highest r_{p_t}) are connected in the design space by a straight line, and a random point on this line is chosen. A "sphere" is constructed about this point with a specified radius (which is significantly smaller than the average dimension of the design space) and a random point is selected within this sphere. This becomes the new design point and is evaluated. Typically, a steady-state replacement strategy is employed wherein one point in the design space is eliminated for each new point created, usually the point closest to the new point (in order to prevent clustering of design points).

Results. The automated design methodology using NIDA achieved an optimal design with a total pressure recovery $r_{p_t} = 0.378$. This represents a 23% improvement compared to a separate design developed independently using NIDA and prior to the automated optimization study, but without automated optimization. This optimal design was confirmed using GASP which obtained $r_{p_t} = 0.375$. The genetic algorithm optimizer was observed to be somewhat more efficient than CFSQP in achieving the optimal design[19], presumably due to the discontinuous structure of the design space [48] as illustrated in the single variable trade study of Fig. 10, wherein r_{p_t} vs x_d is displayed. The gaps in the curve of Fig. 10 represent values of x_d for which NIDA did not return a result.

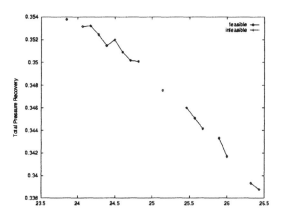

Fig. 10. Trade study of total pressure recovery *vs* axial location of throat (in)

The multi-level design strategy assumes the two flow solvers will have similar trends. Fig. 11 displays the predictions of NIDA and GASP for ten different configurations randomly chosen from the set of designs considered. Notwithstanding some differences among the last two designs[20], there is generally good agreement between the predictions of both codes.

The static pressure contours and streamlines in the vicinity of the inlet throat are shown in Figs. 12 and 13, respectively. For each configuration, the terminal shock system (located at $x \approx 0.60$ to 0.66, depending on the geometry) causes a separation of the boundary layers on the centerbody and/or cowl. The best design d exhibits the smallest separation region on the cowl, and a modest separation on the centerbody.

---

[19] The genetic algorithm also found a design with $r_{p_t} = 0.409$, but a GASP simulation of this configuration yielded $r_{p_t} = 0.374$.

[20] Possibly, these designs were outside the range of validity of the empirical correlations in NIDA.

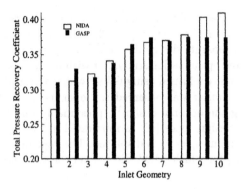

Fig. 11. Comparison of NIDA and GASP predictions

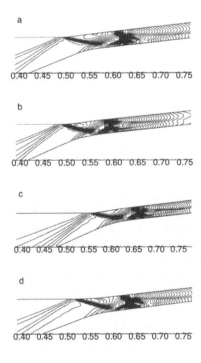

Fig. 12. Static pressure contours for configurations no. 2, 4, 6 and 7

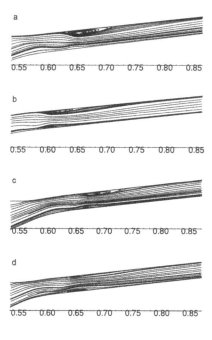

Fig. 13. Streamlines for configurations no. 2, 4, 6 and 7

The supersonic inlet optimization employed approximately 20,000 NIDA evaluations requiring 43 CPU hrs on a DEC ALPHA 4/266 2100 workstation. This figure does not include the CPU time required for the GASP simulations which were performed outside of the automated design optimization to verify and filter the results from NIDA.

5.3 Further Information

Additional information regarding the Propulsion Design Group at Rutgers is available via World Wide Web at http://www.cs.rutgers.edu/hpcd/Area_II.1.

6 Conclusions

CFD is in the midst of a major change. Although used principally to date as an analysis tool, CFD is becoming an essential element of automated optimal design in many disciplines. The new technology of automated design using CFD has clearly demonstrated better designs at lower cost, and industry is consequently adopting this approach. This trend is expected, and reflects both the maturing of

CFD as a discipline (*e.g.*, improved algorithms and models), and the continued advances in high performance computing (*e.g.*, faster processors and parallelism). This trend is not unique, as similar evolution has been seen in other disciplines in computational science and engineering (*e.g.*, Computational Solid Mechanics).

7 Acknowledgments

This research is sponsored in part by the HPCD (Hypercomputing and Design) project which is based at Rutgers University and funded by the Advanced Research Projects Agency of the Department of Defense through contract ARPA-DABT 63-93-C-0064 monitored by Dr. Bob Lucas. The results presented in Section 5 are based on the collaborative efforts of Propulsion Design Group at Rutgers University (Andrew Gelsey, Doyle Knight, Keith Miyake, Khaled Rasheed, Mark Schwabacher, Vijay Shukla, Don Smith, and Gecheng Zha) and Marty Haas (United Technologies Research Center). The contents of this paper do not necessarily reflect the position of the United States government and official endorsement should not be inferred.

References

1. I. Auneau, P. Garnero, and P. Duveau, "Design and Optimization Methods for Scramjet Inlets." AIAA Paper 95-6017, 1995.

2. C. Hirsch, *Numerical Computation of Internal and External Flows, Vols. I and II.* New York: John Wiley & Sons, 1988.

3. J. Moré and S. Wright, *Optimization Software Guide.* Philadelphia: SIAM, 1993.

4. R. Schwartz, *Learning Perl.* Sebastopol, CA: O'Reilly & Associates, Inc., 1993.

5. M. Lores and B. Hinson, *Transonic Design using Computational Aerodynamics*, vol. 81 of *Progress in Astronautics and Aeronautics*, pp. 377–402. American Institute of Aeronautics and Astronautics, 1982.

6. P. Gill, W. Murray, and M. Wright, *Practical Optimization.* New York: Academic Press, 1981.

7. G. N. Vanderplaats, *Numerical Optimization Techniques for Engineering Design : With Applications.* New York: McGraw-Hill, 1984.

8. A. Gelsey, "Intelligent Automated Quality Control for Computational Simulation," Tech. Rep. CAP-TR-21, Department of Computer Science, Rutgers University, Aug. 1994.

9. M. Schwabacher, *The Use of Artificial Intelligence to Improve the Numerical Optimization of Complex Engineering Designs.* PhD thesis, Rutgers University, 1996.

10. S. Obayashi and T. Tsukahara, "Comparison of optimization algorithms for aerodynamic shape design." AIAA Paper No. 96-2394-CP, 1996.

11. W. H. Press, S. A. Teukolsky, W. T. Vetterling, and B. P. Flannery, *Numerical Recipes in C : the Art of Scientific Computing.* Cambridge [England] ; New York: Cambridge University Press, 2nd ed., 1992.

12. A. Jameson, "Aerodynamic Design via Control Theory," *Journal of Scientific Computation*, vol. 3, pp. 33–260, 1988.

13. A. Jameson, "The Present Status, Challenges and Future Developments in Computational Fluid Dynamics," in *AGARD Conference on Progress and Challenges in CFD Methods and Algorithms*, 1995.

14. O. Baysal and M. Eleshaky, "Aerodynamic Design Optimization using Sensitivity Analysis and Computational Fluid Dynamics." AIAA Paper 91-0471, 1991.

15. S. Ta'asan, G. Kuruvila, and M. Salas, "Aerodynamic Design and Optimization in One Shot." AIAA Paper 92-0005, 1992.

16. A. Iollo and M. Salas, "Contribution to the Optimal Shape Design of Two-dimensional Internal Flows with Embedded Shocks." ICASE Report No. 95-20, 1995.

17. N. Metropolis, A. Rosenbluth, M. Rosenbluth, A. Teller, and E. Teller *J. Chemical Physics*, vol. 21, p. 1087, 1953.

18. D. E. Goldberg, *Genetic Algorithms in Search, Optimization, and Machine Learning*. Reading, Mass.: Addison-Wesley, 1989.

19. K. Rasheed and A. Gelsey, "Adaptation of Genetic Algorithms for Continuous Design Space Search," in *Fourth International Conference on Artificial Intelligence in Design: Evolutionary Systems in Design Workshop*, 1996.

20. P. Venkataraman, "Inverse Airfoil Design using Design Optimization." AIAA Paper No. 96-2503-CP, 1996.

21. G. Farin, *Curves and Surfaces for Computer Aided Geometric Design*. Academic Press, 1993.

22. I. Fejtek, T. Barfoot, and G. Lo, "Turboprop Nacelle Optimization using Automated Surface and Grid Generation and Coarse-Grained Parallelization." AIAA Paper 96-2504-CP, 1996.

23. H. Sobieczky and A. Seebass, "Supercritical Airfoil and Wing Design," in *Annual Review of Fluid Mechanics*, vol. 16, pp. 337–363, 1984.

24. M. Siclari, W. V. Nostrand, and F. Austin, "The Design of Transonic Airfoil Sections for an Adaptive Wing Concept using a Stochastic Optimization Method." AIAA Paper 96-0329, 1996.

25. M. Pandya and O. Baysal, "Gradient-Based Aerodynamic Shape Optimization using ADI Method for Large-Scale Problems." AIAA Paper 96-0091, 1996.

26. A. Merchant and M. Drela, "Design and Analysis of Supercritical Suction Airfoils." AIAA Paper No. 96-2397-CP, 1996.

27. S. Eyi, K. Lee, S. Rogers, and D. Kwak, "High-Lift Design Optimization using Navier-Stokes Equations," *Journal of Aircraft*, vol. 33, pp. 499–504, May-June 1996.

28. A. Jameson, "Optimum Aerodynamic Design using CFD and Control Theory." AIAA Paper 95-1729-CP, 1995.

29. N. Naik, S. Krist, R. Campbell, V. Vatsa, P. Buning, and L. Gea, "Inverse Design of Nacelles using Multi-Block Navier-Stokes Codes." AIAA Paper 95-1920, 1995.

30. B. Berkowitz, "Information Age Intelligence," *Foreign Policy*, no. 103, pp. 35–50, 1996.

31. D. Knight, "Parallel Computing in CFD," in *AGARD Conference on Progress and Challenges in CFD Methods and Algorithms*, pp. 3–1 to 3–14, 1995.

32. A. Jameson and J. Alonso, "Automatic Aerodynamic Optimization on Distributed Memory Architectures." AIAA Paper 96-0409, 1996.

33. R. Tanese, "Distributed Genetic Algorithms," in *Proceedings of the 3rd International Conference on Genetic Algorithms*, pp. 434–439, 1989.

34. P. Spiessens and B. Manderick, "A Massively Parallel Genetic Algorithm: Implementation and First Analysis," in *Proceedings of the 4th International Conference on Genetic Algorithms*, pp. 279–286, 1991.

35. W. Seebaugh, R. Doran, and J. DeCarlo, "Detailed Investigation of Flowfields within Large-Scale Hypersonic Inlet Models," Tech. Rep. NASA CR-114305, NASA Ames Research Center, April 1971.

36. A. Gnos, E. Watson, W. Seebaugh, R. Sanator, and J. DeCarlo, "Investigation of Flow Fields Within Large-Scale Hypersonic Inlet Models," Technical Note D-7150, NASA, April 1973.

37. A. Gelsey, D. Knight, S. Gao, and M. Schwabacher, "NPARC Simulation and Redesign of the NASA P2 Hypersonic Inlet." AIAA Paper 95-2760, 1995.

38. V. Shukla, A. Gelsey, M. Schwabacher, D. Smith, and D. Knight, "Automated Redesign of the NASA P8 Hypersonic Inlet Using Numerical Optimization." AIAA Paper 96-2549, 1996.

39. A. Gelsey, "Intelligent Automated Quality Control for Computational Simulation," *Artificial Intelligence for Engineering Design, Analysis and Manufacturing (AI EDAM)*, vol. 9, pp. 387–400, November 1995.

40. A. Gelsey, M. Schwabacher, and D. Smith, "Using Modeling Knowledge to Guide Design Space Search," in *Artificial Intelligence in Design Conference*, Kluwer Academic Publishers, June 1996. To appear.

41. J. Sirbaugh, C. Smith, C. Towne, G. Cooper, R. Jones, and G. Power, *A Users Guide to NPARC Version 2.0*, November 1994.

42. W. McGrory, D. Slack, M. Applebaum, and R. Walters, "GASP Version 2.2: The General Aerodynamic Simulation Program." Aerosoft, Inc, 1993.

43. K.-Y. Chien, "Predictions of Channel and Boundary Layer Flows with a Low Reynolds Number Turbulence Model," *AIAA Journal*, vol. 20, pp. 33–38, January 1982.

44. D. D. Knight, "Numerical Simulation of Realistic High-Speed Inlets Using the Navier-Stokes Equations," *AIAA Journal*, vol. 15, no. 11, pp. 1583–1589, 1977.

45. W. F. Ng, K. Ajmani, and A. Taylor, "Turbulence Modeling in a Hypersonic Inlet," *AIAA Journal*, vol. 27, no. 10, pp. 1354–1360, 1989.

46. K. Kapoor, B. H. Anderson, and R. J. Shaw, "Comparative Study of Turbulence Models in Predicting Hypersonic Inlet Flows," 1992. AIAA Paper 92-3098.

47. C. Lawrence, J. Zhou, and A. Tits, "Users Guide for CFSQP Version 2.3: A C Code for Solving (Large Scale) Constrained Nonlinear (Minimax) Optimization Problems, Generating Iterates Satisfying All Inequality Constraints," Tech. Rep. TR-94-16r1, Institute for Systems Research, University of Maryland, November 1994.

48. G. Zha, D. Smith, M. Schwabacher, K. Rasheed, A. Gelsey, D. Knight, and M. Haas, "High Performance Supersonic Missile Inlet Design Using Automated Optimization." AIAA Paper 96-4142, 1996.

49. M. Haas, R. Elmquist, and D. Sobel, "NAWC Inlet Design and Analysis (NIDA) Code," Tech. Rep. UTRC Report R92-970037-1, United Technologies Research Center, April 1992.

Parallelization of the Discrete Ordinates Method: Two Different Approaches

P. J. COELHO, J. GONÇALVES AND P. NOVO

Instituto Superior Técnico, Technical University of Lisbon
Mechanical Engineering Department
Av. Rovisco Pais, 1096 Lisboa Codex
Portugal

Abstract. One of the most popular methods used in the solution of radiative heat transfer problems is the discrete ordinates method (DOM). The present paper describes two different parallelization strategies of the DOM. One of them is based on angular decomposition and the other one is based on spatial decomposition. In the first case each processor performs the calculations for the whole domain but only deals with a few directions, while in the second case each processor treats all the directions but only for a subdomain. It is shown that the number of iterations is independent of the number of processors in the first parallelization strategy, but increases with the number of processors in the second case. Consequently, higher efficiencies are achievable using the angular decomposition approach. The influence of the order of quadrature, grid size and optical thickness of the medium is also investigated.

1. Introduction

The discrete ordinates method [1,2], DOM, has been widely used to predict radiative heat transfer in combustion chambers. Researchers have used this model as an alternative to the zone and Monte Carlo methods. These methods are seldom used to solve radiation problems coupled with the modelling of a reactive flow because the solution algorithm is rather different from those generally used to solve partial differential equations and also because they are computationally demanding. The DOM is compatible with the CFD algorithms, it is relatively fast and may achieve good accuracy.

The recent development of high performance computers promises to greatly increase the speed and problem capacity of radiation transport calculations. So, in order to exploit high-performance computing capabilities the present paper addresses the parallelization of the DOM. Coupling with CFD algorithms is not addressed here. Two different parallelization strategies are presented and discussed. In the angular decomposition parallelization (ADP) each processor performs the calculations for the whole domain for a certain number of directions. In the spatial domain decomposition parallelization (DDP) the domain is split into subdomains and each processor performs the calculations for one subdomain for the complete set of directions.

Two radiation problems in two-dimensional square enclosures containing an emitting-absorbing medium were studied. The present paper reports the outcome of that study.

2. The Discrete Ordinates Method

The main features of the DOM are described below in order to facilitate the presentation of the parallelization strategies. A complete description of the method may be found elsewhere [1,2]. The DOM is based on the numerical solution of the radiative transfer equation along specified directions. This equation may be written as follows for an emitting-absorbing-scattering grey medium [3]:

$$\frac{dI}{ds} = -\beta I + \kappa I_b + \frac{\sigma_s}{4\pi} \int_0^{4\pi} I\left(\vec{s'}\right) \phi \left(\vec{s'},\vec{s}\right) d\Omega'$$

(1)

The ratio $\phi\left(\vec{s'},\vec{s}\right)/4\pi$ represents the probability that radiation propagating in the direction $\vec{s'}$ and confined within the solid angle $d\Omega'$ is scattered through the angle $\left(\vec{s'},\vec{s}\right)$ into the direction $\vec{s}$ confined within the solid angle $d\Omega$. The radiation intensity was denoted by I and the subscript b refers to a blackbody. The absorption, the scattering and the extinction coefficient of the medium were denoted by κ, β and σ_s, respectively. Equation (1) is a statement of the principle of conservation of energy applied to a pencil of radiation travelling along direction $\vec{s}$.

The present study is restricted to non-scattering grey media. In this case $\sigma_s = 0$ and $\beta = \kappa$, and Eq. (1) can be simplified to

$$\frac{dI}{ds} = -\kappa I + \kappa I_b$$

(2)

The DOM relies on a discrete representation of the directional dependence of the radiation intensity. The radiative transfer equation (2) is solved for a set of n(n+2) directions, yielding the so-called S_n approximation (in two-dimensional problems only one half of the directions needs to be considered due to symmetry). These directions span the total solid angle range of 4π around a point in space, and the integrals over solid angles are approximated using a numerical quadrature rule.

Equation (2) may be written as follows for any discrete direction $\vec{s_i}$:

$$\xi_i \frac{\partial I_i}{\partial x} + \eta_i \frac{\partial I_i}{\partial y} + \mu_i \frac{\partial I_i}{\partial z} = -\kappa I_i + \kappa I_b$$

(3)

where ξ_i, η_i and μ_i are the direction cosines of direction i. This equation is discretized using the finite-volumes approach, yielding a relationship between the volume average intensity, I_{P_i}, and the radiation intensities entering (subscript i) and leaving (subscript e) a control volume:

$$I_{Pi} = \frac{\kappa \, V \, \gamma \, I_{bi} + |\xi_i| \, A_x \, I_{xi,\, i} + |\eta_i| \, A_y I_{yi,\, i} + |\mu_i| \, A_z I_{zi,\, i}}{\kappa \, V \, \gamma + |\xi_i| \, A_x + |\eta_i| \, A_y + |\mu_i| \, A_z}$$

(4)

In this equation V stands for the volume of the cell, and A_x, A_y and A_z denote the areas of the cell faces normal to directions x, y and z, respectively. The parameter γ relates the incoming and outgoing radiation intensities to the volume average intensity according to the following relations:

$$I_{Pi} = \gamma\, I_{x_e,i} + (1-\gamma)\, I_{x_i,i} \qquad\qquad (5a)$$

$$I_{Pi} = \gamma\, I_{y_e,i} + (1-\gamma)\, I_{y_i,i} \qquad\qquad (5b)$$

$$I_{Pi} = \gamma\, I_{z_e,i} + (1-\gamma)\, I_{z_i,i} \qquad\qquad (5c)$$

The most common values of γ are $\gamma = 1/2$ (diamond scheme) and $\gamma = 1$ (step scheme). The step scheme was used in the present calculations.

The boundary conditions may be written as:

$$I_i = \varepsilon_w\, I_{bw} + (1-\varepsilon_w)\frac{q_x^-}{\pi} \qquad\qquad \text{at} \qquad x = x_{min} \qquad\qquad (6a)$$

$$I_i = \varepsilon_w\, I_{bw} + (1-\varepsilon_w)\frac{q_x^+}{\pi} \qquad\qquad \text{at} \qquad x = x_{max} \qquad\qquad (6b)$$

and analogously for the remaining boundaries. The emissivity of the walls was denoted by ε_w. The incident heat fluxes are calculated as

$$q_x^- = \sum_{\substack{j \\ (\xi_j<0)}} w_j\, I_j\, |\xi_j| \qquad\qquad (7a)$$

$$q_x^+ = \sum_{\substack{j \\ (\xi_j>0)}} w_j\, I_j\, |\xi_j| \qquad\qquad (7b)$$

where w_j is the quadrature weight associated with the direction j. In these equations the superscripts + and − identify the boundaries $x=x_{max}$ and $x=x_{min}$, respectively. Similar expressions may be written for the incident heat fluxes on the remaining boundaries.

If the temperature field is not known, it must be determined from the simultaneous solution of the energy conservation equation and the radiative transfer equation. In this paper it is assumed that radiation is the dominant mode of heat transfer, the others being neglected. Therefore, conservation of energy may be expressed as

$$\nabla.q = \kappa\left(4\sigma T^4 - G\right) \qquad\qquad (8)$$

where $\nabla.q$ is the divergence of the radiative heat flux and G is the incident radiation given by

$$G = \int_{0}^{4\pi} I d\Omega \approx \sum_{j} w_j I_j \qquad (9)$$

At each iteration, and for each one of the selected directions, the surface radiosities and the blackbody radiation intensities, I_b, are either known or guessed based on the values computed in the previous iteration. In the first iteration the surface irradiation is neglected. The numerical solution of Eq. (3) is carried out starting from a control volume at one of the corners of the computational domain, which depends on the sign of the direction cosines of the direction under investigation. It is selected the control volume at the corner that permits the calculation of Ip via Eq. (4) using the values of I_{x_i}, I_{y_i} and I_{z_i} available from the boundary conditions. The solution proceeds visiting all the control volumes to compute the radiation intensities according to Eq. (4) and using the auxiliary relations (5a-c). After all the directions have been treated, the surface radiosities may be updated using the boundary conditions. If the temperature field is not prescribed, it is also updated via Eq. (8), after the calculation of the incident radiation in each control volume by means of Eq. (9). The iteration process continues until the convergence criterion has been satisfied.

The convergence criterion demands that one or more of the error measures given below decrease below a prescribed tolerance, typically of the order of 10^{-5}. The error estimation may be based on the maximum absolute value of the normalized difference between the incident heat fluxes on the wall in consecutive iterations:

$$E_1 = \max \left\{ \left| \frac{q_w^m - q_w^{m-1}}{\sigma T_w^4} \right| \right\} \qquad (10)$$

In this equation q_w stands for q_x^-, q_x^+, q_y^-, q_y^+, q_z^- or q_z^+, i.e., the incident heat flux is calculated for all the boundary cell faces, and the maximum difference between the flux at the current iteration, m, and at the previous one, m-1, is taken to compute E_1. The emissive power of one of the walls is used for normalization purposes. Another convergence criterion is based on the change of the radiation intensities between successive iterations:

$$E_2 = \max \left\{ \left| \frac{I_{P_i}^m - I_{P_i}^{m-1}}{\sigma T_w^4} \right| \right\} \qquad (11)$$

All the directions and all the control volumes are tested to determine the maximum value of the quantity into parenthesis in this equation. A disadvantage of this criterion is that it requires the storage of the radiation intensities at the previous iteration, i.e., a four-dimensional array in a 3D problem, which is only used for checking the convergence. To avoid this shortcoming, the incident radiation G may be used instead of the radiation intensity:

$$E_3 = \max \left\{ \left| \frac{G^m - G^{m-1}}{\sigma T_w^4} \right| \right\} \qquad (12)$$

3. Parallelization of the Discrete Ordinates Method

Two different parallelization strategies of the discrete ordinates method have been implemented in a distributed memory computer: angular decomposition and domain decomposition.

3.1 Angular Decomposition

In the angular decomposition approach the total number of directions along which the radiative transfer equation is solved is divided into a number of subsets equal to the number of processors. Each processor performs calculations for the whole domain but treats only a certain subset of directions. In the calculations presented in the next section the number of directions is always a multiple of the number of processors. Hence, load balancing problems are avoided without loss of generality of the parallelization method. The calculation of the incident heat fluxes on a wall involves a summation over all the directions pointing towards the wall (see Eq. 7). Therefore, a processor can only compute a partial value of the incident heat fluxes obtained from a summation over the subset of directions assigned to that processor. The calculation of the total incident heat flux requires that every processor exchanges data with the others. This is accomplished using a binary tree network, as explained below. The communications are synchronized and performed at the end of each iteration.

Taking, as an example, 8 processors (p=8), in the first step processors P_5, P_6, P_7 and P_8 simultaneously send their values of the partial incident heat fluxes on the wall to processors P_1, P_2, P_3 and P_4, respectively. Then, P_1 to P_4 add their partial values to those received from P_5 to P_8. In the second step, P_3 and P_4 simultaneously send their values to P_1 and P_2, respectively, and these last two processors add their partial values to the ones received from P_3 and P_4. In the third step, P_2 sends its value to P_1, and P_1 adds its own contribution to that received from P_2. Therefore, after the third step the total incident heat fluxes on the walls have been computed in P_1. Now, these fluxes must be communicated to all the other processors. Hence, in the fourth step P_1 sends the total fluxes to P_2. In the fifth step P_1 and P_2 simultaneously send this data to P_3 and P_4, respectively, and in the last step P_1, P_2, P_3 and P_4 simultaneously send this information to P_5, P_6, P_7 and P_8, respectively. Therefore, $\log_2 p$ steps are needed to calculate the total heat fluxes in one of the processors, and the same number of steps is required to communicate this information back to all processors. If p is not a power of two, the communication path is similar, but one additional step is required both to be able to compute the total fluxes in one processor and to communicate them to all the others. For example, if p=11 in the first step P_9, P_{10} and P_{11} send data to P_6, P_7 and P_8, and the following steps are as described above.

If a radiative heat source (or sink) is prescribed rather than the temperature field, then the incident radiation at each control volume, given by Eq. (9), also needs to be calculated, and involves a summation over all the directions. Since each processor only deals with a subset of directions, it must exchange data with the others to allow the computation of G in each control volume. This is also done by means of the binary tree network outlined above.

The error measures E_1 and E_3 are used if the radiative heat source is prescribed. They are calculated in all the processors after the computation of the incident heat fluxes on the walls and incident radiation in the control volumes. Therefore, no further communications among the processors are needed to evaluate E_1 and E_3. However if the temperature field is prescribed, the calculation of the incident radiation is not needed. Therefore, the error measure E_2 is used instead of E_3. A processor is only able to determine the maximum in Eq. (11) for the set of directions assigned to that processor. The overall maximum is obtained by exchanging data among processors according to the binary tree network described above. The only difference is that instead of adding the values from the different processors, the maximum of those values is determined.

3.2 Domain Decomposition

In the domain decomposition approach the computational domain is divided into a number of subdomains equal to the number of processors. If the computational domain is mapped using a mesh with NX*NY*NZ grid nodes, and a three-dimensional array with $p_x*p_y*p_z$ processors is used, then the number of grid nodes assigned to each processor will be $(NX/p_x)*(NY/p_y)*(NY/p_z)$. It is assumed here that NX, NY and NZ are multiples of p_x, p_y and p_z, respectively. This avoids load balancing problems caused by different computational loads in different processors, but does not restrict the generality of the parallelization strategy. Each processor performs calculations for a subdomain, solving the equation of radiative heat transfer in that subdomain for all the directions.

Although the subdomains do not overlap, there is a buffer of halo points added to their boundaries, including the virtual boundaries, i.e., boundaries between neighbouring processors. A plane of halo points is added to each subdomain boundary to simplify the exchange of data (radiation intensities) at the virtual boundaries between neighbouring processors. For example, if the i index of the grid nodes in a subdomain ranges from i1 to i2, then a plane of halo points with index i1-1 and another one with index i2+1 are employed. The halo data transfer between neighbouring processors is achieved by a pairwise exchange of data, as explained below.

Suppose that there are p_x processors along the x direction from west to east, and let p_x be even. In the first step processors 1,3....., p_x-1, simultaneously send data to processors 2,4,...., p_x, respectively. In the second step processors 2,4...., p_x-2, simultaneously send data to processors 3,5...., p_x-1. In both steps the data of the control volumes adjacent to the virtual east boundary of a processor are sent to the halo points added to the west boundary of the east neighbour of the processor under consideration (see Fig. 1). The exchanged data consists in the radiation intensities for all the directions $\xi_i > 0$. Then, in the third and fourth steps data is transferred in the reverse direction, as shown in Fig. 1: the data of the control volumes adjacent to the virtual west boundary of a processor are sent to the halo points associated with the east boundary of the west neighbouring processor. The data transferred in these two steps are the radiation intensities for all the directions such that $\xi_i < 0$. This process is easily modified for an odd number of processors. Data exchange along y and z directions is carried out in a similar way.

The transference of the radiation intensities takes place at the end of each iteration. In the first iteration the radiation intensities at the halo points are assumed to be equal to the radiation intensities guessed for the boundaries of the enclosure. The data exchange in this parallelization method influences the convergence rate, i.e., the number of iterations required to achieve convergence. This places an upper limit on the maximum achievable efficiency, as shown in the next section.

The error measures E_1 and E_3 have been used in the calculations reported here. Since the processors are working in parallel, the iterative process must finish simultaneously in all of them, to prevent one processor from trying to communicate with another one that has already finished the calculations. To avoid this problem, E_1 and E_3 are calculated over the whole computational domain, like in the angular decomposition parallelization. Hence, each processor calculates a local maximum of E_1 and E_3 and communicates these maxima to the other processors to find out the overall maximum. This communication proceeds according to the binary tree network described for the angular decomposition parallelization.

4. Results and Discussion

The parallelization strategies described above have been applied to two test cases. In one of them the temperature field is given and in the other one the volumetric radiative heat source is prescribed. The computations were performed in a Parsytec MC3-DE with 112 nodes with the transputer T805 and 4 MBytes of RAM. All the results presented here were obtained taking the number of directions as a multiple of the number of processors in the ADP, or the number of grid nodes along x and y directions as a multiple of the number of processors in those directions in the DDP. In this way the computational load is the same in every processor, and load balancing problems are avoided.

The main objective of parallel computing is to reduce the wall clock time of a particular computation. The performance of parallel computing can be evaluated using the concept of speed-up, a measure of how an algorithm compares with itself on 1 and p processors. The speed-up is defined as $S_p = t_1/t_p$ where t_1 and t_p are the wall clock execution times on 1 and p processors, respectively. It is closely related to the efficiency of the parallel implementation, defined as $E_p = S_p/p$. In the problems studied here the influence of the number of processors, quadrature, grid size and optical thickness of the medium on E_p and S_p was investigated.

4.1 Test Case 1

In the first test case a two-dimensional square enclosure with cold, grey walls ($T_w=0.$, $\varepsilon_w=0.4$) was studied. The enclosure contains a medium maintained at an emissive power of unity. The solution accuracy has been investigated elsewhere [4] and is not addressed here. In all the cases it was verified that the numerical solution obtained is the same regardless of the number of processors, and coincides with the solution computed using the sequential code.

The standard calculations were performed using a grid with 90 x 90 control volumes, a S_{12} quadrature, p = 12 in the ADP and p = 9 in the DDP. The optical thickness of the medium, τ, was taken as 1.0. In the results shown below one of these four parameters (number of processors, quadrature, grid size and optical thickness) was varied, keeping the others unchanged.

The influence of the number of processors is shown in Fig. 2. The efficiency decreases with the number of processors, as expected, but the decrease is relatively slow for the ADP, where $E_p = 74\%$ and $S_p = 62$ for $p = 84$, while it is quite fast for the DDP, where $E_p = 12\%$ and $S_p = 10.6$ for $p = 90$. This discrepancy is explained to some extent by the evolution of the number of iterations required to achieve convergence, n_{iter}, with p. In the ADP n_{iter} is independent of p, while in the DDP there is a marked increase of n_{iter} with the increase of p (see Fig. 2c). In fact, in the DDP the calculations performed in a processor during an iteration require data from the halo points, i.e., radiation intensities calculated at the previous iteration. Hence, during one iteration the boundary radiation intensities cannot travel beyond the physical or virtual boundaries of a processor. Therefore, several iterations are needed to allow this boundary data to spread over the whole domain. On the contrary, both in the sequential algorithm and in the ADP the boundary radiation intensities travel through the whole domain in one iteration.

The efficiency per iteration decreases with p for both parallelization methods, but decreases faster for the ADP. This evolution is directly related to the increase of the ratio of the communication time (t_c) to the execution time (t_e). The reason for this increase is the following. Suppose that p is doubled. Then, the computation time ($t_{cp}=t_e-t_c$) is reduced to approximately one half in the ADP because each processor now only deals with one half of the directions. However, t_c increases, because each processor needs to transfer data to twice as many processors as before. As a consequence, t_c/t_e also increases (see Fig. 2d), with the resultant decrease of the efficiency per iteration.

If the DDP is used, t_{cp} is also reduced to approximately one half when p is doubled, because the number of grid nodes assigned to each processor becomes one half of the initial number. However, t_c also decreases. The time required to exchange the errors E_1 and E_3 among the processors increases with p, but this is only a minor fraction of t_c which is dominated by the time required to transfer the radiation intensities to the halo points. When p increases the number of halo points decreases, justifying the decrease of t_c. Nevertheless, the ratio t_c/t_e still increases, as shown in Fig. 2d, but at a slower rate than in the ADP. This ratio is approximately proportional to the ratio of the number of halo points to the number of grid nodes assigned to a processor, which increases with p. As a consequence, the efficiency per iteration decreases with p.

The influence of the quadrature for a fixed number of processors (p=12 in the ADP, p=9 in the DDP) is shown in Fig.3. Neither n_{iter} nor the ratio t_c/t_e change with the quadrature when the DDP is used. In fact, both t_c and t_{cp} increase linearly with the number of directions along which the radiation transfer equation is solved, but t_c/t_e does not change. As a consequence, both E_p and S_p remain also constant. If the ADP is used, n_{iter} is again independent of the quadrature, as well as t_c. However, t_{cp} increases with the number of directions. Therefore, the ratio t_c/t_e decreases with p, justifying the observed increase of E_p and S_p with the number of directions.

For a fixed number of processors, n_{iter} is independent of the grid size for both parallelization methods, as shown in Fig. 4. If the ADP is used, t_{cp} is proportional to the total number of control volumes, while t_c is proportional to the number of control volumes adjacent to the boundary. Therefore, if the grid size is doubled in both directions, it is expected that t_c/t_{cp} decreases to approximately one half. This is corroborated by the results displayed in Fig. 4d). If the DDP is used the same reasoning can be applied at the processor level, i. e., t_c/t_e has a similar evolution in both ADP and DDP. However, the number of processors used in these calculations is relatively small and, therefore, t_c is a small percentage of t_e. Hence, E_p and S_p show only a slight increase with the grid size.

The influence of τ is illustrated in Fig. 5. In this test case the temperature field is prescribed, i. e., I_b is known. Therefore, when τ increases, the term associated with I_b in Eq. (4) becomes larger compared with the terms associated with the radiation intensities entering a control volume. In other words, the term whose value is known *a priori* becomes larger compared with the other terms which change during the course of the iterative process until convergence is attained. Consequently, n_{iter} decreases with τ for both ADP and DDP. The values of t_c and t_e per iteration do not depend on τ, neither does t_c/t_e.

When the ADP is used, although n_{iter} decreases with τ, n_{iter} is independent of p. Since t_c/t_e is also independent of p, then E_p and S_p do not change with τ. If the DDP is used, the ratio of n_{iter} using p processors to n_{iter} using 1 processor approaches the unity as τ increases. As a consequence, both E_p and S_p increase with τ. However, E_p and S_p per iteration do not change with τ, since t_c/t_e is constant.

4.2 Test Case 2

A two-dimensional square enclosure with cold black walls was studied in the second test case. A volumetric heat source equal to 1.0 Wm^{-3} was prescribed. As in the previous test case, the standard calculations were performed using a grid with 90x90 control volumes, a S_{12} quadrature, p=12 in the ADP and p=9 in the DDP, and $\tau = 1.0$.

In this problem the evolution of E_p and S_p with the number of processors is similar in the two parallelization methods, as shown in Fig. 6. The explanation of these evolutions, based on the influence of p on n_{iter} and on t_c/t_e is identical to that given in test case 1 if the DDP is used. In this case, E_p=21% and S_p=18.6 for p=90. The degradation observed with the increase of p is not as large as in the previous problem, mainly because n_{iter} did not grow as much as before.

If the ADP is employed, n_{iter} is constant, but E_p shows a fast drop with the increase of p, while S_p increases very slowly. In this case, E_p=17% and S_p=14 for p=84. The highest speed-up is obtained for p=42 (S_p=15.7). This means that there is no advantage in using a larger number of processors because the total wall clock time will not decrease further. The reason for this behaviour is the ratio t_c/t_e which increases very fast with p. When p=84, t_c represents 74% of the total wall clock time, while in test case 1 t_c/t_e was only 24.5%. The marked increase of this ratio in the present problem is related to the need to exchange the incident radiation among processors, as explained in section 3.1.

The influence of the quadrature and grid size on the efficiency and speed-up are identical to those discussed in test case 1. The only detail worth of being reported is the marginal influence of the grid size on the ratio t_c/t_e if the ADP is used. In this case, both the incident heat fluxes on the walls and the incident radiation in each control volume need to be communicated to all the processors. The time required to transfer the incident radiation (a two-dimensional array) in each control volume, t_G, is proportional to the number of grid nodes, while the time required to transfer the wall heat fluxes (four one-dimensional arrays), t_H, is proportional to the number of control volumes adjacent to the boundary. Therefore, t_G is larger than t_H. But t_G/t_e does not depend on the grid size, since both t_G and t_e are proportional to the number of control volumes, while t_H/t_e decreases with the grid size, as observed in test case 1. Therefore, $t_c/t_e=(t_G+t_H)/t_e$ exhibits only a small decrease with the grid size, yielding a corresponding marginal increase of the efficiency.

The influence of τ is shown in Fig. 7. The value of n_{iter} increases with τ for both methods, contrary to the evolution reported in test case 1. However, in the present problem the temperature field is not given and, therefore, I_b must be calculated during the course of the iterative procedure via the energy equation. The term of Eq. (4) associated with I_b becomes dominant as τ increases. This explains why in the present problem n_{iter} increases with τ.

The ratio t_c/t_e is not influenced by τ. If the ADP is used, E_p is independent of τ because both t_c/t_e and the ratio of the number of iterations of p to 1 processors are independent of τ. If the DDP is used, the ratio between the number of iterations for p to 1 processors decreases with p, justifying the corresponding increase of E_p and S_p with p. Therefore, E_p and S_p exhibit similar evolutions in the two studied problems.

5. Conclusions

Two different parallelization strategies of the discrete ordinates method were implemented and compared for emitting-absorbing media in two-dimensional enclosures. From the analysis carried out, the following conclusions may be drawn:

i) The number of iterations required to achieve convergence is independent of the number of processors, p, if the ADP is used, and increases with p if the DDP is employed. In both cases the ratio of the communication to the execution time (t_c/t_e) increases with p. As a consequence, the efficiency drops fast and the speed-up increases slowly with p if the DDP is used, or if the ADP is used and a volumetric heat source is prescribed. If the ADP is used and the temperature of the medium is prescribed, high parallelization efficiencies are obtained (74% for p=84).

ii) The number of iterations is independent of the quadrature for both methods, while t_c/t_e is also independent of the quadrature for the DDP, but decreases with the increase of the order of the quadrature for the ADP. Consequently, the efficiency and the speedup increase with the order of the quadrature if the ADP is used, and remain constant otherwise.

iii) The number of iterations is independent of the grid size for both methods, while t_c/t_e decreases with grid refinement. Hence, the efficiency and the speed-up increase with the grid size.

iv) The efficiency and the speed-up increase with the optical thickness of the medium if the DDP is used, because the ratio between the number of iterations required to achieve convergence using p processors and 1 processor becomes closer to one when the optical thickness increases. If the ADP is used there is no influence of the optical thickness on the efficiency and speed-up.

Acknowledgments

This work has been financially supported by the Commission of the European Communities under the ESPRIT project 8114, HP-PIPES - 'High Performance Parallel Computing for Process Engineering Simulation'. The authors are indebted to the Centre of Informatics of the University of Minho, Portugal, for the permission to use their parallel computers.

References

1. Carlson, B.G. and Lathrop, K.D., Transport Theory - The Method of Discrete Ordinates, in *Computing Methods in Reactor Physics*, Gordon & Breach, New York, 1968.
2. Fiveland, W.A., Discrete-Ordinates Solutions of the Radiative Transport Equation for Rectangular Enclosures, *J. Heat Transfer*, Vol. 106, pp. 415-423, 1990.
3. Modest, M.F., *Radiative Heat Transfer*, McGraw-Hill, 1993.
4. Coelho, P.J., Gonçalves, J.M. and Carvalho, M.G., A Comparative Study of Radiation Models for Coupled Fluid Flow / Heat Transfer Problems, in *Numerical Methods in Thermal Problems*, Vol IX, Part 1, pp. 378-389, 1995.

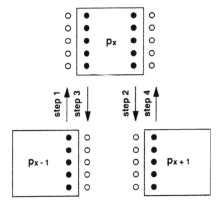

Fig. 1 - Data transfer among processors in the DDP.

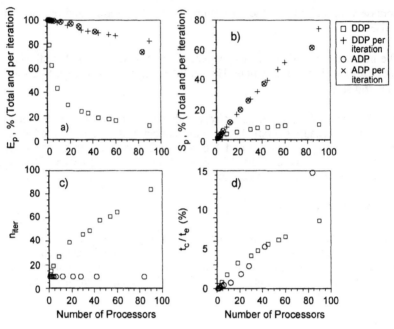

Fig. 2 - Influence of the number of processors on the efficiency, speed-up, number of iterations and ratio of communication to execution time for test case 1.

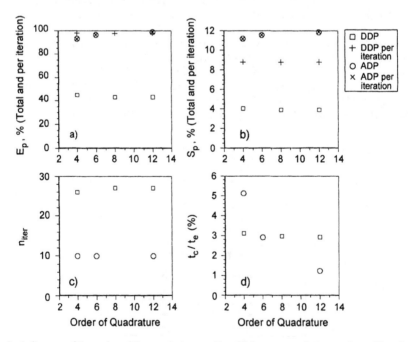

Fig 3 - Influence of the order of the quadrature on the efficiency, speed-up, number of iterations and ratio of communication to execution time for test case 1.

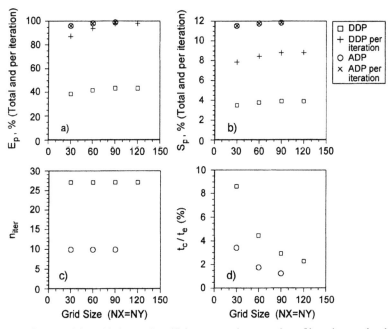

Fig 4 - Influence of the grid size on the efficiency, speed-up, number of iterations and ratio of communication to execution time for test case 1.

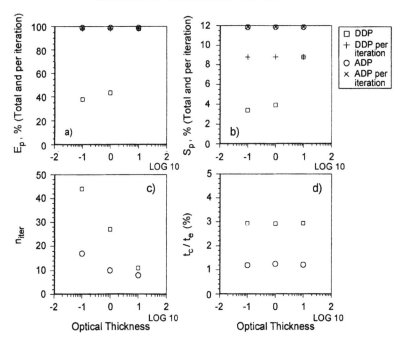

Fig 5 - Influence of the optical thickness on the efficiency, speed-up, number of iterations and ratio of communication to execution time for test case 1.

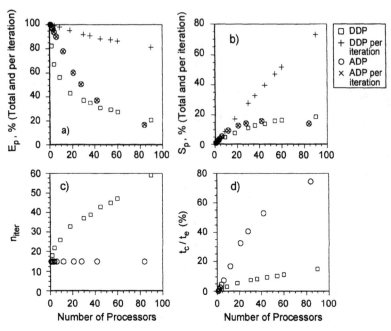

Fig. 6 - Influence of the number of processors on the efficiency, speed-up, number of iterations and ratio of communication to execution time for test case 2.

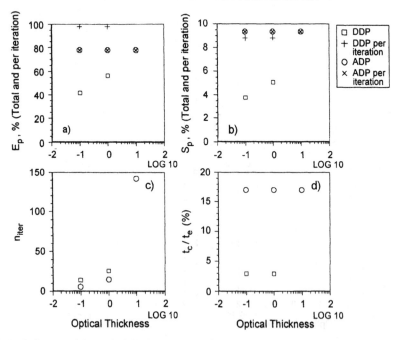

Fig 7 - Influence of the optical thickness on the efficiency, speed-up, number of iterations and ratio of communication to execution time for test case 2.

Experiences with Advanced CFD Algorithms on NEC SX-4

H. van der Ven* and J.J.W. van der Vegt

National Aerospace Laboratory NLR
P.O. Box 90502, 1006 BM Amsterdam
The Netherlands

Abstract. In this paper three topics related to parallel CFD simulations are discussed. The first topic is the shared memory parallelization of the unstructured adaptive flow solver Hexadap. The second topic discusses the performance results of this parallelization on a 16 processor NEC SX-4. The third topic combines the first two and concerns the CFD working environment ISNaS as developed by the National Aerospace Laboratory NLR.

1 Introduction

The National Aerospace Laboratory NLR has a long-standing tradition in the development and use of CFD software. The CFD software includes production solvers using proven technologies, research codes for developing and testing new algorithms and the support of both production and development of CFD software. This paper considers the latter two subjects.

NLR is developing a flow solver based on a discontinuous finite element Galerkin discretization of the Euler/Navier-Stokes equations. The flow solver uses structured hexahedrons as initial mesh, which is followed by solution adaptive, unstructured refinement on a hexahedron by hexahedron basis. Main incentive for the development of this flow solver is time-accurate flow simulation, which will require the utmost of present and future computer hardware. Because of the computational complexity of large scale applications parallelization of the underlying algorithm is required. In June 1996 the NEC SX-3 at NLR has been replaced by a 16 processor NEC SX-4. A first parallelization has been performed on the NEC SX-4/16. This parallelization assesses the SX-4 capabilities and the parallelism of the flux integration scheme, which is the most computational intensive part of the algorithm. The parallelization is based on the shared memory paradigm.

To support the development and use of CFD software NLR has developed a working environment for CFD applications. This working environment appears to the user as a single, virtual computer, even though programs and data are located on different computers in a network. In addition to network transparency,

---

* Partially supported by Dutch Foundation HPCN in the project NICE under contract no. 96009.

the working environment provides facilities for the management of software, data, and documents.

The contents of this paper is as follows. In Section 2 the underlying algorithm of the adaptive, unstructured flow solver will be briefly described. In Section 3 the chosen parallelization strategy will be described. In Section 4 the results of the parallelization will be described and discussed. In Section 5 the working environment will be described. In Section 6 conclusions will be drawn.

2 Algorithm

Computational Fluid Dynamics is used for increasingly complicated problems. Many advanced applications of CFD can only be done with sophisticated grid adaptation algorithms and require significant computer resources. The capture of flow phenomena such as shocks, vortical structures, and time-dependent changes in the flow pattern is still one of the key elements preventing efficient time-accurate simulation of problems in aerospace. Applications requiring time-accurate simulation are structural dynamics, aircraft maneuver, aircraft under large angle of attack, propeller-wing interaction and noise prediction. These applications can only be simulated efficiently with sophisticated grid adaptation techniques and require a numerical scheme which is accurate on highly irregular grids.

This was the main motivation to develop a new, discontinuous Galerkin finite element algorithm for the Euler/Navier-Stokes equations. The discontinuous Galerkin method (DG) uses a local polynomial expansion in each cell which results in a discontinuity at each cell face. This discontinuity can be represented as a Riemann problem which provides a natural way to introduce upwinding into a finite element method. The DG method can therefore be considered as a mixture of an upwind finite volume method and a finite element method. A unique feature of the DG method is that seperate equations for the flow gradients are used, which do not have to be determined from neighbouring cells.

A combination of local grid refinement and the discontinuous Galerkin finite element method is applied in the flow solver Hexadap [3]. This combination is capable of efficiently resolving local phenomena such as shear layers and shocks. This paper will be limited to inviscid flow in order to demonstrate the parallelism in the basic algorithm.

The DG method, combined with a face based data structure, is extremely local in nature and makes it a good candidate for parallel computing.

3 Parallelization strategy

The above described algorithm consists of two parts, namely grid adaptation and flow computation. The grid adaptation part, which consists predominantly of scalar operations, requires a domain decomposition for parallelization and is not considered in this paper. The flow computation has two main components: the calculation of cell face fluxes and a slope limiter.

The flow computation part of the solver consists of three kinds of loops:

face-face: loops over faces, updating face values,
cell-cell: loops over cells, updating cell values,
face-cell: loops over faces, updating cell values.

An example of a face-face loop is the evaluation of the fluxes through cell faces. Examples of cell-cell loops are the loops in the Runge-Kutta scheme where the residuals are added to the flow states. An example of a face-cell loop is the update of the residual of a cell by the flux through one of its faces.

The loops over the faces, both face-face and face-cell, are split into different colours. This is done to be able to vectorize the face-cell loop. Within one colour all faces connect to cells with different cell addresses, hence within one colour there is no data dependency when updating cell values.

The parallelization of the face-cell loops is the most challenging. In general, the loop length of the face-face and face-cell loops is not sufficient for both vectorization and parallelization. The face-face loops can be parallelized over the colours, but this is prohibited for face-cell loops since data dependencies may occur in the residual update.

The face-cell loops in the routine that calculates the residuals can, however, be parallelized by considering the number of variables for which a residual update is needed. For all independent variables, density, momentum components, and total energy, the residual has to be updated, and these updates are independent. For the second order space discretization also the three moment residuals need to be updated (for details see Van der Vegt, [3]). Together with the time step update we thus have $5 + 3 \times 5 + 1 = 21$ independent updates. Since these updates have to be performed for both cells bounding the face, we have 42 independent updates within one colour. For a 16 processor machine this is sufficient parallelism. In Fig. 1 the structure of the flux integration is presented in pseudo-code.

```
                    pardo for all colours
                        do for all faces of a colour
face-face                   calculate flux through face
                        enddo
                    enddo
                    do for all colours
                        pardo for all 42 variables
                            do for all faces of a colour
face-cell                       add fluxes to residuals of cell based variables
                            enddo
                        enddo
                    enddo
```

Fig. 1. Pseudocode representation of the flux integration. Parallel do-loops are written as **pardo**. Also shown are the type of loops.

In order to achieve load balance in the parallel loops over the colours, the number of colours is adjusted to be a multiple of the number of processors, and the vector lengths of all colours are nearly equal.

The above treats the parallelization of the flux integration, the parallelization of the slope limiter is described below. The computation of the slope limiter consists of a number of min-max operations for each cell. The limiter is computed for each cell and for just the five independent variables, and hence the parallelization strategy for the residual updates cannot be followed. For the limiter the loops over the faces within one colour and the five independent variables are collapsed and both vectorized and parallelized. In Fig. 2 the structure of the slope limiter computation is presented in pseudo-code.

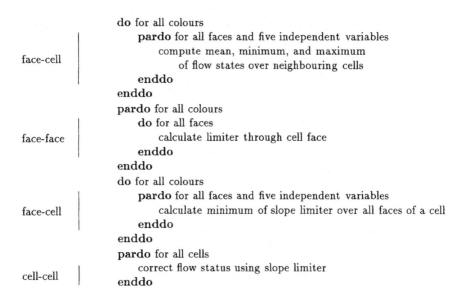

Fig. 2. Representation of slope limiter in pseudo-code. Parallel do-loops are written as **pardo**. Also shown are the type of loops.

The remaining loops in the integration part of the flow solver are cell-cell loops, which are easily both parallelized and vectorized, sometimes collapsing them with the loop over the five independent variables. These loops are split into a multiple of the number of processors to achieve load balance. Eventually, the routines accounting for 99.5% of the integration time are parallelized.

The above parallelization strategy is different from the one proposed in Van der Vegt et al. [4]. The reason for the change is twofold. On the one hand, it was based on the assumption that a critical section was needed for the face-wise update of the cell-based residuals, which restricted the optimal speedup to 5. Using the strategy described above no critical section is needed. On the other hand, the flow solver has been improved significantly and is now better suited for the above parallelization strategy.

Because of memory restrictions on the NEC SX-3 the previous version of the flow solver made extensive use of the extended memory unit (XMU). The XMU is not part of the shared memory, and has to be addressed by explicit I/O. Due to hardware restrictions, these I/O are not parallelizable and the subsequent large serial sections in the code prohibit parallel efficiency. In the present version of the flow solver use of XMU is an option to the user, and for problems that fit into main memory no serial sections in the code occur.

4 Parallelization results and discussion

4.1 NEC SX-4 architecture

The code has been ported to the NLR NEC SX-4, using the above parallelization strategy and compiler directives. The processors in the SX-4 combine a powerful, balanced 2 Gflop/s vector unit with a state-of-the-art superscalar unit. Special synchronization, interprocessor communications and control hardware is implemented to maximize parallel processing efficiency.

The NEC SX-4 installed at NLR has 16 processors, 4 GB main memory and an extended memory unit of 8 GB. Each processor has eight vector pipes of length 256. The main memory consists of 16 modules, each having a bandwidth of 16 GB/s to the processors. Hence the bandwidth scales with the number of processors (with a peak of 256 GB/s), and a better parallel efficieny is expected than on the NEC SX-3. In the acceptance benchmark investigation (reported by Potma et al. [1]) the single processor NEC SX-4 reached a flop rate of 0.618 Gflop/s on the complete program Hexadap, which is an increase of 1.7 with respect to the performance of the NEC SX-3. On the SX-4 a speedup of 9.5 was obtained on 16 processors. These benchmark results were obtained using a previous version of the flow solver Hexadap, the present version has been developed further and is in principle more suited for parallel processing. Moreover, the reported speedups in Potma et al. ([1]) concern only the parallelized routines, where the results in this paper are for the complete flow solver.

4.2 Metrics

The prime metric for parallel processing is elapsed time. Therefore, in the discussion of the results the emphasis will lay on elapsed timings. Other relevant metrics are speedup and scalability.

Speedup is defined as the single processor execution time divided by the multi-processor execution time. For shared memory architectures the single processor time is obtained from the parallel code compiled ignoring the parallelization directives. To be able to measure the performance of parallel algorithms on large problems which do not fit in single processor memory *generalized speedup* ([2]) has been introduced. This is defined as the quotient of sequential speed over parallel speed, where the two speeds may be obtained on different problem sizes.

Of interest to the industry is also the parallelization effort. To measure the effort the required man power is measured and the number of changed lines is counted.

4.3 Results

The parallelization effort has taken 3 man weeks extra in the CFD development effort. A total of 4500 lines has been changed (of the total of 50,000 lines) and 70 compiler directives have been added. The greatest change in the code has taken place in the flux integration routine in the update of the residuals. Smaller changes concern the colouring algorithm, where the number of colours is made a multiple of the number of processors. The loops which are both vectorized and parallelized are split explicitly.

The tests used to assess the parallel performance are constructed from an initial structured mesh around the ONERA M6 wing. Subsequently the mesh has been adapted three times to obtain a series of meshes where each mesh is roughly twice the size of the previous mesh. See Table 1 for a characterization of the meshes. In this table, the work is defined as the number of floating point operations required in the integration part of the solver to advance the flow 100 time steps. The size of the last test case, M6/4, is restricted by memory requirements.

case	# cells	# faces	work [Gflop]	memory [GB]
M6/1	131,072	386,176	492.676	0.398
M6/2	262,190	778,257	1,000.951	0.904
M6/3	517,021	1,554,600	2,000.709	2.029
M6/4	865,623	2,613,670	3,355.267	2.586

Table 1. Mesh characteristics of the four test cases. The work, measured in floating point operations, is defined as the number of floating point operations in the integration part of the solver. Memory is for single processor execution.

In Table 2 the elapsed timings of the flow integration part of the flow solver are presented for the four cases and different number of processors. The larger problems are not executed on a few processors since the timings are not relevant for the discussion. The NEC SX-4 at NLR is part of a multi-user environment and it is not possible to use the SX-4 as a dedicated machine. Even though one can use all 16 processors, other users may use the computer interactively and system processes may run in the background. The performance of the flow solver does not increase on 16 processors, because other processors have to wait for results of the one processor which is (also) used for other processes. Therefore timings results are only presented for up to 14 processors.

case	1	2	4	8	12	14
M6/1	663	341	177	93	67	58
M6/2	1441	726	377	200	146	125
M6/3	-	-	720	375	268	230
M6/4	-	-	-	652	462	397

Table 2. Elapsed timings (in seconds) for the flow integration part of the flow solver.

The (traditional) speedups of case M6/1 and M6/2 are tabulated in Table 3. Using Amdahl's Law and given a parallelization ratio of 99.5% one would predict a speedup of 13.1 on 14 processors. Hence, the experimental speedups are close to the theoretical speedup. Differences most likely are caused by parallel overhead and small serial sections in the parallelized routines.

On 14 processors the performance of the flow solver reaches 8.5, 8.0, 8.7 resp. 8.5 Gflop/s for the respective four cases. Surprising in the timing results is that a n-fold increase in the work, results in roughly an n-fold increase in the elapsed time, even when running on several processors. This implies that for the given algorithm and given architecture traditional and generalized speedup coincide.

In Fig. 3 the obtained speeds for the different cases and number of processors are shown. Here, speed is defined as the quotient of work (see Table 1) over elapsed time (see Table 2). It can be clearly seen that attained speed is nearly independent of the problem size.

The elapsed timings also show that the machine-algorithm combination scales well. An increase in the work and a likewise increase in the number of processors results in comparable elapsed timings.

case	1	2	4	8	12	14
M6/1	1.0	1.94	3.75	7.13	9.89	11.43
M6/2	1.0	1.98	3.82	7.20	9.87	11.53

Table 3. Speedups based on the elapsed timings for the flow integration part of the flow solver.

A parameter in the algorithm determines the number of faces per colour, and hence the vector length in the flux integration and slope limiter routines. This parameter also determines the size of the scratch (or work) arrays in these

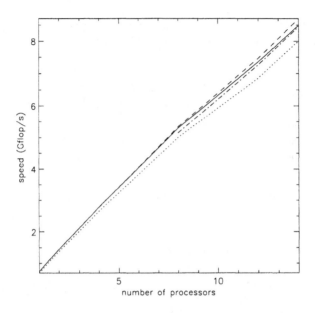

Fig. 3. Attained speeds (in Gflops/s) for the different cases and different number of processors. — M6/1, ··· M6/2, - - - M6/3, - · - M6/4

routines, and therefore the size of scratch memory. In all the above tests the parameter is equal to 12,800. To assess the dependence of the performance on this parameter, test case M6/2 on 8 processors is repeated with two different values of the parameter. Results are tabulated in Table 4. For small values of the parameter the performance degrades, because of decreased vector length. For memory critical runs the parameter may be reduced to limit memory requirements.

As mentioned in Section 3 it is possible to store and retrieve some scratch arrays on the extended memory unit (XMU). Since these stores and retrieves are essentially sequential, the use of XMU will degrade parallel performance. Test case M6/2 is run on 8 processors using XMU. As can be seen in Table 4 execution time is almost doubled. The parameter determining the number of faces per colour is equal to 12,800. Memory use is reduced using XMU, but the memory use is the same as the run with the parameter equal to 4096 not using XMU.

parameter value	4096	12800	25600	using XMU
elapsed time	210	200	200	343
memory use	818	1058	1236	809

Table 4. Timings (in seconds) and memory use (in MB) for different values of the parameter determining the number of faces per colour; and comparison with a run using the extended memory unit (with parameter equal to 12800). All runs for case M6/2 on 8 processors.

4.4 Discussion

The present parallelization efforts have taken a limited time of three manweeks and produced a parallel code which has excellent parallel performance and good scalability. On the 16 processor NEC SX-4 a speedup of 11.5 and a speed of 8.5 Gflop/s is obtained by using 14 processors. The attained speed is independent of the problem size.

The code has been restructured slightly, and the parallel version is part of version pipeline ensuring that future extensions of the functionality will use the parallel structure, which is achieved using the software repository in the CFD working environment.

5   CFD working environment

The parallelization and development of the above algorithm is supported by (tools in) the working environment ISNaS [5] for CFD applications. Within the NICE project this working environment serves as a prototype for the Dutch HPCN Center for Flow Simulation, HFS. HFS supports cooperative work across the network consisting of the combined networks of the partners with the NLR NEC SX-4 in its center.

CFD development and use takes place on such diverse systems as mainframes, parallel and/or vector computers and graphical work stations. ISNaS is designed to make the network transparent to the user. Together with the management of data and documentation, processing of CFD codes in a production environment is greatly facilitated by ISNaS.

ISNaS also supports the developers of CFD software during the development phase. In large projects where several disciplines cooperate, software configuration management and information exchange support the software engineers. The use of the working environment enforces quality control. The Informatics division of NLR is certified for ISO-9001.

To support the development of parallel codes the parallelization tools of the NEC SX-4 are integrated in ISNaS. Two main support tools for parallelization are the parallelizer and analyzer tool. The integrated tools allow users easy access to this tools without the need to read all details in the manuals. Options of the tools which are considered to be generic to all problems are presented to the user as clear text. For the analyzer tool, for example, the user can choose between static analysis, execution time analysis of the entire program and do-loop analysis of specified loops. In this way, the tools are made accessible in a user-friendly way and it is expected that the parallel use of the NEC SX-4 will increase. Moreover, job control and assignment of processors are handled by the working environment.

ISNaS is a so-called instantiation of SPINE, the general tool developed at NLR to create application area specific working environments. ISNaS or related products can be installed on any UNIX network.

6 Conclusions

In this paper the use of the NEC SX-4 for advanced CFD processing is described.

First, a hexahedron based, flow solver with unstructured grid adaptation is parallelized on the 16 processor NEC SX-4 using the shared memory paradigm. Roughly 10% of the code is restructured to achieve an efficient algorithm which reaches 8.5 Gflop/s on 14 processors. The obtained speed is independent of the problem size. This proves the capabilities of the NEX SX-4 as a high performance computing platform. The present parallelization only concerns the flow solver part of the algorithm, for the parallelization of the adaptation part a grid partitioning is required. This constitutes future work.

Second, the efficient use of the NEC SX-4 is supported by the working environment ISNaS for CFD applications. Support tools for the analysis and shared memory parallelization of algorithms have been integrated in the working environment. This integration supplies easy access to the tools for first-time users. In the production phase, ISNaS supports users of the CFD software in job control, version consistency and assignment of processors.

Summarizing, the NEC SX-4 allows for good parallel efficiency using the easy-to-use shared memory paradigm. Application to production codes will reduce computing time and costs.

Acknowledgment The authors want to thank both G.A. van der Velde (NEC), for his idea of independent residual updates, and K. Potma and M.C.Z. Schoemaker (NLR) for their support in the use of the NEC SX-4.

References

1. K. Potma, G.J. Hameetman, W. Loeve and G. Poppinga, *Early benchmark results on the NEC SX-4*, NLR Technical Publication TP96464L, 1996, presented at Parallel CFD'96, Capri.
2. Xian-He Sun and J. Gustafson, *Toward a better parallel performance metric*, Parallel Computing 17 (1991) 1093-1109.

3. J.J.W. van der Vegt, *Anisotropic grid refinement using an unstructured discontinuous Galerkin method for the three-dimensional Euler equations of gas dynamics*, AIAA Paper 95-1657, 1995.
4. J.J.W. van der Vegt and H. van der Ven, *Hexahedron based grid adaptation for future Large Eddy Simulation*, AGARD symposium 'Progress and challenges in CFD methods and algorithms', Seville, October 1995.
5. M.E.S. Vogels and W. Loeve, *Development of ISNaS: an information system for flow simulation in design*, NLR TP89025, 1989. Also see *http://www.nlr.nl/public/fac/fac-isna/isnas.html*

Parallelization of CFD Code Using PVM and Domain Decomposition Techniques

L.M.R. Carvalho and J.M.L.M. Palma[1]

Faculdade de Engenharia da Universidade do Porto
Rua dos Bragas
4099 Porto, Portugal
(e-mail: *jpalma@fe.up.pt*)

Abstract. The present work discusses the results of parallelization of a conventional CFD code, initially developed for sequential machines. The authors used the domain decomposition technique with overlapping and the software package PVM. The results showed parallel efficiency in excess of 100%, attributed to the favourable effect of the domain splitting on the numerical efficiency of the algorithm and cache hit ratio.

1   Introduction

As in many other scientific areas, parallel computing has a great impact on CFD (Computational Fluid Dynamics), see for instance Pelz *et al.*(1993) and Ecer *et al.*(1995), and different strategies have been followed to take advantage of emerging technologies (either hardware or software).

The present work describes the parallelization of a conventional CFD code based on the SIMPLE algorithm, initially developed for sequential machines by Patankar and Spalding(1972). This is a well known algorithm within the CFD community. It has been used for over 20 years, and there is accumulated experience on its numerical behaviour and convergence characteristics (e.g. McGuirk and Palma(1993), Wanik and Schnell(1989), van Doormaal and Raithby(1984)).

For parallelization the authors used the domain decomposition technique with overlapping and the software package PVM (Geist *et al.*(1994)). PVM is a software library allowing the use of a network of heterogeneous UNIX computers.

The paper is made-up of 3 main sections. The mathematical model is included in Section 2. The results are discussed in Section 3. Section 4 summarizes the main conclusions of the work.

2   Mathematical Model

The fluid flow equations for a cartesian grid system, assuming incompressible, time-independent and newtonian flow are the following:

$$\frac{\partial U_i}{\partial x_i} = 0 \tag{1}$$

$$\rho\frac{\partial U_i U_j}{\partial x_j} = -\frac{\partial P}{\partial x_i} + \frac{\partial}{\partial x_j}\mu\left(\frac{\partial U_i}{\partial x_j} + \frac{\partial U_j}{\partial x_i}\right) \tag{2}$$

where P is the pressure, U_i is the velocity along the direction x_i, and ρ and μ are the density and fluid dynamic viscosity, respectively.

SIMPLE is a sequential algorithm, where the momentum equations and continuity equation (a pressure correction equation, strictly speaking) are solved one after another, as if they were independent, until both the velocity and pressure fields fulfill the requirements of mass and momentum conservation. This is called a global iteration, opposed to the inner iterations of the iterative method ("solver") used to solve each of the system of equations, made up of the algebraic form of the conservation equations referred above.

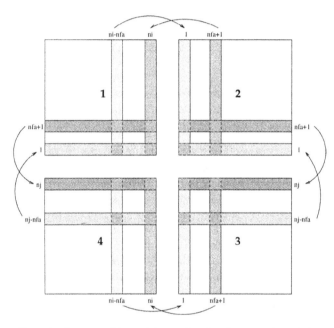

Fig. 1. Domain decomposition with overlapping in case of 4 subdomains

2.1 Strategy of Parallelization

The integration domain was split into subdomains, with overlapping (Fig. 1). The thickness of the overlapping region (*nfa*) was an input variable and there is a number of *nfa* lines evaluated simultaneously in 2 or 4 subdomains, in case of the nodes inside the inner square of size *nfa* × *nfa*. A non-staggered grid (Rhie and Chow(1983)) has been used, and the overlapping occurred always between nodes of similar type; i.e. a grid node in the centre of the control volume in subdomain 1 overlaps with a similar node type in subdomain 2. The overlapping between a grid node in the centre of the control volume and the face of a control volume in another subdomain was tried in an initial stage of the work and proved to be unsuccessful.

In each subdomain a maximum number of global iterations of SIMPLE was performed ($niter$), before interchanging the data in the overlapping region. This included the value of all variables (U, V and P) along the lines as indicated in Fig. 1. The subdomains were all of identical size and this alleviated the need for a load balancing methodology, although even under these conditions it has been found that the convergence criterion in some subdomains could be reached earlier.

The strategy being followed here is not the only one possible. For instance, Lewis and Brent(1993) parallelized the SIMPLE algorithm using separate processors for each transport equation. This, however, set a limit to the number of processors being used (3 in the present case). Furthermore, and compared to our strategy, this increases the number of communications and requires a load balancing methodology, given the difficulties of convergence of the pressure compared to the momentum equations. Many authors have concentrated only on the parallelization of the solver, because this is for large grid sizes the most time consuming part of the algorithm. This is also a very active topic of research in numerical analysis.

3   Discussion of Results

The discussion of results is organized into 2 subsections. The objective in subsection 3.1 is to assess the influence of the various numerical parameters on the convergence of the algorithm. Subsection 3.2 is restricted to a single set of numerical parameters, but calculations have been performed for a wider range of numerical grids to see the effects of the problem size.

The test case geometry was the usual lid driven cavity. Results have been obtained for a single Reynolds number ($Re = \rho U L/\mu = 100$). The numerical grid is non-staggered and uniformly distributed. The tri-diagonal matrix algorithm was used for solving the system of algebraic linearized equations; the under-relaxation factors were 0.7 and 0.3, for velocities and pressure, respectively. The number of sweeps was fixed and set at 2 for U and V velocities, and 4 for pressure.

The parallel efficiency was defined by:

$$\epsilon = \frac{T_S}{p\,T_P} \tag{3}$$

where T_S stands for the execution time of the sequential, and T_P is the execution time of the parallel version of the code. The number of PVM processes is given by p, which can be larger than the number of processors. This definition here differs from the standard where it is commonly used the number of processors, rather than the number of processes. Also the *"speed-up"*, S_n, was defined by

$$S_n = \frac{T_S}{T_P} \tag{4}$$

The times being quoted for each run were obtained via the UNIX command *time*. In case of the sequential version, the number of iterations is a global iteration of SIMPLE, as defined in section 2. The number of iterations of the parallel version (in Tables 2 to 4) is the total number of global iterations, i.e. the sum of the global iterations in every subdomain. Note that the maximum number of global iterations in each subdomain could change; the values of *niter* in Tables 2 to 4 were the maximum allowed, and occurred only if a reduction to 10% of the initial residual were not reached before.

The results were obtained either on a workstation cluster with 2 Hewlett Packard (models 715/50 and 715/75) connected by standard Ethernet and running PVM, or in case of the results in subsection 3.2, on a single workstation (HP 715/75) running PVM also.

3.1 The Convergence of the Parallel Algorithm

Results are shown here for 3 grid sizes (20×20, 40×40 and 80×80), with different width of the overlapping region between subdomains (*nfa*) and variable number of global iterations of SIMPLE per each subdomain (*niter*) before interchanging data located on the boundaries. Table 1 shows the results obtained with the sequential version of the code.

Grid size	Elapsed time	N. Iter.
20×20	16.1	79
40×40	124.0	314
60×60	539.0	704
80×80	1748.9	1039
100×100	3670.6	1944
120×120	7333.6	2816

Table 1. Algorithm performance in sequential calculations (Results obtained on an HP 715/75)

Before proceeding with the discussion of the results it is important to show that the flowfield obtained with either the sequential or the parallel version are identical. This is the objective of Fig. 2.

Tables 2 to 4 show the results for each of the 3 grid sizes tested. The number of iterations in these tables are not directly comparable with those for the sequential version (in Table 1), because the number of iterations per subdomain in the parallel version could vary between 1 and the value set by *niter* in Tables 2, 3 and 4.

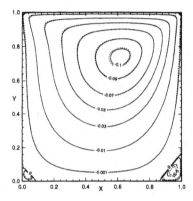

Fig. 2. Streamline patterns obtained with the sequential (solid line) and parallel version (dotted line) using 2 subdomains, $nfa = 3$ and $niter = 3$

• 2 subdomains

	$niter = 3$	5	7	9
$nfa = 3$	7.7 (419)	9.0 (577)	9.5 (720)	47.3 (5901)
5	5.2 (233)	4.6 (286)	5.2 (355)	5.3 (447)
7	4.8 (191)	4.1 (213)	4.0 (237)	3.0 (268)

• 4 subdomains

	$niter = 3$	5	7	9
$nfa = 3$	10.7 (1143)	—	—	—
5	6.3 (512)	6.2 (747)	6.9 (1222)	6.5 (1102)
7	5.1 (435)	4.9 (506)	4.8 (604)	4.9 (722)

• Sequential version: 16.1 (79)

Table 2. Algorithm performance in parallel calculations (20×20 grid) as a function of subdomain overlapping, internal iterations and number of subdomains for 2 and 4 PVM processes running on 2 workstations. (Elapsed time in seconds and total number of global iterations between parenthesis)

For all cases in Tables 2 to 4 the parallel was always faster than the corresponding sequential run (in Table 1). The few exceptions to this rule occurred for high values of $niter$; for instance, in Table 2 for $nfa = 3$ and $niter = 9$, with 2 subdomains. The best result of the parallel runs led to speed-up of $5.4, 6.1$ and 8.7 for grids 20×20, 40×40 and 80×80, respectively. In general, the number of global iterations per subdomain is higher than the number of iterations of the sequential version; the results obtained are possible only because one iteration in a subdomain is faster than one iteration in the whole domain of the sequential version.

• 2 subdomains		
$niter = 3$	5	7
$nfa = 3$ 73.6 (1407)	81.1 (3790)	—
5 35.8 (1114)	35.9 (1426)	49.3 (2687)
7 27.5 (823)	25.0 (953)	23.5 (1097)
9 22.6 (733)	21.6 (795)	20.4 (869)

• 4 subdomains		
$niter = 3$	5	7
$nfa = 3$ 97.9 (6189)	105.7 (9242)	—
5 42.7 (2940)	44.9 (3816)	—
7 29.9 (1977)	30.3 (2369)	32.1 (2781)
9 27.4 (1656)	29.0 (1856)	26.0 (2071)

• *Sequential version:* 124.0 (314)

Table 3. Algorithm performance in parallel calculations (40×40 grid) as a function of subdomain overlapping, internal iterations and number of subdomains for 2 and 4 PVM processes running on 2 workstations. (Elapsed time in seconds and total number of global iterations between parenthesis)

The splitting into 2 subdomains is the most favourable situation. The case with 4 subdomains reduces the number of combinations of nfa and $niter$ for which convergence could be achieved and the calculations took longer compared to the 2 subdomains case. Only the 80×80 grid was favoured by the splitting into 4 subdomains; the fastest run required 201.3 s compared to 270.7 s, in case of 2 subdomains.

For a given level of overlapping (nfa), results become worse with increasing $niter$. For a given $niter$, the increase of nfa improves convergence. This is a behaviour already expected and consistent with the elliptic nature of this problem. The domain decomposition splits the initial one single problem in 4 apparent independent problems. The increased thickness of the overlapping region reinforces the coupling between the 4 subdomains, whereas the increased number of iterations in each subdomain and solving the equations to higher accuracy before interchanging the values reinforces the decoupling between the 4 subdomains and may even lead to divergence.

These results are displayed in Fig. 3, in terms of the parallel efficiency for grid 80×80. Values of efficiency in excess of 100% have been reported in the literature. For instance, Michl et al.(195)) attributed this behaviour to the cache size. The subdomain calculations, opposed to a full domain calculation can fit in the processor memory, decreasing the cache miss ratio and therefore reducing execution time. This is a machine dependent feature, influenced not only by the

● 2 subdomains

niter = 3	5	7	9
nfa = 3 1345.8 (15246)	1847.8 (21783)	—	—
5 532.8 (6054)	687.1 (8185)	868.1 (10537)	1040.4 (12821)
7 289.7 (3919)	389.1 (4783)	394.0 (5736)	461.7 (6814)
9 276.4 (3222)	297.8 (3604)	330.3 (4102)	368.3 (4627)
11 270.7 (2953)	277.7 (3152)	293.0 (3385)	306.0 (3692)

● 4 subdomains

niter = 3	5	7	9
nfa = 3 1625.5 (43133)	—	—	—
5 806.9 (16958)	1011.9 (22900)	—	—
7 398.5 (10018)	760.2 (12593)	539.3 (13523)	—
9 332.7 (7627)	357.0 (8902)	201.3 (10397)	—
11 302.2 (6623)	312.5 (7339)	333.6 (8103)	361.2 (9023)

● Sequential version: 1748.9 (1039)

Table 4. Algorithm performance in parallel calculations (80×80 grid) as a function of subdomain overlapping, internal iterations and number of subdomains for 2 and 4 PVM processes running on 2 workstations. (Elapsed time in seconds and total number of global iterations between parenthesis)

cache size and cache organization but also by the bandwidth of both cache and main memory. Quantitative results obtained here are expected to be processor dependent; however, the conclusions are most likely to be valid for any cache-base RISC architecture.

3.2   The Influence of the Problem Size

The results in this subsection have been obtained for 2 and 4 PVM processes (subdomains) running on a single machine. The number of global iterations per cycle in each subdomain (*niter*) and *nfa* have been kept constant and equal to 3. Results have been obtained for a wider range of grid sizes. Here there has been no use of the criterion based on the reduction to 10% of the initial residual, and we call cycle the updating of the boundary values; which may be compared to the number of global iterations of the sequential version of the code (in Table 1). The number of cycles is obtained from the number of iterations divided by (*niter* × *number of subdomains*) and it is included in Tables 5 and 6, to facilitate the comparison to the number of iterations in Table 1.

Fig. 4 shows the results obtained for the sequential version of the code compared to the cases with 4 and 2 subdomains. The results are plotted in log scales and show the elapsed time as a function of the equivalent number of nodes. The version with 4 subdomains is faster than the sequential version for grid sizes below 10000 nodes (see also values in Tables 1 and 6). The case with 2 subdo-

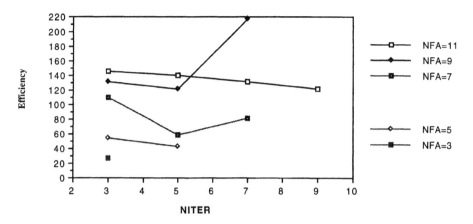

Fig. 3. Efficiency of parallelization as a function of number of internal iterations (*niter*) and width of the overlapping region between subdomains (*nfa*).

mains and running 2 PVM processes simultaneously is between 4.8 and 12.2× faster than the sequential version. The slopes of the curves for sequential and 4 subdomains are similar and steeper than the curve for 2 subdomains. The scale-up between the case with 2 processes and the case with 4 processes increases with the grid size, reaching 19.6 for an equivalent grid of 190 × 190. This value was made possible mainly, because the number of cycles has been reduced 28.2 times. These are even lower than those obtained for the sequential version.

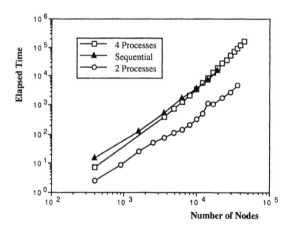

Fig. 4. Elapsed time as a function of number of nodes, for sequential version and parallel versions with 2 and 4 PVM processes on 1 single workstation

The domain splitting led to favourable effects on the numerical convergence

N.	$NI \times NJ$	$(\equiv NI \times NJ)$	T_{2P} (s)	N.Iter	Cyles	T_S/T_{2P}	T_{4P}/T_{2P}	$\frac{N.Cyc.}{N.Cyc_{4P}}$
1	12×20	(20×20)	2.6	408	68	6.1	2.8	1.7
2	17×30	(30×30)	9.2	948	158	—	—	—
3	22×40	(40×40)	25.6	1686	281	4.8	2.7	—
4	27×50	(50×50)	51.1	2316	386	—	—	—
5	32×60	(60×60)	78.4	2568	428	6.9	4.9	4.1
6	37×70	(70×70)	108.7	2568	428	—	6.7	6.3
7	42×80	(80×80)	142.5	2622	437	12.2	8.9	8.8
8	47×90	(90×90)	205.0	2778	463	—	10.4	11.5
9	52×100	(100×100)	322.8	3366	561	11.4	11.0	12.4
10	57×110	(110×110)	500.8	4110	685	—	11.4	13.0
11	62×120	(120×120)	1083.3	4644	774	6.8	7.8	14.5
12	67×130	(130×130)	1032.9	5124	854	—	12.6	16.3
13	77×150	(150×150)	1719.6	6120	1020	—	15.9	19.9
14	87×170	(170×170)	2815.6	7098	1183	—	18.0	23.9
15	97×190	(190×190)	4659.3	8088	1348	—	19.6	28.2

Table 5. Algorithm performance of parallel version with 2 subdomains (2 PVM processes in one workstation). T_{4P} and T_{2P} are the execution time for 2 and 4 PVM processes. ($\equiv NI \times NJ$) is the efective size of the whole domain.

of the algorithm. This was a very surprising result for which we have not found yet a convincing reason. There was the possibility that the parallel version could converge to a different flowfield, but the observation of the streamline pattern for many of the calculations showed results identical to those displayed in Fig. 2.

Even when using 1 single workstation running various PVM processes the execution time is reduced, compared to the sequential version. For instance, the sequential version for a grid 80×80 (in Table 6) takes 1749 s and 1039 iterations, compared to the parallel version with 4 subdomains, which although performing 3826 iterations, requires 1267 s only. This can be explained by cache missing problems which will be less for smaller domain sizes. The grid size indicated for a parallel calculation (between parenthesis) is the equivalent grid size; the grid 80×80 was obtained from calculations in 4 subdomains of size 42×42.

4 Conclusions

The work shows results of parallelization of a conventional sequential algorithm used in Computational Fluid Dynamics. Results were obtained for a wide range of grid sizes. The main conclusions of these work were the following.

1. The best results showed parallel efficiencies above 200%: Calculations with very fine numerical grids even showed speed-up factor of 6 approximately.

N.	$NI \times NJ$	$(\equiv NI \times NJ)$	CPU [s]	N.Iter	N.Cyles
1	12×12	(20×20)	7.3	1368	456
2	22×22	(40×40)	69.7	1806	602
3	32×32	(60×60)	382.3	21288	1774
4	37×37	(70×70)	733.1	32136	2678
5	42×42	(80×80)	1267.7	45912	3826
6	47×47	(90×90)	2128.8	62868	5239
7	52×52	(100×100)	3558.1	83184	6932
8	57×57	(110×110)	5726.4	107172	8931
9	62×62	(120×120)	8459.4	135060	11255
10	67×67	(130×130)	12974.0	166956	13913
11	72×72	(140×140)	18759.8	203232	16936
12	77×77	(150×150)	27365.4	243732	20311
13	82×82	(160×160)	36970.7	289188	24099
14	87×87	(170×170)	50798.9	339516	28293
15	92×92	(180×180)	70216.0	394788	32899
16	97×97	(190×190)	91484.6	113841	37947
17	102×102	(200×200)	118659.6	521940	43495
18	107×107	(210×210)	156955.2	593268	49439

Table 6. Algorithm performance of parallel version with 4 subdomains (4 PVM processes in one workstation). ($\equiv NI \times NJ$) is the efective size of the whole domain.

This was mainly the consequence of the favourable effect introduced by the domain splitting on the characteristics of convergence of the algorithm.

2. The convergence of the parallel algorithm improves with the increased overlapping between subdomains and is reduced if a tighter convergence is imposed in each subdomain before interchanging information on the boundaries.

Acknowledgments. The authors are very grateful to their colleagues F.A. Castro and A. Silva Lopes, for their help during all stages of this work.

References

Ecer, A., Haüser, J., Leca, P., and Periaux, J., editors (1995). *Parallel Computational Fluid Dynamics'95*. North-Holland, Amsterdam. Proceedings of the Parallel CFD'93 Conference, Paris-France (May 10–12, 1993).

Geist, A., Beguelin, A., Dongarra, J., Jiang, W., Manchek, R., and Sunderam, V. (1994). PVM3 user's guide and reference manual. Technical report, ORNL/TM-12187, Oak Ridge National Laboratory.

Lewis, A. and Brent, A. (1993). A comparison of coarse and fine grain parallelization strategies for the SIMPLE pressure correction algorithm. *International Journal for Numerical Methods in Fluids*, **16**, 891–914.

McGuirk, J. and Palma, J. (1993). The efficiency of alternative pressure-correction formulations for incompressible turbulent flow problems. *Computers & Fluids*, **22**(1), 77–87.

Michl, T., Wagner, S., Lenke, M., and Bode, A. (195). Dataparallel implicit Navier-Stokes solver on different multiprocessors. In Ecer *et al.*(1995), pages 133–140. Proceedings of the Parallel CFD'93 Conference, Paris-France (May 10–12, 1993).

Patankar, S. and Spalding, D. (1972). A calculation procedure for heat, mass and momentum transfer in three-dimensional parabolic flows. *International Journal of Heat and Mass Transfer*, **15**, 1787–1806.

Pelz, R., Ecer, A., and Haüser, J., editors (1993). *Parallel Computational Fluid Dynamics'92*. North-Holland, Amsterdam.

Rhie, C. and Chow, W. (1983). Numerical study of the turbulent flow past an airfoil with trailing edge separation. *AIAA Journal*, **21**(11), 1525–1532.

van Doormaal, J. and Raithby, G. (1984). Enhancements of the SIMPLE method for predicting incompressible fluid flows. *Numerical Heat Transfer*, **17**, 147–163.

Wanik, A. and Schnell, U. (1989). Some remarks on the PISO and SIMPLE algoritms for steady turbulent flow problems. *Computers & Fluids*, **17**(4), 555–570.

Possibilities of Parallel Computing in the Finite Element Analysis of Industrial Forming Processes

Eugenio Oñate

International Center for Numerical Methods in Engineering (CIMNE)
Universidad Politécnica de Cataluña
Gran Capitán s/n, Campus Norte UPC, 08034 Barcelona, Spain

Abstract. The paper presents an overview of the possibilities of parallel computing for the analysis of industrial forming processes using the finite element method. The theoretical and computational aspects of the various finite element formulations are presented in some detail as well as the different strategies for parallelization of the solver, the mesh generation, the error simulation and the mesh adaption modules. Some examples of parallel analysis of powder compaction and sheet stamping processes using parallel finite element codes developed at CIMNE are finally presented.

1 Introduction

In the last decade, computer applications aimed at facilitating and improving forming processes in the manufacturing industry have been honed from tools used only by specialised reseachers to everyday devices for use from the shopfloor to the designers desk [1-7]. The nature of these new computer-based tools and methodologies is such that the software programs themselves and the expertise needed to use them are not specific to a single forming process.

The potential applications and technical benefits of computer based systems for the forming industry have been explicitly recognised in the report "Computer-Aided Engineering for Metal Forming" published by the EC in 1995 [8]. In this report the state of the research in Europe, USA and the rest of the world is analysed together with a study of the market size and a description of the main companies and software products.

It should not be however forgotten that the development of reliable computational procedures to predict the behaviour of deformation processes during forming operations has encountered many serious obstacles. Together with the nonlinearity of the material, other important effects like the unsteady nature of the process, the large magnitude of the strains involved and the importance of contact and frictional effects at the tool-material interfaces make the study of forming processes so complex that its analysis justifies the use of sophisticated numerical techniques such as the finite element method [6] and usually leads to huge computer requirements.

The ultimate aim of computational procedures is to advance in the design and manufacturing of metal, plastic and glass based products, hereby bringing about quantitative improvements in the competitive state of manufacturing enterprises. The fundamental numerical methods developed form only a part of the simulation requirements of the manufacturing industry and a comprehensive support system would need additional facilities. In particular, the important aspects of a CAD link and access to a material and process database are necessary for computational tools to be considered useful for routine industrial applications. Additionally, software modules which assess the "fitness-for purpose" of the industrial codes are also essential.

It is also important that such computational tools can be implemented on relatively low cost hardware platforms (such as the PC networks) which is considered essential for acceptance of the methodology within general manufacturing industries.

Industrial forming processes such as sheet stamping, rolling, extrusion, molding, etc., involve the continuous deformation of the material in time. The equations of motion of this transient problem can be written in terms of the displacements of the deformed material points measured from an appropriate reference configuration (displacement approach), or else in terms of their velocities at each moving configuration (flow approach).

Both, displacement and flow approaches can make use of sophisticated elastoplastic/viscoplastic material models. Alternatively by neglecting elastic effects simpler rigid-plastic/viscoplastic models can be employed. Finally the equations of motion can be of quasi-static type leading to the so called *implicit* finite element codes or else incorporate dynamic inertia effects, this being the basis of the *explicit dynamic* codes.

Despite recent advances in enhancing the overall efficiency of implicit and explicit finite element codes, the numerical simulation of forming processes of industrial size is still today a formidable task which generally requires excessive computing effort to be efficiently applicable as an everyday design tool. The need for improving the efficiency of existing serial codes via parallelization is therefore obvious. The opportunity and necessity of such a development is precisely the goal of much current research which aims to fill this gap and release parallelized codes for analysis and design of industrial forming problems in clusters of PC's and workstations affordable by SME's in the manufacturing sector.

This paper aims at presenting an overview of the possibilities of enhancing current finite element based computational procedures for analysis and design of industrial forming processes by means of parallel computing techniques. The layout of the paper is the following. In next section the crucial steps in the parallelization of a finite element code for analysis of industrial forming processes are listed. The following sections are devoted to the description of the quasi-static and explicit dynamic finite element solution techniques. Here the possibilities of their parallelization in different computer architectures are discussed. The paper follows by describing different alternatives for parallel mesh generation and parallel mesh adaption using error estimation procedures. In the last part of the paper some examples of application of implicit and explicit finite element parallel codes developed at CIMNE for analysis of industrial forming problems such as metal powder compaction and sheet stamping are presented.

2 Basic concepts in the finite element analysis of industrial forming processes

2.1 Objectives of finite element analysis

The main objective of finite element numerical simulation of industrial forming processes is to accurately predict the deformation of the material during forming so as to evaluate that the final product has the desired shape and properties. The analysis should also provide accurate results of internal variables such as strains and stresses at any stage of the process, as well as the residual stresses in the formed part. Finally, a prediction of the possible deffects (tearing, wrinkling, shrinking, fracture, etc.) induced by the large deformations occuring in the process is also essential in order to assess the quality of the final product. Computer simulation, therefore, plays the role of a *virtual testing laboratory* to be used by process design engineers for evaluating the suitability of the geometry of dies and tools, the mechanical properties of materials and the process parameters chosen for a particular forming operation. These results of the analysis are used to introduce the necessary corrections in the design variables until the prescribed requirements for the final product are satisfied.

Figure 1 shows a typical example of the finite element analysis of a sheet stamping operation using the explicit code STAMPACK developed at CIMNE [76]. The case studied is the forming of a kitchen sink. Figure 1a shows the finite element discretization of the different parts involved in the stamping process, i.e. the metal sheet, the punch, the die and the blank holder used to prevent lifting of the sheet while stamping. Figure 1b shows the geometry of the kitchen sink at the end of the process. The contours plotted on the sheet indicate the thickness changes in the different zones of the sink. Note that the minimum thickness occurs at the right bottom angle where fracture initiates. Onset of fracture can be detected by using the forming limit diagram (Figure 1c) where the major and minor strains in all sheets points are plotted. Strain values over the critical straight lines in this figure indicate that the material has exceeded its forming capacity and fracture appears, as shown in Figure 1d. These predictions were used to modify the die shape until a final fracture-free sink was obtained [74]. The full analysis with the serial version of STAMPACK [76] took some 32 hours in a single processor Silicon Graphic Power Challenge (R4000) workstation. This time was reduced to 13 hours using a three 486 PC network using the parallel version of the code [80].

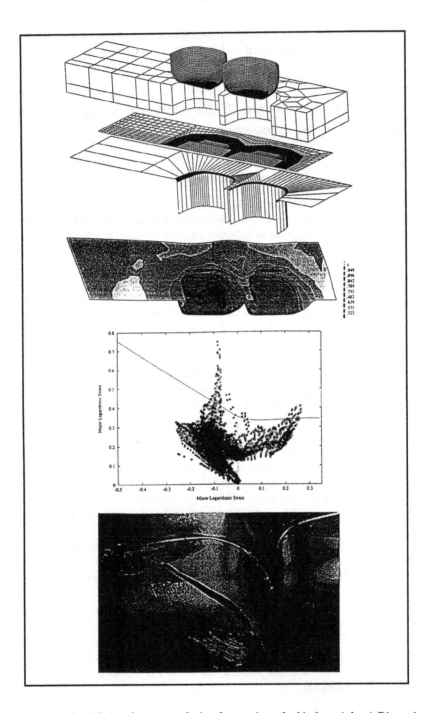

Fig. 1. Example of finite element analysis of stamping of a kitchen sink. a) Discretization of tools and sheet. b) Final deformation of sheet and thickness contours. c) Limit forming diagram indicating onset of fracture. d) Detail of fracture zone.

2.2 Steps in the finite element analysis

The solution of practical industrial forming problems by the finite element method typically involves three basic steps.

Step 1: Mesh generation. The finite element mesh must be generated using information from geometrical data. This step can consume a considerable amount of time and effort if an adequate CAD description of the analysis geometry is not available. Also access to an efficient three dimensional surface and volume unstructured mesh generator is essential for dealing with complex industrial shapes in acceptable times [6], [60-71], [77], [78]. Figure 2 shows the CAD description of a car roof and the finite element discretization of the roof surface using triangular shell elements for the purpose of performing a sheet stamping analysis. Note the finer density of elements in zones where higher curratures in the sheet geometry occur [77], [78].

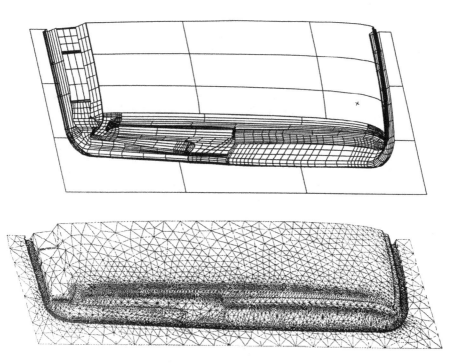

Fig. 2. From CAD to the finite element mesh. CAD description of a car roof. Discretization into shell triangles for sheet stamping analysis.

Step 2: Finite element solver. Two distinct classes of finite element solvers are found in practice. Following the terminology of Section 1, these will be named as implicit and explicit solvers. In an implicit solver the solution for the basic nodal unkowns (i.e. displacements or velocities) are obtained by solving a non linear system of equations iteratively. This implies solving for each iteration a system of algebraic equations of the form

$$^{t+\Delta t}\mathbf{H}^k \Delta \mathbf{a}^k = {}^{t+\Delta t}\mathbf{r}^k \tag{1}$$

where $\mathbf{H}$ is an adequate iteration matrix, $\mathbf{r}$ is a vector of nodal forces which vanishes once equilibrium is reached and $\Delta \mathbf{a}$ is a vector containing the increment of the basic nodal unknowns. The superindex $t + \Delta t$ denotes values at time $t + \Delta t$ whereas k is an iteration counter.

The solution of eq. (1) can be performed directly or else by using iteration techniques such as conjugate gradient, GMRES, etc. A more detailed description of the equation solution process is presented in Sections 2.5 and 3.

Accounting for all the requirements for the simulation of industrial forming processes, the analysis of a moderate size forming problem such as the stamping of a metal sheet using a quasi-static implicit code demands significant memory requirements and computer power, with solution of the linear equations system dominating the overall cost. That is the case of the so called *implicit codes* such as ABAQUS, MARC, ANSYS, etc. An interesting alternative has emerged in last years where by introducing dynamic inertia effects the transient problem can be solved using an explicit (central differences) scheme. The so called *explicit dynamic* codes, such as DYNA 3D, LS-DYNA, ABAQUS-EXP, PAMP-STAMP, ELFEN, RADIOSS, OPTRIS and STAMPACK, have distinct advantages over implicit solution based packages and this explains their popularity for solving industrial sheet stamping problems. Among the advantages of explicit codes we can mention first that the central memory requirements are one order of magnitude less than in the implicit case as the solution of a system of equations is not longer necessary. Secondly, the treatment of non-linear effects and frictional contact conditions is considerably simpler than in the implicit case.

In an explicit dynamic solver the vector of basic nodal unknowns is found directly by an expression such as

$$^{t+\Delta t}\mathbf{a} = \mathbf{M}_D^{-1}{}^t\mathbf{f} \tag{2}$$

where $\mathbf{M}_D$ is a diagonal matrix and $\mathbf{f}$ is a nodal force vector. Note that the solution of a system of equation is not required in this case. Further details are given in Section 4.

The main drawback of explicit codes is the stability constraint which typically demands time steps of 10^{-7} - 10^{-8} seconds. This usually requires 10^5 - 10^6 explicit solutions in time for the analysis of a standard sheet stamping problem. Taking into account that the average CPU time to solve a single time step for a mid-size industrial problem such as the stamping of a car roof (involving some 10.000 unknowns) is 2.6 seconds, it is immediately deduced that a single stamping analysis would easily take 20-30 hours of computing time. The need for improving the computational efficiency of these codes via parallelization is therefore obvious.

Step 3: Error estimation and mesh adaption. The accuracy of the finite element solution depends highly on the quality of the mesh used. The optimal mesh for each problem can be found by using error estimation and mesh adaption techniques. This implies that the original mesh is progressively enriched or derefined in some zones in a more or less automatic manner on the basis of adequate error parameters computed *a posteriori* [6,72]. This topic is further discussed in Section 5.

Indeed the performance of each of above steps can be enhanced by using parallel computing techniques. Figure 3 shows the flow chart of a fully parallelized finite element package where the different analysis steps are performed in parallel. Details of the insight of some of these operations are given in the next sections.

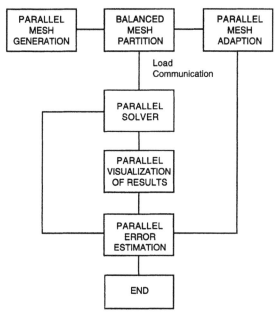

Fig. 3. Flow chart of a fully parallelized finite element package for analysis of industrial forming process.

3 Description of the quasistatic (implicit) finite element formulation

The term quasistatic stems from the fact that, although a transient problem is solved, the effect of dynamic inertia terms is neglected in the deformation process. Time can be therefore seen as equivalent to a fictitious load parameter and the equilibrium equations at any instant of the process are analogous to those governing a standard static problem. Within the quasistatic formulation two conceptually different approaches are possible: the displacement approach (also known as "solid" approach) and the flow approach.

3.1 Displacement approach

This approach uses a total or updated description of the motion. The basic variables are the displacements, **u**, of the deforming material and these are related to the strains, ε, by standard non linear kinematic expressions [6]. On the other hand, the constitutive equations relating the apropriate stress measures, σ, and the strains are usually written in a rate (incremental) objective form to allow for large strain computations. Here both elasto-plastic and elasto-viscoplastic constitutive models have been used with success for different forming problems [1-8]. Finally the equilibrium equations can be written point-wise through the adequate differential equations to be satisfied in the reference geometry configuration or (what is more usual) in a global sense through the principle of virtual work (PVW) [6]. Box 1 presents in a schematic form the basic equations of the displacement approach.

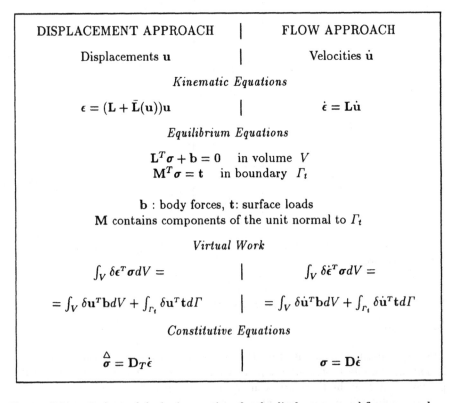

Box 1. Schematic form of the basic equations for the displacement and flow approaches

Note that in Box 1 σ and ε represent adequate conjugate stress and strain measures. The form of the PVW as written in Box 1 corresponds to the use of 2d Piola-Kirchhoff stresses and Green-Lagrange strains for σ and ε, respectively. The expressions of the PVW for other stress and strain definitions (i.e. Cauchy stresses and Almansi strains, etc.) can be found in many text books [9].

3.2 Flow approach

This approach is typical of fluid mechanics problems where a fixed Eulerian reference frame defining a control volume through which the material flows is generally used. This method appears to be more natural for bulk forming problems like rolling, extrusion, mould filling, etc. However, it can be also applied to forging and sheet forming problems in a straightforward manner, simply by identifying the control volume with the material geometry at each deforming step [1-8, 10, 11].

Box 1 shows also the basic equations of the flow approach. The main variables are now the *velocities* of the deforming body, $\dot{u}$, and these are linearly related to the rates of deformation, $\dot{\varepsilon}$, in a linear manner. The equilibrium and PVW equations are written in terms of the Cauchy stresses, σ, showing a clear analogy with the corresponding equations for the solid approach. Note, however that V and Γ_t denote now the body volume and the traction prescribed surface in the *current* deforming configuration.

The numerical solution of the flow problem can be substantially simplified if a direct relationship between Cauchy stresses and strain rates can be obtained in the form

$$\sigma = \mathbf{D}\,\dot{\varepsilon} \qquad (3)$$

Eq.(1) is typical of fluid mechanic problems where $\mathbf{D}$ in an apropriate constitutive matrix depending on the flow viscosity only [6], [10].

It can be shown that eq.(3) is readily obtained for rigid-plastic and rigid-viscoplastic materials in which elastic effects have been neglected. In the isotropic case matrix $\mathbf{D}$ is a function of a single flow viscosity parameter, μ, given for a rigid-plastic Von-Mises material by [10]

$$\mu = \frac{\sigma_y}{3\bar{\dot{\varepsilon}}} \qquad (4)$$

where σ_y is the Von-Mises yield stress and $\bar{\dot{\varepsilon}} = \left(\frac{2}{3}\dot{\varepsilon}_{ij}\dot{\varepsilon}_{ij}\right)^{1/2}$.

Remark 1. Eq.(4) defines a non linear viscosity thus implying a non-Newtonian type of flow. The expression of μ for viscoplastic materials including the effect of microscopic voids can be found in [12, 13, 19]. Also note that rigid zones are characterised by $\bar{\dot{\varepsilon}} = 0$ which leads to $\mu = \infty$. Therefore, a cut off value of μ should be used in these zones to prevent singularity.

Remark 2. The form of the constitutive equation (3) for Von-Mises metals defines an incompressible flow problem (i.e. $\dot{\varepsilon}_{ii} = 0$). This introduces serious difficulties if the finite element solution is based on "solid" elements. However, the incompressibility condition can be simply imposed in "shell type" elements by setting the Poisson's ratio equal to 0.5 and then updating the element thickness making use of the plane stress condition [4, 11, 13].

Remark 3. It is interesting to note that the overall equations of the flow approach as written in Box 1 are analogous to those of standard infinitesimal (incompressible) elasticity. This analogy can be exploited to simplify further the computational procedure by directly using standard finite element codes written for the elasticity case simply replacing displacements and strains by velocities and strain rates, respectively, and the shear modulus by the (non linear) flow viscosity [4, 6, 10, 11, 13, 19].

3.3 Finite element models

The discretization of the forming material can be made using continuum elements (in the case of bulk forming processes) or else by means of shell type elements. The later these are mainly useful for sheet stamping operations.

Continuum elements are based on the (non linear) equations for a 2D/3D deformable body. The finite element formulation is straightforward and uses standard C_o 2D or 3D solid elements for the axisymmetric and the three dimensional cases, respectively [6]. Continuum elements allow to treat two- face contact conditions and can also be used in sheet stamping processes. However, the accurate modelling of bending effects in presence of plasticity requires the use of 3-5 elements across the sheet thickness in these cases. Some drawbacks of solid elements are the difficulties for dealing with plastic incompressibility conditions in presence of large strains, the large number of variables involved and the pre and postprocessing.

Shell elements require the selection of an adequate shell formulation. The most popular shell elements are based on the so called degenerated continuum theory [6] which does not require an explicit integration across the thickness. Extensive work has been carried out in recent years to formulate new shell elements adequate for the non linear computations typical of sheet stamping problems involving thickness changes [14, 15, 74].

Shell elements can be extremely simplified if bending effects are neglected. The resulting membrane formulation can be used to model many sheet forming problems where stretching effects are dominant. Several authors have attempted to introduce bending effects in the membrane approach so that the accuracy of the bending solution is retained at a lower cost [4, 13, 16, 19].

3.4 Treatment of frictional contact

In the last years, much attention has been devoted to the numerical analysis of frictional contact problems. Despite that, contact mechanics is not so well developed as continuum mechanics and further work in this area is still needed.

Contact and friction appear as a consequence of the interaction between different bodies. Such interaction is typical of forming problems where the material is formed by means of a punch or gas pressure. During the forming process the material interacts with the tools, adding a new source of complexity to the numerical simulation due to the nonlinear nature of the boundary conditions. The numerical treatment of frictional contact problems involves two main steps. First, a contact search procedure must be performed in order to detect the penetrations (kinematic incompatibilities) between the different bodies involved in the analysis. Second, the penetrations detected must be canceled and the kinematic compatibility constraints must be satisfied.

3.4.1 Contact search

Using the slave-master terminology, it is imposed that the slave nodes do not penetrate the master surface [17].

In order to detect if penetration has taken place a search procedure for each slave node must be performed. Here different node to segment and node to patch contact search algorithms have been proposed [17]. If the search is done using the elements normals, special care must be taken in order to treat different situations arising from the discontinuity of the normal field.

An alternative has been proposed by Dalin and Oñate [18] and later modified by Agelet de Saracibar [19] in which a continuous normal field based on the normal at the nodes is used. The development of efficient parallel contact search algorithms is a challenging research topic nowadays. A possiblity is to use a parallel node to patch contact search strategy where one (or more) element patches are assigned to each slave processor. The main problems here are the need to update the candidate contact patches along the deformation process and the efficient combination of the data structure for the contact search with that of the global mesh partition scheme chosen (see Section 3.2.2).

3.4.2 Frictional contact formulations

Different formulations for the numerical analysis of frictional contact problems have been proposed. In the penalty method a penalized functional is added to the standard functional of the unconstrained problem [17,20]. Frictional effects between sliding surfaces are usually accounted using a Coulomb type law. The main drawback of this method is that the strains are exactly satisfied for infinite values of the penalty parameter only, which leads to an infinite ill-conditioning of the tangent operator. Otherwise, this is a very simple way to enforce the constraints and it is quite easy to implement.

Penalty frictional contact models can be described using a plasticity theory framework where the penalty and the frictional parameters may be viewed as constitutive parameters [17].

In the Lagrange multipliers technique a new field (the multipliers) is introduced by means of a contact functional. This leads to an increase of the number of the unknowns and of the system of equations to be solved. Furthermore the tangent operator is indefinite (zero diagonal block associated with the multipliers) and special care must be taken during the solution process. Its main advantage is that the constraints are satisfied exactly [17].

Using the perturbed Lagrange multipliers method one can bypass this draw-back as the tangent operator is definite. With this approach both the penalty and Lagrange multipliers methods can be formulated in an unified manner [17].

In the augmented Lagrangian method, traditionally used in conjunction with Uzawa's algorithm, the constraints are satisfied exactly at finite values of the penalty parameter [21,22]. This overcomes the problems associated to the choice of the penalty parameter and the ill-conditioning of the tangent operator early mentioned. However, no increase of number of the equations to be solved is pro-duced and the multipliers are simply updated after each converged equilibrium step (nested Uzawa's algorigthm) or after each equilibrium iteration (simultane-ous Uzawa's algorithm) [17], [21]. In the first case an outer loop is needed but otherwise quadratic rate of convergence can be expected if consistent tangent operators have been used. In the later, no extra loops are needed but the update of the multipliers destroys the quadratic rate of asymptotic convergence of the consistent Newton-Raphson scheme [21].

In the context of frictionless contact problems a formulation bsed on a three-field Hu-Washizu type functional has been proposed by Papadopoulus and Taylor [23]. In such a formulation contact between elements rather than between node and elements are postulated, introducing an assumed gap function that is taken as an independent variable in the formulation. A similar procedure was proposed by Wriggers using a two-field functional [17].

3.5   Computational aspects

For both solid and flow approaches the resulting non linear equilibrium equations can be written after finite element discretization in the form (see Box 2)

$$\mathbf{r}(\mathbf{a}, \mathbf{x}, t) = \mathbf{p}(\mathbf{a}, \mathbf{x}) - \mathbf{f}(t, \mathbf{x}) = 0 \tag{5}$$

where $\mathbf{r}$, $\mathbf{p}$ and $\mathbf{f}$ stand for the vectors of residual forces, internal forces and external forces, respectively, $\mathbf{a}$ are displacements and velocities in the displace-ment and flow approaches respectively, $\mathbf{x}$ is the cartesian coordinate vector and, t represents the time in the flow approach and the load increment in the solid approach.

Equation (5) can be iteratively solved for the values of vector $\mathbf{a}$ at $t + \Delta t$. For the kth iteration we have

$$\Delta \mathbf{a}^k = - \left[{}^{t+\Delta t}\mathbf{H}^k \right]^{-1} {}^{t+\Delta t}\mathbf{r}^k \tag{6}$$

where $\mathbf{H}$ in an adequate iteration matrix. Vector $\mathbf{a}$ is subsequently updated as

$$ {}^{t+\Delta t}\mathbf{a}^{k+1} = {}^{t+\Delta t}\mathbf{a}^k + \Delta \mathbf{a}^k \tag{7}$$

The next step is to compute the new stress field. In the rigid/plastic-viscoplastic flow approach the stresses are *directly* obtained from the updated velocity field (see Box 2). In the displacement approach the computation of stresses requires the integration of the rate constitutive equations.

DISPLACEMENT APPROACH	FLOW APPROACH

Basic Discretization

$$\mathbf{u} = \mathbf{N}\mathbf{a} \qquad\qquad \dot{\mathbf{u}} = \mathbf{N}\mathbf{a}$$

$$\epsilon = \mathbf{B}_s\mathbf{a} \qquad\qquad \dot{\epsilon} = \mathbf{B}_f\mathbf{a}$$

$$\overset{\triangle}{\sigma} = \mathbf{D}_T\dot{\epsilon} = \mathbf{D}_T\bar{\mathbf{B}}_s\mathbf{a} \qquad\qquad \sigma = \mathbf{D}\dot{\epsilon} = \mathbf{D}\mathbf{B}_f\mathbf{a}$$

Loop over load cases or time increments
LOOP   ILOAD = 1,NLOAD
Equilibrium Equation

$$\mathbf{r} = \mathbf{p} - \mathbf{f} = 0$$

$$\mathbf{p} = \int_V \bar{\mathbf{B}}_s^T\sigma dV \qquad\qquad \mathbf{p} = \left(\int_V \mathbf{B}_f^T\mathbf{D}\mathbf{B}_f dV\right)\mathbf{a} = \mathbf{K}\mathbf{a}$$

$$\mathbf{f} = \int_V \mathbf{N}^T\mathbf{b}dV + \int_s \mathbf{N}^T\mathbf{t}ds$$

Solution Procedure

$$^{t+\Delta t}\mathbf{a}^0 = 0 \qquad\qquad ^{t+\Delta t}\mathbf{a}^0 = {}^t\mathbf{a}$$

$$^{t+\Delta t}\mathbf{x}^0 = {}^t\mathbf{x} \qquad\qquad ^{t+\Delta t}\mathbf{x}^0 = {}^t\mathbf{x} + {}^t\mathbf{a}\Delta t$$

LOOP   k=1,NITER

$$\Delta\mathbf{a}^k = -[^{t+\Delta t}\mathbf{H}^k]^{-1}(^{t+\Delta t}\mathbf{r}^k)$$

$$^{t+\Delta t}\mathbf{a}^{k+1} = {}^{t+\Delta t}\mathbf{a}^k + \Delta\mathbf{a}^k$$

$^{t+\Delta t}\sigma^k$ are computed by integrating the constitutive equations $\qquad ^{t+\Delta t}\sigma^{k+1} = \mathbf{D}\mathbf{B}_f{}^{t+\Delta t}\mathbf{a}^{k+1}$

Geometry Updating

$$^{t+\Delta t}\mathbf{x}^{k+1} = {}^{t+\Delta t}\mathbf{x}^k + \Delta\mathbf{a}^k \qquad ^{t+\Delta t}\mathbf{x}^{k+1} = {}^t\mathbf{x} + {}^t\mathbf{a}(1-\theta)\Delta t +$$

$$+{}^{t+\Delta t}\mathbf{a}^{k+1}\theta\Delta t$$

$$\theta = 0 \text{ explicit solution}$$
$$\theta \neq 0 \text{ implicit solution}$$

CHECK CONTACT

Compute $^{t+\Delta t}\mathbf{r}^{k+1}$ and $^{t+\Delta t}\mathbf{f}^{k+1}$

Compute error norm $E = \frac{\|^{t+\Delta t}\mathbf{r}^{k+1}\|}{\|^{t+\Delta t}\mathbf{f}^{k+1}\|}$

$E.GT.TOL. \overset{yes}{\to} \quad$ continue iterations
$k = k + 1$

$\downarrow$ no

INITIATE NEW LOAD CASE OR TIME INCREMENT

Box 2. Quasi-static finite element solution algorithm for displacement and flow approaches

The final step is to update the material geometry. This can be simply done from the displacement and velocity fields as shown in Box 2. At this stage the mechanical properties are also updated as well as some specific geometrical parameters, like the thickness in sheet stamping problems, and the contact and friction conditions are checked.

The updated geometrical and mechanical parameters of the deformed material are used to compute the new increments Δa^k of the next iteration from eq.(6). The process is continued until convergence is achieved. This is usually measured by the satisfaction of eq. (5) using a mean quadratic norm for the residual forces. A summary of the solution process for both the displacement and flow approaches is shown in Box 2.

The detailed discussion of the computational aspects of the finite element solution falls outside the scope of this paper. Details of the different matrices and vectors appearing in Box 2 can be found in [6]. Nevertheless the following remarks should be noted at this point.

a) The iteration matrix $\mathbf{H}$ in the solid approach is usually taken as the tangent stiffness matrix computed as $\mathbf{H} = \frac{\partial \mathbf{p}}{\partial \mathbf{a}}$. The solution algorithm coincides in this case with the standard Newton-Raphson iteration scheme [6].

b) For the flow approach the selection of matrix $\mathbf{H}$ can be made on the basis of *secant* or *tangent* iteration procedures. The expression of $\mathbf{H}$ for the secant case coincides with that of the finite element stiffness matrix $\mathbf{K}$ for standard infinitesimal elasticity [3]. The form of the exact tangent matrix is complex and generally non-symmetric, and approximate simpler forms are used [19].

c) The use of an explicit geometry updating scheme ($\theta = 0$) in the flow approach results in an iterative algorithm in which the geometry at $t + \Delta t$ is known "a priori" and kept fixed during the iterations and the velocities at $t + \Delta t$ are the only possible unknowns. However, in the implicit case ($\theta \neq 0$) the material coordinates at $t + \Delta t$ change during the iteration process. This would allow to formulate the problem in terms of the velocities at $t + \Delta t$, or in terms of the displacement increments between the two configurations at t and $t + \Delta t$, by noting that the geometry updating equation can be written in the form

$$\Delta \mathbf{u} = ({}^t\mathbf{a} + \Delta \mathbf{a}\theta)\Delta t \tag{8}$$

where

$$\Delta \mathbf{u} = {}^{t+\Delta t}\mathbf{x}^{k+1} - {}^t\mathbf{x}$$
$$\Delta \mathbf{a} = {}^{t+\Delta t}\mathbf{a}^{k+1} - {}^t\mathbf{a} \tag{9}$$

are the displacement and velocity increment vectors, respectively. This alternative has not been fully exploited in practice and it opens new possibilities for research.

d) The terms "explicit" and "implicit" in Box 2 refer to the integration scheme chosen for updating the geometry in the flow approach. This should not be mixed up with the so called "explicit dynamic" methods based on the solution of the full dynamic equations with inclusion of inertia terms using an explicit backwards integration scheme. This possibility is discussed later in Section 4.

Due to the high non linearity of the problem the number of loading or time increments NLOAD often has to be relatively large in order to guarantee convergence. Also the number of iterations within each loading or time increment (NITER) may be quite large in some cases and it can of course vary from one increment to another. Values of NLOAD = 30 and NITER = 5 are common, thus leading to 150 solutions for a single problem. This implies that even for relatively small-sized finite element meshes the computational cost of the solution may be important. It is therefore obvious that the term large scale computation does not necessarily refer in this context to the actual size of the finite element mesh.

A typical time distribution of a quasistatic FE solution in a serial machine is as follows. For small to medium size problems (up to 2000-3000 degrees of freedom) the three main contributions come to equal parts from the computation of the iteration matrix H and the external force vector $\mathbf{f}$ including the evaluation of frictional contact forces, and from the resolution of the linearized problem. The other contributions (evaluation of strains, stresses and the internal force vector $\mathbf{p}$) are relatively less costly (10%). However, when the size of the problem increases (over 5000 degrees of freedom) the solver becomes the dominant part of the solution reaching over 90 % of the total time.

3.6   Direct versus iterative methods

The use of direct methods such as Gauss-Jordan technique, for solving the linear system of equations $\mathbf{H}\Delta\mathbf{a} = -\mathbf{r}$ for each iteration is not the most suitable for parallelization purposes as it requires the assembly of the iteration matrix $\mathbf{H}$ and the residual force vector $\mathbf{r}$ prior to the solution process. This involves a large amount of global communications among processes and the overall parallel efficiency is drastically reduced.

For this purpose iterative solvers such as conjugate gradient (CG) or generalized minimal residual (GMRES) are more suitable as the assembly of the global iteration matrix is never required. CG solvers are applicable when the iteration matrix $\mathbf{H}$ is symmetric. Unfortunately, in many practical forming processes the symmetry is lost due to the appearance of non symmetric terms contributed by the frictional forces or by non-associated plasticity/viscoplasticity rules [6]. In these cases the GMRES algorithm should be used. Note that the GMRES method coincides with the GC technique for symmetric systems. Therefore it is concluded that the GMRES algorithm has the desired vesatility to be applicable in all situations appearing in the analysis of practical industrial forming processes.

The main operations involved in the GMRES method are vector operations, scalar products, matrix vector products and a minimization using a QR decomposition. It can be verified that the cost of the matrix vector products takes over 90 % of the total cost in serial computations [32] and, therefore, it is the crucial operation in the implementation of this iterative solution method in a parallel computer.

4. On the parallelization of quasistatic (implicit) finite element codes for analysis of industrial forming processes

The possibilities of parallelizing a quasistatic finite element code for analysis of forming processes in fine grain SIMD computer such as the CM 200 and in heterogeneous distributed environments using message passing PVM and MPI tools are described next.

4.1 Implementation of implicit finite element codes on a fine grain SIMD parallel computer

Finite element codes are well suited for implentation in fine grain SIMD computers such as the CM 200. Despite the current low popularity of this parallel architecture, the main features of the parallelization process are given below for reference purposes as this architecture has been found well suited for large scale finite element computations.

4.1.1 Data structures

A key issue in the implementation of a finite element is the data structure. Clearly most of the success of the implementation lays in the efficient exploitation of the capacity of the machine to perform parallel computations with a minimum of communication among processors. In an implicit finite element code most of the operations performed at a single element level are repeated for all the elements in the mesh. Also the exchange of information is performed among neighbouring elements only. For this reason a data structure at elemental level emerges here as the natural approach [24-31].

This type of structure associates each element in the mesh to a single physical (or virtual) processor. Note that the term virtual processor in the CM refers to its capacity of dividing each physical processor in as many (2^n) virtual processors as necessary for enlarging the capacity of storing a parallel variable. All the required element data are stored in the local memory of the corresponding processor. In this way no communication is needed for operations performed at element level.

Unfortunately, the element data structure is not suitable for the optimal parallelization of the matrix vector operations appearing in the GMRES solver. Here the information contained in the vectors refers to the nodes only. An element - data storage would require to store the information for a particular node as many times as elements share this node. The optimal solution to this problem is a mixed element - node data structure which contains an element data structure for storing the element stiffness matrix terms and a node data structure for storing the information of the vectors containing nodal information. This type of data structure requires using two communication routines: one for transferring the information from the degrees of freedom (nodes) to the elements (gather) and another one for the inverse (scatter) operations (element information to nodes) for assembly purposes. Further details can be found in [32], [37], [38].

4.1.2 Communication systems

The use of non structured meshes is very popular in the finite element analysis of complex geometries such as these occuring in forging or sheet stamping processes. Non structured meshes are also essential for the successful implementation of general adaptive refinement procedures [6]. These meshes introduce a certain degree of randomness in the exchange of messages among processors. This invariably imposes the need of using the communication systems known as ROUTER [33], [34].

In case of structured meshes a more efficient communication system called NEWS can be used [34]. This is based on mapping the structured distribution of elements into a regular grid of processors. Belytschko and Plaskacz [28, 35] have extended this communication system to partially structured meshes. Each subdomain containing an structured distribution of elements is treated with the NEWS system, whereas the interfaces between the subdomains are treated with the ROUTER system [32].

Within the ROUTER system the CMSSL library of the CM offers two sets of communication routines for performing gather and scatter operations known as CPS (Communication Primitive Set) and CC (Communication Compiler) [32], [33]. In references [32] and [35] a study of all these communication routines for the analysis of a range of finite element problems of different sizes was performed. The conclusion was that for a large number of scattering operations (in meshes of over 10.000 elements) the so called Fast Graph method of the CC routine showed better performance. Similarly, for gathering operations in large meshes the NOP method (also of the CC routine) was the best option.

References [32], [38-42] show comparative results of the finite element analysis of powder compaction processes using serial Silicon Graphics Indigo workstations available at CIMNE and the CM2 with 4096 processors provided by the European Center of Parallelism of Barcelona (CEPBA). Speeds up to 15 where obtained for relatively medium size problems (up to 15.000 degrees of freedom). It was noted in [42] that the efficiency of the CM increases very rapidly with the problem size and, therefore, higher speed up can be expected for larger scale problems of this kind.

4.2 Implementation of implicit finite element codes on coarse grain parallel computer environments

The industrial forming sector is mostly formed by SMEs basically equiped with PC's and (in some exceptional cases) with high-end workstations. This situation makes it mandatory to search for an efficient parallel implementation of existing forming finite element codes in heterogeneous distributed computer environments combining clusters of PC's and workstations using PVM and MPI tools. The obvious advantage of such a challenging development is that the resulting parallelized codes can also automatically run in a quasi-efficient manner in more sophisticated message passing and shared memory multiprocessor computers such as the Silicon Graphics Power Challenge Series, the IBM SP-2 and the Convex C-4 among others.

Most of the recently proposed computational methods for solving partial differential equations with the finite element method on multiprocessor architectures stem from the divide and conquer paradigm and involve some form of domain decomposition (DD) [43]. Generally these methods are motivated by the fact that DD provides a natural route to parallelism. Some of the proposed methodologies correspond to a divide and conquer implementation of existing serial algorithms and, therefore, exploit mainly a spacial type of parallelisms. Others are genuine DD methods, that is, methods which can be interesting even as serial algorithms in some situations [44]. In all cases, these methods require a careful partitioning of the underlying computational mesh into a specified number of submeshes, generally equal to the number of available processors. A mapping procedure is then invoked to assign a submesh to a processor. If a regular mesh decomposition is possible, then defining an appropriate mesh partition may be a trivial task. However, for complex domains and irregular (unstructured) meshes this process is more challenging [45, 46].

It is, in general, well accepted that a mesh partition scheme should meet at least three basic requirements [45]: (1) it must handle irregular geometrics and arbitrary discretization in order to be general purpose; (2) it should create subdomains which allow the overall computational cost to be as evenly distributed as possible among the processors, and (3) it must minimize the amount of interface nodes in order to minimize intersubdomain communications and/or synchronisation overhead. Numerical experience shows that several other issues in addition to load balancing and communication costs need to be addressed. These are usually related to the particular case of computational method chosen as discussed in [45]. Further discussion on this topic is presented in Section 4.2.2.

4.2.1 The Schur complement method

The Schur complement method is the most popular technique for solving a linear system of equations using subdomain techniques. The method will be applied here to the solution of each of the linear equation systems $H\Delta a = -r$ resulting from the incremental iterative solution of an industrial forming problem using an implicit finite element formulation (see Box 2). For simplicity, let us rename the system of equations to be solved as

$$\mathbf{K}\mathbf{a} = \mathbf{f} \qquad (10)$$

where $\mathbf{K}$ is the iteration matrix, $\mathbf{f}$ is (minus) the residual force vector $\mathbf{r}$ and $\mathbf{a}$ is the vector containing the unknown nodal variables.

The first step in Schur's method is the partition of the finite element mesh into N submeshes (subdomains) that can be treated independently, each one possibly being assigned to a single processor. Let us assume that the numbering of the degrees of freedom in the mesh leads to the following pattern for the system of equations (10)

$$\begin{bmatrix} \mathbf{K}_{11} & 0 & \cdots & 0 & \cdots & 0 & \mathbf{K}_{1I} \\ 0 & \mathbf{K}_{22} & \cdots & 0 & \cdots & 0 & \mathbf{K}_{2I} \\ \vdots & \vdots & \ddots & \vdots & \cdots & \vdots & \vdots \\ 0 & 0 & \cdots & \mathbf{K}_{ii} & \cdots & 0 & \mathbf{K}_{iI} \\ \vdots & \vdots & \cdots & \vdots & \ddots & \vdots & \vdots \\ 0 & 0 & \cdots & 0 & \cdots & \mathbf{K}_{NN} & \mathbf{K}_{NI} \\ \mathbf{K}_{I1} & \mathbf{K}_{I2} & \cdots & \mathbf{K}_{Ii} & \cdots & \mathbf{K}_{IN} & \mathbf{K}_{II} \end{bmatrix} \begin{Bmatrix} \mathbf{a}_1 \\ \mathbf{a}_2 \\ \vdots \\ \mathbf{a}_i \\ \vdots \\ \mathbf{a}_N \\ \mathbf{a}_I \end{Bmatrix} = \begin{Bmatrix} \mathbf{f}_1 \\ \mathbf{f}_2 \\ \vdots \\ \mathbf{f}_i \\ \vdots \\ \mathbf{f}_N \\ \mathbf{f}_I \end{Bmatrix} \qquad (11)$$

where index I denotes the interfaces between domains, $\mathbf{a}_I$ and $\mathbf{f}_I$ are the unknown variables and the nodal forces at the subdomain interfaces and $\mathbf{a}_i$, $\mathbf{f}_i$ $i = 1, N$ are the unknown variables and the nodal forces at the interior of each subdomains. The different matrices $\mathbf{K}_{ij}$ in eq.(11) are formed by assembling the stiffness matrices $\mathbf{K}^{(s)}$ contributed by each subdomain s.

The next step is eliminating the variables at the subdomain interiors $\mathbf{a}_i$ by solving

$$\mathbf{K}_{jj}\mathbf{a}_j = \mathbf{f}_j - \mathbf{K}_{jI}\mathbf{a}_I , \quad j = 1, N \qquad (12)$$

Substituting the value of $\mathbf{a}_j$ obtained from (12) into each row of eq. (11) gives the final equation system for the interface problem

$$\left(\mathbf{K}_{II} - \sum_{j=1}^{N} \mathbf{K}_{Ij}\mathbf{K}_{jj}^{-1}\mathbf{K}_{jI} \right) \mathbf{a}_I = \mathbf{f}_I - \sum_{j=1}^{N} \mathbf{K}_{Ij}\mathbf{K}_{jj}^{-1}\mathbf{f}_j \qquad (13)$$

Eq. (13) can be rewritten in compact form as

$$\bar{\mathbf{K}}_I\mathbf{a}_I = \bar{\mathbf{f}}_I \qquad (14)$$

where

$$\bar{\mathbf{f}}_I = \mathbf{f}_I - \sum_{j=1}^{N} \mathbf{K}_{Ij}\mathbf{K}_{jj}^{-1}\mathbf{f}_j \qquad (15)$$

$$\bar{\mathbf{K}}_I = \mathbf{K}_{II} - \sum_{j=1}^{N} \mathbf{K}_{Ij}\mathbf{K}_{jj}^{-1}\mathbf{K}_{jI} \qquad (16)$$

Matrix $\bar{\mathbf{K}}_I$ is called the *Schur complement*.

In practice, each subdomain matrix $\mathbf{K}^{(s)}$ and the corresponding vectors $\mathbf{a}^{(s)}$ and $\mathbf{f}^{(s)}$ are individually stored in the following form

$$\mathbf{K}^{(s)} = \begin{bmatrix} \mathbf{K}_{s,s} & \mathbf{K}_{I_s} \\ \mathbf{K}_{I_s,s} & \mathbf{K}_{I_s,I_s} \end{bmatrix} ; \quad \mathbf{a}^{(s)} = \begin{Bmatrix} \mathbf{a}_s \\ \mathbf{a}_{I_s} \end{Bmatrix} ; \quad \mathbf{f}^{(s)} = \begin{Bmatrix} \mathbf{f}_s \\ \mathbf{f}_{I_s} \end{Bmatrix} \qquad (17)$$

where I_s is the projection of the complete interface I onto the subdomain s. Note that $\mathbf{K}^{(s)}$ will generally be a banded matrix.

With this notation, the interface problem (13) can be reformulated as

$$\left[\sum_{s=1}^{N}(\mathbf{K}_{I_s I_s} - \mathbf{K}_{I_s s}\mathbf{K}_{ss}^{-1}\mathbf{K}_{sI_s})\right]\mathbf{a}_I = \sum_{s=1}^{N}(\mathbf{f}_{I_s} - \mathbf{K}_{I_s s}\mathbf{K}_{ss}^{-1}\mathbf{f}_s) \tag{18}$$

where $\sum$ denotes the assembly operator.

The Schur complement $\bar{\mathbf{K}}_I$ and the interface force vector $\bar{\mathbf{f}}_I$ are defined now as

$$\bar{\mathbf{K}}_I = \sum_{s=1}^{N}\mathbf{K}_I^{(s)} \quad ; \quad \bar{\mathbf{f}}_I = \sum_{s=1}^{N}\mathbf{f}_I^{(s)} \tag{19}$$

with

$$\mathbf{K}_I^{(s)} = \mathbf{K}_{I_s I_s} - \mathbf{K}_{I_s s}\mathbf{K}_{ss}^{-1}\mathbf{K}_{sI_s} \tag{20}$$

$$\mathbf{f}_I^{(s)} = \mathbf{f}_{I_s} - \mathbf{K}_{I_s s}\mathbf{K}_{ss}^{-1}\mathbf{f}_s \tag{21}$$

The solution process goes as follows: First the interface variables $\mathbf{a}_I$ are obtained by solving the linear equation system (14) with $\bar{\mathbf{K}}_I$ and $\bar{\mathbf{f}}_I$ defined by eq.(19). Then the unknown variables at the subdomain interiors $\mathbf{a}_s$ are obtained by solving the following linear systems

$$\mathbf{K}_{ss}\mathbf{a}_s = \mathbf{f}_s - \mathbf{K}_{sI_s}\mathbf{a}_{I_s} , \quad s = 1, N \tag{22}$$

Note that the size of systems (22) is smaller than that of the original system (12).

Four classes of computational strategies can be distinguished for the practical implementation of the Schur complement method [45]. They are labelled as iterative/iterative, iterative/direct, direct/iterative and direct/direct where the first qualifier relates to the subdomain computations (eq.(22)) and the second qualifier relates to the interface ones (eq.(14)). A solution scheme that requires any factorization process is labelled here as "direct". A scheme that involves only vector, matrix and matrix-vector sums or products (i.e. iterative methods such as Jacobi, CG and GMRES) is labelled as "iterative".

For instance, an iterative/iterative strategy will use iterative methods for solving for both the interface and the subdomain variables. Conversely, in a direct/direct strategy both the interface and the subdomain variables are solved with a factorization solution scheme. In all cases the inverses of the $\mathbf{K}_{ss}$ matrices are needed. For this purpose a **LU** factorization is performed which helps to considerable savings in memory. A more detailed discussion of the four computational strategies mentioned can be found in [45]. Experience shows that the most frequent computations in DD methods follow a direct/iterative strategy [45, 47, 48], i.e. the subdomain problems are solved using a factorization scheme such as the LU technique and the interface problem is solved utilizing some variant of the preconditioned CG or GMRES algorithms. In this case the necessary matrix-vector products appearing in the solution of the interface problem are carried out in the following three steps:

- The part of the global vector **a** corresponding to the interfaces, $\mathbf{a}_I$, is *gathered* in subvectors $\mathbf{a}_{I_s}$ containing the variables corresponding to the interfaces of the different subdomains I_s
- The products $\bar{\mathbf{K}}_I^{(s)} \mathbf{a}_{I_s} (= \mathbf{q}_{I_s})$ necessary for the CG and GMRES schemes are computed in parallel in each subdomain. For this purpose, the LU factorization of the $\mathbf{K}_{ss}^{-1}$ matrices is used.
- The resulting subvectors $\mathbf{q}_{I_s}$ are *scattered* into an assembled vector $\mathbf{q}_I$.

In a standard master - slave parallel environment the master program will be assigned the task of solving the interface problem (using either a direct or an iterative strategy) and will assemble the necessary variables. Each slave processor will be responsible for managing (at least) one subdomain and will carry out all the computations at subdomain level including the evaluation of the internal variables via eq. (22).

Figure 4 shows the organization of computations and the necessary messages in a master - slave architecture using PVM tools. Recall that the computation of the products $\bar{\mathbf{K}}_I \mathbf{a}_I$ in the slave processors is only necessary if the interface problem is solved using an iterative scheme.

An evaluation of the communication costs for the implementation of the computational scheme shown in Figure 4 in a PC network using the PVM message passing library can be found in [49]. The test problem was the analysis of the compaction of a metal powder prismatic part. Maximum theoretical speeds up of 2.66 in a 3 PC network where obtained for a mesh with 10.082 degrees of freedom when the interface problem was solved with an iterative scheme. The speed up rose to 2.96 for the same mesh when a direct solver was used for the interface problem. However, the total CPU time of the solution was greater (almost double) in this case.

4.2.2 About mesh partitioning techiques

Given a mesh and a number of processors one would like to automatically partition the mesh into a number of submeshes (subdomains) such that: (a) the subproblems associated to each submesh have about the same complexity, and (b) the amount of communication between the processors assigned to the submeshes (ideally one processor is assigned to each submesh) is minimized.

As quoted by Fahrat and Simon [50] "the various forms of mesh partitioning algorithms that try to achieve load balancing and minimum communication are as different as the number of researchers working on the problem". In general, the recently published mesh decomposition algorithms can be grouped into three categories: engineering, optimization and graph theory based heuristics [51-59]. Among these we can list the Greedy algorithm, the Reverse Cuthill-McNee (RCM) algorithm, the Recursive RCM algorithm, the Principal Inertia algorithm, the Recursive Principal Inertia algorithm, the Recursive Graph Bisection algorithm, the 1D Topological Frontal algorithm and the Recursive Spectral Bisection algorithm. Details of these algorithms can be found in [51-59]. A comprehensive and concise description can be found in [50].

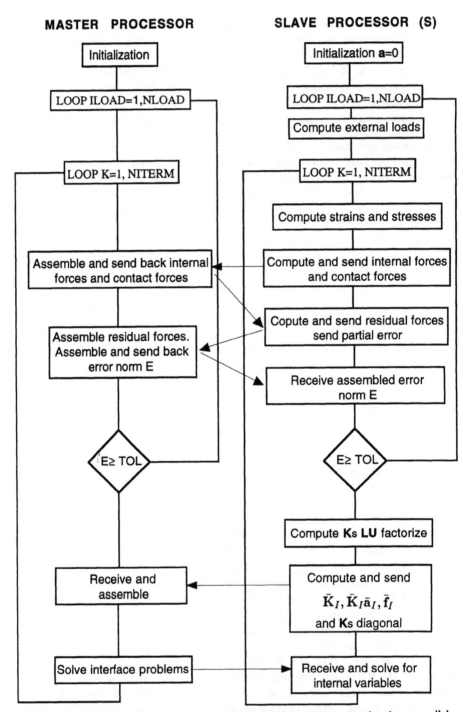

Fig. 4. Organization of a non linear implicit finite element computation in a parallel master - slave architecture. All slave processors will perform identically.

Deciding which specific algorithms or category of algorithms is best is as difficult as defining what exactly constitutes an "efficient" partition. The answers to both questions are mesh, problem and machine dependent. Generally the analyst should decide himself which partitioning algorithm is best for his/her problem. Some of the partitioning atributes that can be evaluated for a specific problem and a particular parallel arquitecture are:

Interface size. The total number of nodes or edges on the subdomain interfaces is today the most popular criteria for assessing a mesh partition. Because of interprocessor cost considerations, it is believed that the mesh partition that has the smallest interface size is the best one, even though this is not necessarily true [50], [57].

Memory requirements. On a distributed memory parallel processor environment, memory limitations can be imposed either by the maximum memory requirements of a given subdomain or by the total memory requirements of all subdomains.

Other important issues for accepting or rejecting a mesh partition are the *load balance* factors, the *communication pattern and network traffic* and the *communication costs*. For more details see [50].

5 Parallelization of explicit dynamic finite element codes for analysis of industrial forming problems

As mentioned earlier explicit dynamic methods have recently become very popular in the context of the displacement (solid) approach, as they do not require the solution of a system of equations. The basis of the method is the solution of the dynamic equilibrium equations at time t using an explicit time integration scheme with a diagonal mass matrix (see Box 3). The advantages of this approach is that the iteration matrix does not need to be formed for finding the solution of the kinematic variables (displacements, velocities and accelerations) and the strain and stress fields at each time step in the transient dynamic integration process. This naturally leads to considerable savings in memory and allows to solve large industrial forming problems in relatively small memory PC's and workstations. In addition contact conditions are easily modelled due to the small time steps required for stability reasons [72, 73, 74].

Box 3 Solution steps in an explicit dynamic finite element code (displacement approach)

Looking at Box 3 we can note the most expensive operations performed in the explicit dynamic solution. These are:

1) Computation and assembly of the internal nodal force vector **p** (70-75 % of total CPU time). This requires:

   - Computation of the element stresses. This typically involves matrix-vector multiplications at element level.

EXPLICIT DYNAMIC ALGORITHM

Discretized dynamic equilibrium equation

$$\mathbf{M\ddot{a}} + \mathbf{p(a)} = \mathbf{f}$$

$\mathbf{M}$: mass matrix

$\mathbf{p}(e)$: internal force vector $= \displaystyle\int_v \mathbf{B}^T \boldsymbol{\sigma}\, d\mathrm{V}$

$\mathbf{f}$: external force vector

Explicit time integration

$n + 1$ time step

Data: $\mathbf{a}^n, \mathbf{a}^{n-1/2}, \varepsilon^n, \sigma^n, \mathbf{f}^n, \mathbf{p}^n$

1) Compute kinematic variables

Accelerations: $\qquad \ddot{\mathbf{a}}^n = \mathbf{M}_D^{-1}[\mathbf{f} - \mathbf{p(a)}]^n$

Velocities: $\qquad \dot{\mathbf{a}}^{n+1/2} = \dot{\mathbf{a}}^{n-1/2} + \ddot{\mathbf{a}}^n \Delta t^{n+1/2}$

Displacements: $\qquad \mathbf{a}^{n+1} = \mathbf{a}^n + \dot{\mathbf{a}}^{n+1/2}\Delta t^{n+1}$

$$\Delta t^{n+1/2} = \tfrac{1}{2}[\Delta t^n + \Delta t^{n+1}]$$

$$\mathbf{M}_D = \text{diag. } \mathbf{M}$$

2) Compute strains and stresses $\varepsilon^{n+1}, \sigma^{n+1}$
3) Compute internal force vector

$$\mathbf{p}^{n+1} = \int_v \mathbf{B}^T \sigma^{n+1}\, d\mathrm{V}$$

4) Check frictional contact conditions
5) Compute external force vector $\mathbf{f}^{n+1}$
6) Go back to 1) and repeat the process for the next time step

Box 3. Solution steps in an explicit dynamic finite element code (displacement approach).

- Matrix-vector multiplications and numerical integration at element level to obtain the internal force vector.
- Assembly of the element contributions $\mathbf{p}^{(e)}$ to form the global internal force vector $\mathbf{p}$.

2) Computation of the "external" force vector $\mathbf{f}$ (15-20 % of total CPU time). This requires:

- Computation of the frictional contact forces between the tools and the deforming elements. This typically involves expensive search algorithms to identify the elements in contact. A number of fast search strategies based on node to element patch techniques can be implemented to speed up the contact search process.
- Assembly of the elementary contact forces $\mathbf{f}^{(e)}$ in the global force vector $\mathbf{f}$.

3) Computation of the kinematic variables (displacements, velocities and accelerations (5-10 % of total CPU time). This is a scalable operation involving a sum of vectors and a simple product of a vector $(\mathbf{f}\text{-}\mathbf{p})$ by a diagonal matrix $\mathbf{M}_D$.

It is clear that parallelization can substantially alliviate the cost associated to above operations. In particualar:

a) The implementation of a parallel domain decomposition structure will scale down the cost of computation of the kinematic variables. The need for data transfer across domain boundaries in explicit dynamic computations is minimum and includes only the percentage of mass and the forces assigned to each boundary node. This allows to use any of the well known domain decomposition techniques available in the literature.

b) Contact computations can also be speeded up using parallel contact search algorithms based on enhanced mesh partition techniques and selective node to segment and node to element patch contact search strategies. Much research in this field is in progress [17].

c) The computation of the element internal force vector and the correponding global assembly are typical gather-scatter operations which can be effectively parallelized and also involve a minimum of inter- processor communications.

The drawbacks of the explicit dynamic solution are the small time steps (10^{-6} - 10^{-8} seconds) required for stability reasons, the difficulties for predicting local instabilities such as wrinkling in sheet stamping operations (these instabilities are typically detected by the singularity of the consistent tangent matrix which is not required in this case). Also, the computation of elastic recovery effects (spring back) in some forming problems, like forging or sheet stamping, requires the addition of "ad-hoc" damping terms and the relaxation process to recover the unloaded stage is often too costly for practical purposes. An increasingly popular was out to this problem is to couple the explicit dynamic code with an implicit code for the purpose of predicting local instabilities and elastic spring back effects only.

6 About parallel mesh generation, parallel error estimation and mesh adaption

This review would not be complete without mentioning, although briefly, some of the current trends in such important topics as the generation of unstructured meshes, the estimation of the solution error and the techniques for mesh enhancement in a parallel computing enviroment.

6.1 Parallel mesh generation

Indeed, mesh generation itself can be the bottleneck in large scale finite element computations involving complex geometries [60, 78]. The main difficulties here lay in the efficient interfacing between the geometrical CAD data and the mesh generator algorithm chosen. Very typically the CAD data is inconsistent and need some "repairs" which can take considerable amount of time and effort [77].

Assuming that all the necessary data is available, the speed of existing codes for efficiently generating a mesh of tetrahedral elements in a serial computer varies from 2500 - 3000 tetrahedra per minute. The CPU time to generate a 3D mesh for a complex forging problem requiring some 200 - 300.000 tetrahedra would therefore rise to 2-3 hours. The gains to be achieved by efficient parallel mesh generations are obvious.

Most unstructured mesh generators are based in Advancing Front Techniques [62-66] and the so-called generalized Voronoi algorithms [68–71, 78]. A typical sequence of operations for parallel mesh generations using the Advancing Front method is the following [60],[61]:

Step 1. Subdivide the global domain into subdomains using the background grid.

Step 2. Generate and unstructured mesh for each subdomain separately.

Step 3. Grid up consistently the intersubdomain region.

Step 4. Reassemble result.

A schematic example of this procedure is shown in Figure 5.

Clearly the mesh generations process can be combined with the partition strategy chosen for the subsequent application of the DD solution method. Therefore the outcome of the parallel mesh generation process should ideally be the required optimum mesh partition sought. Much research work is in progress in this direcction and the interested reader can consult references [8] and [60, 61].

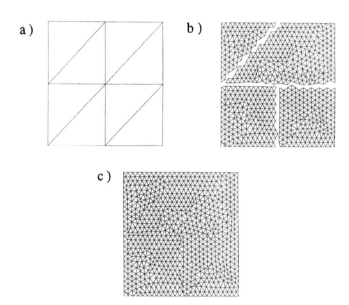

Fig. 5. Example of parallel mesh generation [60]. a) background grid. b) gridded sub-domains. c) final assembled mesh.

6.2 Parallel error estimation and mesh enrichment

Adaptive mesh enhancement strategies are becoming more and more popular for the accurate solution of practical industrial forming problems. The overall process follows the scheme shown in Figure 3.

The numerical results provided by the finite element code are used to compute "a posteriori" global or local estimation of the solution error. These error measures allow to predict the new element sizes by using a certain "mesh" optimality criteria [72], so that a new enriched mesh can be appropriately defined. The process is repeated until a certain norm of the global or local error values is satisfied [6], [72], [78].

The evaluation of the solution error is a typical parallelizable operation. The global error norm is obtained by adding up the contributions from the different individual elements. These contributions can be readily computed in parallel in the different slave processors.

The mesh adaption process requires information on the new element sizes. These are computed using information provided by the actual global error norm, the prescribed global error tolerance and the individual error contributed by each element [72]. This is, therefore, an element by element operation which can be easily distributed among the different slave processors once the global error norm has been obtained.

The new adapted mesh can be obtained as an enrichment of the actual mesh (i.e. some elements are splitted into smaller elements and others are merged to build up a larger element) or else it can be completely regenerated using the new element sizes [6], [78]. Both options can be parallelized as follows. The regeneration process is, in fact, equivalent to a brand new mesh generation and the parallelization techniques previously mentioned can be used. Here the actual mesh is commonly used as the background grid if the Advancing Front Technique is used. Naturally a new partition of the regenerated mesh is usually required as the element distribution can change considerably within the original subdomains due to the adaption process.

The case of mesh enrichment is simpler to parallelize as the enrichment process can be performed in parallel for the different subdomains. A posteriori remeshing of the subdomain boundaries is usually needed to ensure element compatibility at the subdomain interfaces in the new mesh, while maintaining the prescribed size requirements at the interfaces (see Section 4.2.2).

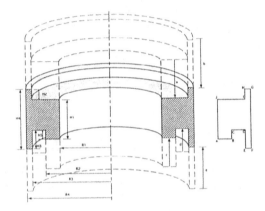

Fig. 6. Powder compaction of a cylindrical part. Definition of the geometry and the compaction process [32].

7 Examples

The first example taken from reference [32] is the analysis of the powder compaction process of an axisymmetric part including the unloading and the mould extraction stages. Frictional contact effects at the mould-part interfaces where also taken into account. Figure 6 shows the geometry of part and the definition of the compaction process. Futher details on the geometrical, material and process parameters of this problem can be found in [32].

The analysis was performed using the implicit non-linear code POWCOM [49,73] developed joinly by CIMNE and the Universidad Politécnica de Cataluña in the framework of ESPRIT PACOS project "PARCOM" [79] for the parallel analysis of powder compaction process in PC networks using PVM tools. A mesh of 488 four node axisymmetric quadrilateral elements was used. The parallel computations where performed in a network of 3 PC's. Average speed ups of

2.65 where obtained for all the different loading cases studied. Figure 7 shows the changes in the original mesh defining the powder material during the compaction process. The evolution of the final compacted part during the extraction process is shown in Figure 8. Further details of this example can be found in reference [32].

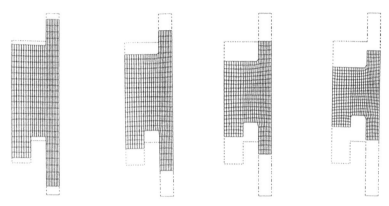

Fig. 7. Powder compaction of a cylindrical part. Deformed geometry of the finite element mesh at different stages of the compaction process [32].

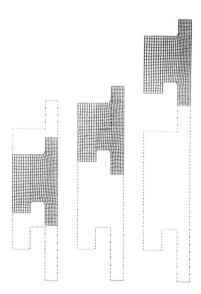

Fig. 8. Powder compaction of a cylindrical part. Evolution of the specimen during the extraction process [32].

The second example is the stamping of a rectangular pan. Figure 9 shows the geometrical description of the problem. Futher details can be found in [75].

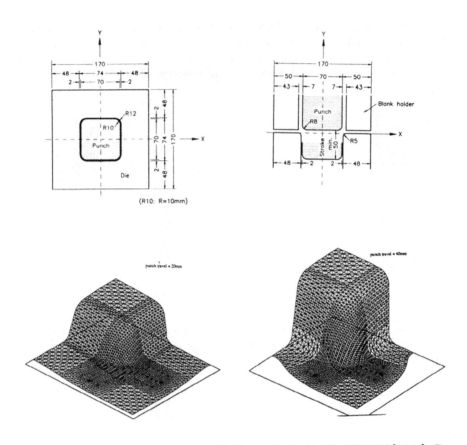

Fig. 9. Stamping of rectangular pan with the parallel code STAMPACK [75,76]. Geometry of tools and sheet and deformed mesh for different punch travels.

The analysis was performed with the explicit dynamic parallel finite element code STAMPACK [76] developed at CIMNE in cooperation with the European Center for Parallelism of Barcelona (CEPBA) in the context of ESPRIT PACOS project STAMPAR [80]. The parallel computations were performed in this case using a cluster of three Silicon Graphic workstations and PVM tools. Figure 9 shows the mesh of 540 triangular shell elements used for this problem as well as some results of the numerical analysis. An average speed up to 2.53 was obtained in the different stamping cases analyzed for this problem.

Figures 10 - 12 show some results for a number of industrial sheet stamping problems analyzed with the code STAMPACK. These include the stamping of an oil pan (Figure 10), the stamping of a curved rail (Figure 11) and the stamping of a car roof (Figure 12). Further information on these examples can be found in references [76, 80].

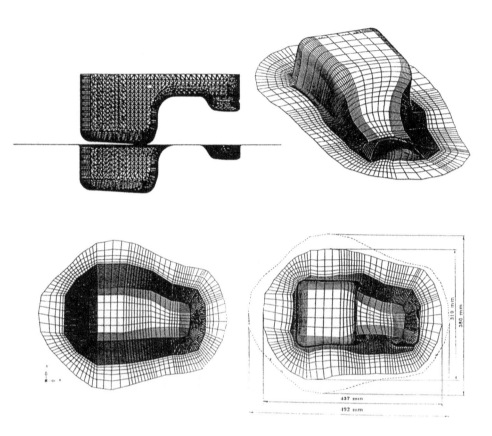

Fig. 10. Some results from the finite element analysis of an oil pan with the sheet stamping code STAMPACK [76].

Concluding remarks

This paper has presented an overview of the posibilities of parallel computing for the finite element analysis of complex non linear problems in continuum mechanics such as those occuring in the industrial forming process industry. The content of the paper has covered a wide range of topics from the basic theoretical and computational concepts of the non linear finite element solution, to the description of some parallel computing strategies, as well as some examples of application to powder compaction and sheet stamping problems. It is expected that the reader will recognize the opportunities of parallel computing in a challenging field of major industrial relevance.

Important contributions from the research community are expected in next coming years in order to bring closer the enhanced numerical simulation methods and parallel computing tools to the practical needs of the industrial forming sector.

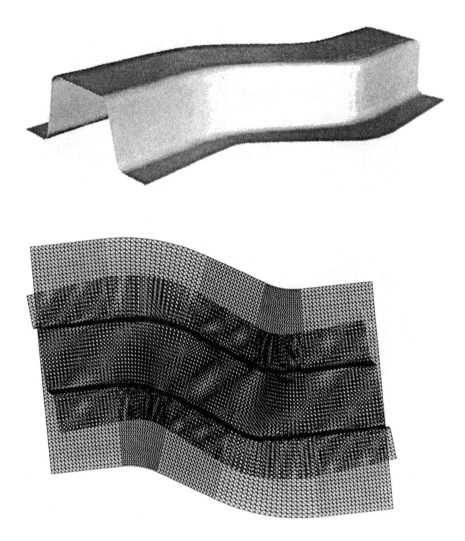

Fig. 11. Finite element analysis of a curved rail with the sheet stamping code STAM-PACK [76]. Geometry and original and deformed meshes.

Acknowledgements

Results for the powder compaction example were gently provided by Drs. J.C. Cante, J. Oliver and S. Oller. The parallel simulations with STAMPACK were performed by Dr. L. Neamtu. Thanks are also given to Drs. J. Rojek and P. Cendoya and to Mrs. Y. Jovicevic and Mr. O. Fruitos for their help in providing results for the sheet stamping examples shown.

The author wishes to acknowledge several fruitful discussions with the previously named scientists as well as with Drs. G. Duffet and R. Löhner.

This research was partially supported by ESPRIT PACOS project STAMPAR (ESPRIT PCI project N. P21037). The support received from the European Center of Parallelism of Barcelona (CEPBA) is also acknowledged.

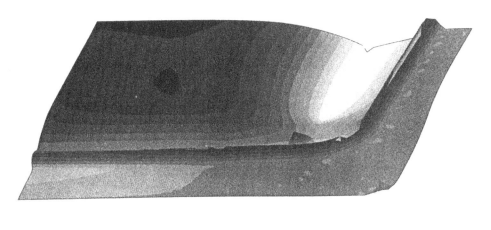

Fig. 12. Finite element analysis is a car roof with the sheet stamping code STAMPACK [76]. Deformed shape of one quarter of the roof at 70% punch travel. Full deformed shape of half geometry.

References

1. Chenot, J.L, Wood, R.D. and Zienkiewicz, O.C. (Eds.) "Numerical Methods in Industrial Forming Process". *Proceedings of NUMIFORM'92*, Balkema, 1992.

2. Chenot, J.L. and Oñate, E. (Eds.), *Modelling of Metal Forming Processes*, Kluwer Academic Publishers, 1988.

3. Oñate, E. and Agelet de Saracibar, C. "Alternatives for finite element analysis of sheet metal forming processes", in *Numerical Methods in Industrial Forming Processes*. J.L. Chenot, R.D. Wood and O.C. Zienkiewicz (Eds.), Balkema 1992.

4. Oñate, E. and Agelet Saracibar, C. "Numerical modelling of sheet metal forming forming problems", in *Numerical Modelling of Material Definition Processes*, C. Sturgess *et al.* (Eds.), Spring-Verlag, London 1991.

5. P. Dawson et. al. (Eds.), *"Numerical Methods in Industrial Forming Processes"*. Proceedings of NUMIFORM 95, Balkema, 1995

6. Zienkiewicz, O.C. and Taylor, R.C., *The finite element method*, Mc Graw Hill, Vol. 1, 1989, Vol. **2**, 1991.

7. Oñate, E. "Possibilities of finite element methods for analysis of industrial methods for analysis of industrial forming processes", (in Spanish). Publication CIMNE, Barcelona, 1996.

8. "Computer-aided engineering for metal forming", written by Eurotech Data and published by EC (DG XVIII - Credit et Investissements), February 1995.

9. Malvern, L.E., *"Introduction to the mechanics of a continuous medium"*. Prentice Hall, 1969.

10. Zienkiewicz, O.C., Jain, P.C. and Oñate, E., "Flow of solids during forming and extrusion. Some aspects of numerical solution", *Int. J. Solids Struct.*, 14, 14-28, 1978.

11. Oñate, E. and Zienkiewicz, O.C., "A viscous shell formulation for analysis of thin sheet metal forming", *Int. J. Mech. Sci.*, **25**, 305-35, 1983.

12. Agelet de Saracibar, C. and Oñate, E., "Plasticity models for porous metals", in *Computational Plasticity*, D.R.J. Owen, E. Hinton and E. Oñate, (Eds.), Pineridge Press 1989.

13. Oñate, E. and Agelet de Saracibar, C., "Analysis of sheet metal forming problems using a selective bending-membrane formulation", *Int. J. Num. Mech. Engng.*, **30**, 1577-93, 1990.

14. Simo, J.C., Rifai, M.S. and Fox, D.D., "On a stress resultant geometrically exact shell model. Part IV. Variable thickness shell with through the thickness stretching", *Comp. Meth. Appl. Mech. Engng.*, **81**, 91-126, 1990.

15. Oñate, E., Zarate, F. and Flores, F., "A simple triangular element for thick and thin plate and shell analysis", *Int. J. Num. Meth. Engng.*, **37**, 2569–82, 1994.

16. Oñate, E. and Zarate, F., "New shell element with translationed degrees of freedom". Publications CIMNE, Barcelona, 1996.

17. Wriggers, P., "Finite element algorithms for contact problems", in *Archives of Computational Methods in Engineering*, Vol. **2**, 4, 1–49, 1995.

18. Dalin, J.B. and Oñate, E., "An automatic algorithm for contact problems". J.Thompson *et al.* (Eds.), *Numerical Simulation of Industrial Forming Processes*, Balkema 1989.

19. Agelet de Saracibar, C., "Finite element analysis of sheet metal forming process", (in Spanish). Ph.D. Thesis, Univ. Politécnica de Cataluña, Barcelona, Spain, 1990.

20. Agelet de Saracibar, C., "A nex frictional time integration algorithm for large slip multi-body frictional contact problems", Publication CIMNE, Barcelona, Spain 1995.

21. Simo, J.C. and Laursen, T.A., "An augmented lagrangian treatment of problems involving frictions", *Computers and Structures*, 42, 97 - 116

22. Alart, A. and Currier, A., "A mixed formulation for frictional contact problems for Newton like solution methods", *Comp. Meth. Appl. Mech. Engng.*, **92**, 353-75, 1991.

23. Papadopoulous, P. and Taylor, R.L., "A mixed formulation for the finite element solution of contact problems", *Comp. Meth. Appl. Mech. Engng.*, 94, 373 - 89, 1992.

24. Farhat, C., "Which parallel finite element algorithm for which architecture and which problem?", *Eng. Comput.*, 1990, Vol. **7**, Sept. 1990.

25. Farhat, C., "On the mapping of masively parallel processors onto finite element graphs", *Computers and Structures*, Vol. **32**, N. 2, pp. 347-353, 1989.

26. Farhat, C., Sobh, N. and Park, K.C., "Transient finite element computations on 65536 processors: the Connection Machine", *Int. Journal for Numerical Methods in Engineering*, Vol. **30**, pp. 27-55, 1990.

27. Johnsson, S.L. and Mathur, K.K., "Data structures and algorithms for the finite element method on a data parallel supercomputer", *Int. Journal for Numerical Methods in Engineering*, Vol. **29**, pp. 881-908, 1990.

28. Belytschko, T., Plaskacz, E.J., Kennedy, J.M. and Greenwell, D.L., "Finite element analysis in the Connexion Machine", *Comp. Meth. Appl. Mech. Engng.*, **81**, 229-54, 1990.

29. Belytschko, T. and Plaskacz, E.J., "SIMD Implementation of a non-linear transient shell program with partially structured meshes", *Int. Journal for Numerical Methods in Engineering*, Vol. **33**, pp. 997-1026, 1992.

30. Mathur, K.K., "On the use of randomized address maps in unstructured three dimensional finite element simulations", *Thinking Machines Corporation*, Technical Report TR-37 CS90-5, 1990.

31. Johan, Z., Hughes, T.J.R., Mathur, K.K. and Johnsson, S.L., "A data parallel finite element method for computational fluid dynamics on the Connection Machine System", *Comp. Methods. Appl. Mech. Engrg.*, Vol. **99**, pp. 113-134, 1992.

32. Cante, J.C., "Numerical simulations of powder compaction processes. Applications of parallel processing techniques". Ph.D. Thesis (in Spanish), Univ. Politécnica de Cataluña, Barcelona, 1995.

33. Thinking Machines Corp. Programming in Fortran. Version 1.0., January 1991, Cambridge, Massachusetts.

34. Fischer, T., Oñate, E. and Miquel, J., "Finite element analysis of high speed viscous flows in a massive parallel computer", C. Wagner *et al.* (Eds.), ECCOMAS *Conference on Computational Fluid Dynamics*, J. Wiley, 1994.

35. Belytschko, T., Plaskacz, E.J., Kennedy, J.M. and Greenwell, D.L., "Finite element analysis on the Connection Machine", *Comp. Methods. Appl. Mech. Engng.*, Vol. **81**, pp. 229-254, 1990.

36. Cante, J.C., Ilpide, E., Oliver, J. and Oller, S., "Experiences on massive parallel computations in a Finite Element context". Publication CIMNE, Barcelona, Spain 1993.

37. Cante, J.C., Ilpide, E., Oliver, J. and Oller, S., "A study of the GMRES algorithm in a non linear finite element context", Publication CIMNE, Barcelona, Spain, 1993.

38. Cante, J.C., Ilpide, E., Oliver, J. and Oller, S., "Optimizing gathering and scattering costs on the Connection Machine in a finite element context", Publication CIMNE, Barcelona, Spain, 1993.

39. Cante, J.C., Ilpide, E., Oliver, J. and Oller, S., "Comparisons of the serial and parallel versions of a non linear finite element program", Publication CIMNE, Barcelona, Spain, 1995.

40. Cante, J.C., Ilpide, E., Oliver, J. and Oller, S., "On the structure of a finite element code for its implementation on a massive parallel SIMD computer", Publication CIMNE, Barcelona, Spain, 1994.

41. Cante, J.C., Ilpide, E., Oliver, J. and Oller, S., "Towards large scale parallel computations using a powder compaction simulation program", Publication CIMNE, Barcelona, Spain, 1994.

42. Ilpide, E., "Experiences in the use of parallel computing in non linear solid mechanics", (in Spanish). Publication CIMNE, Barcelona, Spain, 1994.

43. Chan, T., Glowinski, R., Periaux, J. and Widlund, O. (Eds.), *Domain Decomposition Methods.* SIAM Philadelphia, 1989.

44. Keyes, D.E. and Gropp, W.D., "A comparison of domain decomposition techniques for elliptic partial differential equations and their parallel implementations", *SIAM J. Sci. Stat. Comp.*, 8, S166-S202 (1987).

45. Farhat, C. and Lesoinne, M., "Automatic partitioning of unstructured meshes for the parallel solution of problems in computational mechanics", *Int. J. Num. Meth. Engng.*, 36, 745-64, 1993.

46. Farhat, C., Maman, N. and Brown, G., "Mesh partitioning for implicit computations via iterative domain decomposition. Impact and optimization of the subdomain aspect ratio", *Int. J. Num. Meth. Engng.*, 38, 989-1000, 1995.

47. Farhat, C. and Roux, F.X., "A method of finite element tearing and interconnecting and its parallel solution algorithm", *Int. J. Num. Methods Engng.*, 32, 1205-1227, 1991.

48. Nour-Omid, B. and Park, K.C., "Solving structural mechnics problems on the CALTECH hypercube machine", *Comp. Methods Appl. Mechn. Eng.*, 61, 161-176, 1987.

49. Cante, J.C. Ilpide, E., Oliver J. and Oller S., "Development of a parallel software for industrial powder compaction processes", Publication CIMNE, Barcelona, Spain, 1996.

50. Fahrat, C. and Simon, H.D., "TOP/DOMDEC. A software tool for mesh partition and parallel processing", *Report of Center for Space Structures and Control*, CU-CSSC-93-11, Univ. of Colorado, Boulder, USA, May, 1993.

51. Flower, J., Otto, S. and Salama, M., "Optional mapping of irregular finite element domains to parallel processors", A.K. Noor, ed., in *Parallel Computations and their Impact on Mechanics*, The American Society of Mechanical Engineers, AMD Vol 86, 239-252, 1987.

52. Nour-Omid, B., Raefsky, A. and Lyzenga, G., "Solving finite element equations on concurrent computers", A. K. Noor, ed., in *Parallel Computations and their Impact on Mechanics*, The American Society of Mechanical Engineers, AMD- Vol. 86, (1987) 209-228.

53. Farhat, C., "A simple and efficient automatic FEM domain decomposer", *Comp. and Struct.*, Vol. 28, N. 5, (1988) 579-602.

54. Malone, J.G., "Automated mesh decomposition and concurrent finite element analysis for hypercube multiprocessors computers", *Comp. Meth. Appl. Mech. Eng.*, Vol. 70, N. 1, 27-58, 1988.

55. Pothen, A., Simon, H. and Liou, K.P., "Partitioning sparse matrices with eigenvectors of graphs", *SIAM J. Mat. Anal. Appl.*, Vol. 11, N.3, 430-452, 1990.

56. Simon, H., "Partitioning of unstructured problems for parallel processing", *Comput. Sys. Engng.*, Vol. 2, N. 3, 135-148, 1991.

57. Farhat, C. and Lesoinne, M., "Automatic partitioning of unstructured meshes for the parallel solution of problems in computational mechanics", *Internat. J. Numer. Meth. Engng.*, Vol. 36, N. 5, 745-764, 1993.

58. Farhat, C., "A simple and efficient automatic FEM domains decomposer", *Comp. Struct.*, 28, 579-60, 1988.

59. Farhat, C. and Crivelli, L. "A general approach to nonlinear FE computations on shared memory multiprocessors", *Comp. Meth. Appl. Mech. Engng.*, **72**, 153-72, 1989.

60. Löhner, R., Camberos, J. and Merriam, M., "Parallel unstructured mesh generations", *Comp. Meth. Appl. Mech. Engng.*, **95**, 343-57, 1992.

61. Mestres, J.C., Bugeda, G. and Oñate, E., "On parallel mesh generations", Publication CIMNE, Barcelona, Spain, 1996.

62. Lo, S.H., "A new mesh generation scheme for arbitrary planar domains", *Intern. J. Num. Meth. Engng.*, **21**, 1403-1426, 1985.

63. van Phai, N., "Automatic mesh generation with tetrahedron elements", *Int. J. for Num. Meth. in Engng.*, **18**, 237-289, 1982.

64. Peraire, J., Peiro, J., Formaggia, L., Morgan, K. and Zienkiewicz, O.C., "Finite element Euler computations in three dimensions", AIAA-88-0032, 1988.

65. Löhner, R., "Some useful data structure for the generation of unstructured grids", *Com. Appl. Numer. Methods*, **4**, 123-135 ,1988.

66. Löhner, R. and Parikh, P., "Three-dimensional grid generation by the advancing front method", *Int. J. Num. Methods Fluids*, **8**, 1135-1149 1988.

67. Watson, D.F., "Computing the N-dimensional Delaunay tesselation with application to Voronoi polytopes", *Comput. J.*, **24**, (2), 167-172, 1981.

68. Tanemura, M., Ogawa, T. and Ogita, N., "A new algorithm for three-dimensional Voronoi tessellation", *J. Comput. Phys.*, **51**, 191-207, 1983.

69. Yerry, M.A. and Shepard, M.S., "Automatic three-dimensional mesh generation by the modified-octree technique", *Internat. J. Num. Meth. Engng.*, **20**, 1965-1990, 1984.

70. Cavendish, J.C., Field, D.A. and Frey, W.H., "An approach to automatic three-dimensional finite element mesh generation", *Int. J. Num. Meth. Engng.*, **21**, 329-347, 1985.

71. Baker, T.J., "Three-dimensional mesh generation by triangulation of arbitrary point sets", AIAA-87-1124-CP, 1987.

72. Oñate, E. and Bugeda, G., "Mesh optimality criteria in adaptive finite element computations", *Engng. Comp.*, Vol. **3**, 307–321, 1994.

73. User Guide to POWCOM. "A parallel finite element code for analysis of powder compaction processes". Publication CIMNE, Barcelona, Spain 1996.

74. Rojek, J., Jovisevic Y. and Oñate, E., "New finite elements for analysis of industrial sheet stamping process using an explicit dynamic code", Publication CIMNE, Barcelona, Spain 1996.

75. Oñate, E., Rojek, J. and Garcia, C., "NUMISTAMP. A research project for assessment of finite element models for stamping processes", *Journal of Material Processing Tech.*, **50**, 17–38, 1995.

76. User Guide of STAMPACK. "An explicit dynamic program for analysis of industrial sheet stamping problems". Publication CIMNE, Barcelona, Spain 1996.

77. "GID. A customizable system for geometric interfacing and generation of data for finite element analysis", Publication CIMNE, Barcelona, Spain 1996.

78. George, P.L., *Automatic Mesh Generation*, J. Wiley, 1991.

79. "Development of parallel computer technique for the finite element analysis of powder compaction processes", Project PARCOM of the ESPRIT PACOS Programme of EC, Project no. P9601, 1996.

80. "Development of parallel computation techniques for finite element analysis of sheet stamping processes", Project STAMPAR of the ESPRIT PACOS Programme of EC. Project no. P21037, 1996.

Preconditioners for Nonsymmetric Linear Systems in Domain Decomposition Applied to a Coupled Discretization of Navier-Stokes Equations

Filomena D. d'Almeida[1] and Paulo B. Vasconcelos[2]

[1] Faculdade de Engenharia da Universidade do Porto
4099 Porto codex, Portugal
falmeida@fe.up.pt
[2] Faculdade de Economia da Universidade do Porto
Rua Dr Roberto Frias, 4200 Porto, Portugal
pjv@fe.up.pt

Abstract. We will consider the linear system generated by a coupled discretization and linearization method for the Navier-Stokes equations. This method consists of a discretization of the momentum equations to obtain the velocities and pressure at the faces of a finite volume, in terms of the values of these variables at the grid points followed by the integration of the momentum and continuity equations in the finite volumes.

This integration leads to equations where the values of the variables at the cell faces are to be replaced by the expressions obtained at the previous stage.

The linear system to be solved at each nonlinear iteration connects values of velocities and pressure at each grid point in each equation. The coefficient matrix is large, nonsymmetric, sparse, with non-null entries on the diagonal. The characteristics of these linear systems indicate the use of nonstationary iterative methods, for instance preconditioned GMRES, for their solution. The application of preconditioners based on the incomplete factorization of the system matrix will be analyzed.

When nonoverlapping domain decomposition is used to parallelize this method, the system matrix at each outer iteration is in block-partitioned form. The solution of $Ax = b$ is then equivalent to the solution of the Schur complement reduced system (corresponding to the solution on the interfaces separating subdomains) followed by the update of the solution in the subdomains. Incomplete factorization preconditioners applied to S (Schur complement) are very expensive as they need the explicit construction of S.

The use of other type of preconditioners based on probing techniques or approximate inverses will be considered. In the first case the preconditioner M will be a narrow band approximation of S obtained by the use of matrix-vector products only. In the second case, approximate inverse techniques allow the approximation of S by a less expensive sparse matrix to be used as a preconditioner.

1 Introduction

A nonoverlapping domain decomposition strategy and the GMRES method were used to parallelize a given code for coupled discretization of Navier-Stokes equations [9]. That code yields a linear system to be solved at each nonlinear iteration whose coefficient matrix has some special characteristics such as nonsymmetry, sparsity, nonsingularity of diagonal blocks. In this work some block preconditioners for GMRES that can fit the domain decomposition and the characteristics of this code are studied and tested on a cluster of 4 workstations DEC ALPHA connected by FDDI (Fibber Distributed Data Interface) and a GigaSwitch, using PVM (Parallel Virtual Machine).

This work is organized as follows.

In Sect. 2 the test problem is described, in Sect. 3 some algorithmic details are given. Section 4 explains the application of domain decomposition to parallelize this problem and Sect. 5 describes the preconditioners used. Some numerical results of the parallelization are reported in Sect. 6.

2 Test Problem Specification

The test problem to be considered is the steady two dimensional incompressible laminar flow in a square lid-driven cavity. The lid slides across the cavity inducing a zone of recirculatory fluid and for fine grids small counter-rotating eddies in the corners can be expected .

The problem specification is represented in Fig. 1, where ρ stands for the density, μ for the dynamic viscosity, Re for the Reynolds number, U_i for the velocities and P for the static pressure.

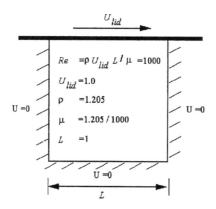

Fig. 1. Test problem

The Navier-Stokes equations governing the flow in this case are

$$div(U) = 0 \qquad (1)$$

$$\rho(D_x U)U = -gradP + \mu \Delta U \qquad (2)$$

the continuity equation (mass conservation) and the momentum equations, respectively. Where $D_x U$ stands for the Jacobian matrix of U, $gradP$ for the column matrix of the partial derivatives of P and ΔU for the Laplacian operator of U.

The method used, as opposed to the classical decoupling of the equations, solves the fluid flow equations as a coupled system. It uses a numerical model where the cell face velocities are predicted using the momentum equations, assuming mass conservation and constant viscosity.

The equations were discretized in the solution domain using a finite volume cells formulation where the fluid flow equations are solved as a completely coupled system ([9], [2]).

This method starts with the discretization of the momentum equations (assuming mass conservation and constant viscosity) using 9-point stencil finite differences approximations (Fig. 2) to the derivatives in

$$\rho(gradU_j)^T U = -\frac{\partial P}{\partial x_j} + \mu \, \Delta U_j , \qquad (3)$$

where $j = 1, 2$ for the two-dimensional case, U stands for U_i, $i = 1, 2$.

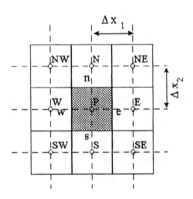

Fig. 2. 9-point stencil discretization

To be more specific, the left hand side can be written as, when $U = U_1$:

$$\rho(\frac{\partial U}{\partial x_1}U_1 + \frac{\partial U}{\partial x_2}U_2) = 2\rho u_e{}^{old}\frac{u_e - U_P}{\Delta x_1} + \rho v_e{}^{old}\frac{2u_e - U_S - U_{SE}}{2\Delta x_2} , \qquad (4)$$

where u_α and v_α denote the velocities U_1 and U_2 , respectively, at cell face α ($\alpha = e, w, n, s$).

For the right hand side of (3) a second order finite differences scheme is used

$$\mu \frac{\partial^2 U}{\partial x_1^2} = 4\mu \frac{U_P - 2u_e + U_E}{\Delta x_1^2} \tag{5}$$

$$\mu \frac{\partial^2 U}{\partial x_2^2} = \mu/2 \left(\frac{U_N - 2U_P + U_S}{\Delta x_2^2} + \frac{U_{NE} - 2U_E + U_{SE}}{\Delta x_2^2} \right) \tag{6}$$

and for the pressure

$$-\frac{\partial P}{\partial x_1} = -\frac{P_E - P_P}{\Delta x_1} \; ; \; -\frac{\partial P}{\partial x_2} = \frac{P_S - P_P}{\Delta x_2} . \tag{7}$$

When $U = U_2$, corresponding formulae are used.

Then (3) is solved to obtain the velocities and pressure at the faces of the finite volume in terms of the values of these variables at the grid points (see Fig. 2).

The mass conservation equation is then integrated over the control volume in terms of the cell faces velocities, i.e.

$$\int div(U) \, dV = (u_e - u_w)(\Delta x_2) + (v_n - v_s)(\Delta x_1). \tag{8}$$

For the solution of the momentum equations the same procedure was followed.

The resulting square coefficient matrix is large, sparse, block tridiagonal and unsymmetric. The matrix size is N , $N = 3mn$, for a mxn grid. Each outer iteration involves the solution of such a linear system.

3 Algorithm

Since the method described above yields a large nonsymmetric linear problem to be solved at each outer iteration, it is crucial to use the other characteristics of the matrices involved to make this competitive with the classical (decoupled discretization) approach.

These characteristics are the sparsity inside the band profile of the coefficient matrix and the fact that the solution of the linear system need not to be very accurate since the nonlinear iteration will further improve the coefficients of the matrix.

This suggests the use of an iterative linear solver [3] which can be stopped whenever the accuracy is good enough. The coefficient matrix will only be used on matrix-vector multiplications which can be done using sparse techniques for storage and computations [14].

The nonsymmetry of the system leads to the choice of the GMRES method [19], [3]. GMRES (Generalized Minimum RESidual) is a method based on an orthonormal basis of a Krylov subspace. To solve $Ax = b$ the kth approximate solution is expressed as $x_k = x_0 + z_k$, where x_0 is an initial solution with residual $r_0 = b - Ax_0$. Then z_k is computed such that its residual projected onto the Krylov subspace generated by r_0 is minimized.

Iterative nonstationary methods, namely GMRES, require appropriate preconditioners [3], [23].

This linear iterative method is included in the outer iteration described in Sect. 2 and the procedure stops when the 1-norm of the residual falls down to some pre-defined tolerance for all variables; for velocity, the residual is weighted by $(\rho U_{lid}^2)^{-1}$ and for pressure by $(\rho U_{lid})^{-1}$.

Algorithm (sequential)

1. generate grid and A_k and b_k as in Sect. 2

2. $k = 0$

3. repeat until convergence

   (a) apply the linear solver to $A_k x_k = b_k$

   (b) $k = k + 1$

   (c) rebuild matrix A_k and b_k

   (d) compute the residual

In the tests reported in the following sections, the residual of the nonlinear iteration is required to be less than 10^{-5} and the tolerance on the precision for the GMRES method is $eps = 5 * 10^{-6}$. GMRES is not restarted since it proved to be enough to use a Krylov space of dimension at most $nsave = 30$ to have convergence of the outer iteration. However some tests were made with smaller $nsave$ and greater eps and the results will be referred to in Sect. 5.1.

4   Parallel Approach and Domain Decomposition

Since the nonlinear method and the linear solver are iterative, the best option is to exploit the physical parallelism of the problem by domain decomposition techniques. A nonoverlapping one-way dissection strategy was used in this work.

The basic idea of this technique (see Fig. 3) is to divide the domain of the original problem into disjoint subdomains, Ω_i, connected by interfaces Γ_j, solve the original problem corresponding to each subdomain in parallel, and then concatenate the partial solutions. This concatenation, in the case of the nonoverlapping domain decomposition, is performed via Schur complement.

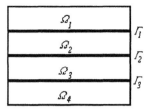

Fig. 3. Domain decomposition, associated partitioned matrix and Schur complement

If the nodes are numbered consecutively first in each subdomain and then in the interfaces, the corresponding matrix has the following structure, for p subdomains and $p-1$ interfaces :

$$\begin{bmatrix} A_{11} & & & & A_{1p+1} \\ & A_{22} & & & A_{2p+1} \\ & & \ddots & & \vdots \\ & & & A_{pp} & A_{pp+1} \\ A_{p+11} & A_{p+12} & \cdots & A_{p+1p} & A_{p+1p+1} \end{bmatrix}$$

If we want to separate subdomain and interface blocks, A can be represented as

$$\begin{bmatrix} B & F \\ E & C \end{bmatrix}.$$

The interface matrix C and connections between matrices and subdomains are very sparse and have a special pattern: C is block diagonal with $p-1$ blocks (for one-way dissection), and E and F^T have the following profile:

$$\begin{bmatrix} 0000xx & xx00000 & 000000 & 000000 \\ 000000 & 0000xx & xx0000 & 000000 \\ 000000 & 000000 & 0000xx & xx0000 \end{bmatrix}.$$

The Schur complement S is obtained by block Gaussian elimination of the system

$$\begin{bmatrix} B & F \\ E & C \end{bmatrix} \begin{bmatrix} x_d \\ x_I \end{bmatrix} = \begin{bmatrix} b_d \\ b_I \end{bmatrix} \tag{9}$$

thus yielding

$$\begin{bmatrix} B & F \\ 0 & S \end{bmatrix} \begin{bmatrix} x_d \\ x_I \end{bmatrix} = \begin{bmatrix} b_d \\ f_I \end{bmatrix} \tag{10}$$

where $S = C - EB^{-1}F$, $f_I = b_I - EB^{-1}b_d$.

The computations involving B^{-1} correspond to the solution of p linear systems whose coefficient matrices are $A_{11}, ..., A_{pp}$, that can be done in parallel, as well as the calculation of the coefficients of submatrices $A_{ii}, A_{ip+1}, i = 1, ..., p$. The coefficients of submatrices $A_{p+1j}, j = 1, ..., p+1$ are computed in another processor.

5 Preconditioners

When solving $Ax = b$ by nonstationary iterative methods it is necessary to use a preconditioner in order to improve the rate of convergence.

A preconditioner is a nonsingular matrix M such that the equivalent system $M^{-1}Ax = M^{-1}b$ has a better condition number.

A right preconditioner can also be used: $AM^{-1}y = b$ with $x = M^{-1}y$.

The choice between left or right preconditioners is dictated by the iterative method [19].

The matrix M may be obtained in two different ways:

1. M is an approximation to A
2. M^{-1} is an approximation to A^{-1}.

As GMRES only needs the coefficient matrix for matrix-vector products the explicit computation of M^{-1} is not needed. We only need its action on a vector in each iteration of GMRES.

In the first case the application of the preconditioner requires the solution of a linear system (to obtain $M^{-1}(Ax)$). In the second case we have approximate-inverse preconditioners and as M^{-1} is known, preconditioning reduces to the computation of matrix-vector products.

A large number of interface preconditioners has been proposed in the literature such as exact eigendecomposition of S [6], [11] Neumann-Dirichlet preconditioner [6], [8] [17], Neumann-Neumann preconditioner [7], probing preconditioner [10],[16],[12], multilevel preconditioner [20].

5.1 Incomplete LU Factorization

A detailed study of several options of incomplete LU factorization (ILU) recommended ILUT(30) as a good preconditioner for this problem in the sequential case (as described in [22], [23], [18]).

Factorizations ILU(0), MILU0, ILU(k), $k = 1, ..., 4$, ILUT($lfil$), from SPARSKIT library [18],[3] were tested.

The simplest incomplete LU factorization is ILU(0) where no fill-in is allowed during the Gaussian elimination. In ILU(k) only fills of level k are allowed. Fill levels are defined [3] by calling an element of level k if it is caused by elements at least one of which is of level $k - 1$. Level 1 fills are caused by the elements of the original matrix. MILU0 is similar to ILU(0) but factors are constructed to have the same row sums as the original matrix.

The ILUT(*lfil*) is an incomplete factorization that uses a dual threshold strategy for dropping elements:

1. any element whose size is less than a prescribed tolerance, *tol*, is dropped and
2. each of L and U rows will have a maximum of *lfil* nonzeros .

The storage overhead is controlled by *lfil* but the cost of this factorization is considerable. However the fact of applying it only in the subdomains where the systems are smaller (in parallel) and to the Schur complement makes it very efficient for this problem when S is known.

The pattern of the LU factors obtained from the ILUT preconditioner for a 20x20 grid with $eps = tol = 5 * 10^{-3}$ is plotted in Fig. 4.

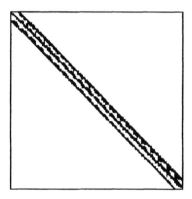

Fig. 4. LU factors from ILUT on a 20x20 grid

With domain decomposition, the initial system is equivalent to (10). The ILU preconditioner is then a block diagonal matrix M such that

$$M = \begin{bmatrix} L_d U_d & 0 \\ 0 & L_S U_S \end{bmatrix}$$

where $L_d U_d$ is a block diagonal matrix containing the ILU factorization of each diagonal block of B and $L_S U_S$ is the ILU factorization of the Schur complement S. So, the application of this preconditioner is equivalent to solve the subsystems corresponding to the subdomains and interfaces by GMRES with ILU local preconditioner.

In fact, (10) preconditioned on the right with M is equivalent to

$$\begin{bmatrix} B & F \\ 0 & S \end{bmatrix} M^{-1} \begin{bmatrix} y_d \\ y_I \end{bmatrix} = \begin{bmatrix} b_d \\ f_I \end{bmatrix} ; \begin{bmatrix} x_d \\ x_I \end{bmatrix} = M^{-1} \begin{bmatrix} y_d \\ y_I \end{bmatrix} \tag{11}$$

or

$$\left[\begin{array}{c} B(L_dU_d)^{-1}y_d + F(L_SU_S)^{-1}y_I = b_d \\ S(L_SU_S)^{-1}y_I = f_I \end{array} \right]. \tag{12}$$

This is done in 3 steps:

1. Compute approximation $\tilde{x}_d$ to x_d by

$$B(L_dU_d)^{-1}\tilde{y}_d = b_d \; ; \; \tilde{x}_d = (L_dU_d)^{-1}\tilde{y}_d \tag{13}$$

and matrix $X_d = B^{-1}F$ by

$$B(L_dU_d)^{-1}\tilde{Y}_d = F \; ; \; \tilde{X}_d = (L_dU_d)^{-1}\tilde{Y}_d \tag{14}$$

(diagonal blocks of B are distributed among different processors so this is done in parallel).

2. Use $\tilde{x}_d$ and X_d to compute $S = C - EX_d$ and $f_I = b_I - E\tilde{x}_d$. Then solve

$$S(L_SU_S)^{-1}y_I = f_I \; ; \; x_I = (L_SU_S)^{-1}y_I. \tag{15}$$

3. Update the solutions in the subdomains

$$B(L_dU_d)^{-1}x_d = b_d - Fx_I \tag{16}$$

(this again can be done in parallel, each subdomain correspond to a process).

After Steps 1 and 2 the results have to be sent to the interface process and to the subdomains processes, respectively.

Note that in the computation of S the part $A_{ii}^{-1}A_{ip+1}$ is done by GMRES preconditioned with ILU factors of block A_{ii}. This computation is done in the processor that contains the ith diagonal block and passed to the interface processor column by column in order to allow the interface processor to go on working while the subdomain processor computes the next column.

The computation of L_SU_S can only be done when the Schur complement S is explicitly known. This is possible when the interfaces are not many.

The advantage of this approach is that it is very stable and leads to a number of outer iterations that is not influenced by the grid size and number of interfaces. It also improves the condition number, as it can be seen in Table 1. The disadvantage is that it is very expensive to compute S explicitly.

The condition numbers were estimated by LAPACK routines using the 1-norm. The time per iteration reflect the effect of preconditioning both the Schur complement and the diagonal blocks of B.

Table 1. Performance of ILU preconditioner

grid	32 x 32			64 x 64		
N	3072			12288		
nb. subdomains	2	3	5	2	3	5
nb. rows of S	192	288	480	384	576	960
$eps = 5*10^{-6}$						
nb. of outer iter.	15	14	14	16	15	15
$\kappa(S)$	1093	2008	3891	714	1302	2645
$\kappa(S*(L_S U_S)^{-1})$	52	97	434	55	278	1040
time per iter.	35.3	18.3	17.6	480.0	324.3	266.7

grid	32 x 32	
nb. subdomains	2	3
$eps = 5*10^{-4}$		
nb. of outer iter.	15	14
$\kappa(S)$	1093	2008
$\kappa(S*(L_S U_S)^{-1})$	52	98

5.2 Implicit Use of S

As the explicit computation of S is too expensive the alternative is to compute S implicitly, that is to compute its action on a vector v:

1. Compute (in parallel, in the processors corresponding to the blocks of B and F)
   $t = Fv$;
   solve $Bu = t$.
2. Compute (in the interface processor)
   $w = Cv - Eu$.

Experience shows that the systems appearing in Step 1, involving B, (distributed among the processors) have to be solved with a good accuracy. That is done using GMRES preconditioned by ILUT(30).

The product $w = Sv$ is the basis for an implicit GMRES that uses it to compute the residual r_0 and the Krylov basis $\{Sr_0, S^2 r_0, ..., S^{nsave-1} r_0\}$.

This implicit GMRES cannot be preconditioned by ILU factorization of S since this is not known explicitly.

As it is known, the Schur complement has a better condition number than A, so the first heuristic proposed is not to precondition S and only precondition by ILU factors the solves performed in the subdomain processors including the ones required by the computation of Sv that is in Step 1.

The corresponding global preconditioner is

$$M = \begin{bmatrix} L_S U_S & 0 \\ 0 & I \end{bmatrix}.$$

The results obtained are reported in Table 3.

Probe Preconditioner The next test done was the use of probing preconditioner. The probing technique which was developed by Chan and Resasco [10] and Keyes and Gropp [16], is an algebraic technique for the computation of interface preconditioner in domain decomposition algorithms. The basic idea is to approximate the interface matrix S by a matrix having a specific sparsity pattern, for example a band matrix.

The preconditioner, M_d, is constructed by multiplying S by $2 \times d + 1$ selected probe vectors. For instance, if S is tridiagonal it can be exactly recovered when $Sv_i, i = 1, 2, 3$ is known, for $v_1 = (1, 0, 0, 1, 0, 0, 1, ...)^T$, $v_2 = (0, 1, 0, 0, 1, 0, 0, ...)^T$, $v_3 = (0, 0, 1, 0, 0, 1, 0, ...)^T$. So if we want a tridiagonal preconditioner M_1 then $M_1 v_i = Sv_i, i = 1, 2, 3$. When S is unknown but there is a procedure to compute $Sv_i, i = 1, 2, 3$ such as the one described in Sect. 5.2, we can recover a tridiagonal M_1 from $[Sv_1 Sv_2 Sv_3]$ (see [12]):

$$\begin{bmatrix} m_{11} & m_{12} & 0 & 0 & ... \\ m_{21} & m_{22} & m_{23} & 0 & ... \\ 0 & m_{32} & m_{33} & m_{34} & ... \\ 0 & 0 & m_{43} & m_{44} & ... \\ & & ... & & \end{bmatrix} [v_1 v_2 v_3] = \begin{bmatrix} m_{11} & m_{12} & 0 \\ m_{21} & m_{22} & m_{23} \\ m_{34} & m_{32} & m_{33} \\ m_{44} & m_{45} & m_{43} \\ & ... & \end{bmatrix} \tag{17}$$

If the elements of S outside the main tridiagonal are small, M_1 can be a good approximation to S, that contains some influence of the other elements, since the column Sv_1 is the sum of columns 1,4,7,... of S, Sv_2 is the sum of columns 2,5,8,... of S, Sv_3 is the sum of columns 3,6,9,... of S.

It is well adapted to large variations in the coefficients, and to changes of the relative size of the subdomains but it is not optimally adapted to increasing numbers of grid points. It has very good performance when S is weakly connected globally and strongly connected locally because then the elements of S decay rapidly away from the diagonal.

In our case, the fact that 9-point finite differences are being used and that the partial differential equations are solved in a coupled way gives S a structure that is not as strongly decaying as one would wish for this preconditioner.

Fig. 5 shows the profile of S for 3 subdomains. Fig. 6, 7, 8 represent the horizontal profiles at columns 90, 91, 92, for a 32x32 grid (S has 288 rows).

The decay property is not verified very clearly as we can see in Fig. 8 and so many subdiagonals have to be considered to improve the condition number. In Fig. 9 the over all effect of these preconditioners is shown by plotting the number of outer iterations against d (bandwidth of M_d) for a 32x32 grid with 3 subdomains. As we can see the best case, in terms of number of iterations is

32x32 grid, Schur profile 2 interfaces

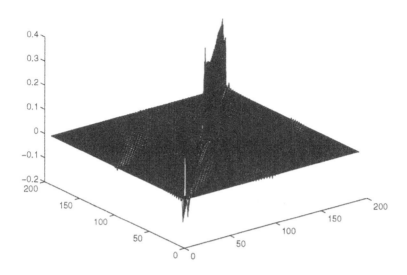

Fig. 5. Profile of the Schur complement

when we take $d = 48$ or $d = 27$, for $nsave = 20$ or 30, respectively, but then the computation and factorization of M_d are very expensive.

For a small number of subdomains these values of d correspond also to a large improvement in the condition number, for instance, with 2 subdomains and a 32x32 grid, for $d = 48$ the condition number is 220 and for $d = 52$ it is 36.

Approximate Inverse Preconditioner A number of techniques have recently been developed to construct a sparse approximate inverses to a matrix to be used as a preconditioner. They approximate each row or column independently by minimizing

$$min\|e_j - Ax\|_2 \,, \tag{18}$$

where $j = 1, 2, \ldots$ and e_j is the jth column of the identity matrix.

Chow and Saad [13] proposed an algorithm that minimizes (4) with a reduction of the norm at each step of GMRES: beginning with a sparse initial guess x with a residual r, the iteration proceeds by taking a search direction d obtained by dropping entries in the direction r. The sparsity pattern of x is controlled in this way. If the number of nonzeros is less than a given $lfil$ one entry is added by chosing the step length $\alpha = (r, Ad)/(Ad, Ad)$. Then x and r are updated.

Benzi and Tuma [4], [5] describe a biconjugation algorithm [15] to be used as a preconditioner. Two sets of vectors A- biconjugate are built: $\{z_i\}_{i=1}^{n}$, $\{w_i\}_{i=1}^{n}$,

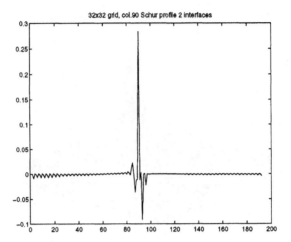

Fig. 6. Profile of the Schur complement - column 90

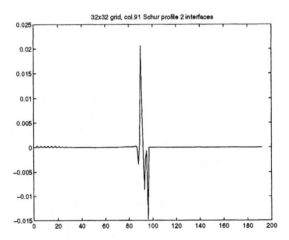

Fig. 7. Profile of the Schur complement - column 91

$w_i^T A z_j = 0, i \neq j$. If $Z = [z_1 \, z_2 \dots z_n]$ and $W = [w_1 \, w_2 \dots w_n]$, $W^T A Z = D = diag(d_1, ..., d_n)$. Then $A^{-1} = Z D^{-1} W^T = \sum_{i=1}^{n} (z_i \, w_i^T)/d_i$.

Here we propose another approach based on defect correction [21]. Let G be an affine operator, approximate inverse to the problem $Ax = b$ as defined in [1], then there is a norm for which $\|x - G(Ax - b)\| < 1$, for all x in the neighbourhood of the solution and the iteration

$$x_0 = G(0) \tag{19}$$

$$x_k = x_0 + x_{k-1} - G(Ax_{k-1} - b), \tag{20}$$

converges to the solution of the linear system.

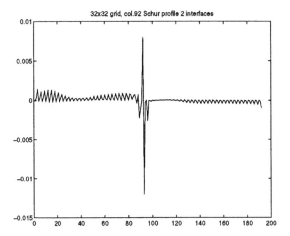

Fig. 8. Profile of the Schur complement - column 92

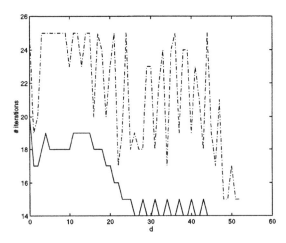

Fig. 9. Number of outer iterations vs probe bandwidth, for 32x32 grid, with 3 subdomains, $nsave = 20$ - dashed line, $nsave = 30$ - solid line

For the block form of the system (9) let us choose the approximate inverse G to be

$$G(x) = \begin{bmatrix} \tilde{B}^{-1} & 0 \\ 0 & \tilde{C}^{-1} \end{bmatrix} \begin{bmatrix} x_d \\ x_I \end{bmatrix} + \begin{bmatrix} \tilde{B}^{-1}b_d \\ \tilde{C}^{-1}b_I \end{bmatrix} . \tag{21}$$

$\tilde{B}^{-1}x_d$ and $\tilde{C}^{-1}x_I$ may be computed by GMRES.

The iteration

$$x_0 = \begin{bmatrix} \tilde{B}^{-1} b_d \\ \tilde{C}^{-1} b_I \end{bmatrix}$$

$$x_k = x_0 + (x_{k-1} - \begin{bmatrix} \tilde{B}^{-1} B & 0 \\ 0 & \tilde{C}^{-1} C \end{bmatrix} x_{k-1}) - \begin{bmatrix} \tilde{B}^{-1} & 0 \\ 0 & \tilde{C}^{-1} \end{bmatrix} \begin{bmatrix} 0 & F \\ E & 0 \end{bmatrix} x_{k-1} \qquad (22)$$

converges to the solution if $\tilde{B}^{-1}$ and $\tilde{C}^{-1}$ are good approximations to B^{-1} and C^{-1}, respectively, and if the spectral radius of

$$\begin{bmatrix} 0 & \tilde{B}^{-1} F \\ \tilde{C}^{-1} E & 0 \end{bmatrix}$$

is less than 1.

If we want to use this to precondition the Schur complement, when it is not known explicitly, we can use the same iteration. Since (9) is equivalent to (10) if the iteration limit is the solution of (9), then the inferior block is the solution of $S x_I = f_I$.

To use this in conjunction with implicit GMRES, whenever there is the need to compute Sv we use the algorithm described in Sect. 5.2 after preconditioning (on the right) with this iterative algorithm:

In the subdomain processes

- $x_d^0 = B^{-1} 0$

- repeat until convergence

  1. send $x_d{}^{k-1}$

  2. receive $x_I{}^{k-1}$

  3. $t = F x_I{}^{k-1}$

  4. $w = B^{-1} t$

  5. $x_d{}^k = -w$

  6. $k = k + 1$

In the interface process

- $x_I^0 = C^{-1} v$

- repeat until convergence

1. receive $x_d{}^{k-1}$

2. send $x_I{}^{k-1}$

3. $t = E x_d{}^{k-1}$

4. $w = C^{-1} t$

5. $x_I{}^k = x_I{}^0 - w$

6. $k = k + 1$

This preconditioner for S is well suited for block distributed computation. It can also be used as preconditioner for A and in that case, the initial vector would not be zero in the subdomains. That corresponds to the block-Jacobi preconditioner.

Some tests were done with the Laplacian test problem where this algorithm works very well, but unfortunatelly in our test problem it fails to converge since the spectral radius of $B^{-1}F$ and $B^{-1}E$ are not less than 1.

6 Numerical Results and Conclusions

The numerical examples were run on a farm of 4 workstations DEC ALPHA connected by FDDI and a GigaSwitch using PVM.

Tables 2 and 3 report CPU and elapsed times in seconds for 32x32, 64x64, 96x96, 128x128 grids with 2, 3 and 5 subdomains. The last column shows the relative speedup compared to the case of 3 processes (1 Master and 2 Slaves).

When using 2 or 3 subdomains, each process correspond to a processor. For 5 subdomains, 2 processes have to share the same machine, but this leads to a better load balance since 2 of the machines (the Master processor and one Slave processor) are slower.

In conclusion we may say that ILU factorization of the Schur complement is very expensive but efficient and feasible when the grid size is not large. The implicit use of the Schur complement allows the parallel solution of the problem with smaller elapsed times and with a small increase in the number of outer iterations. For larger numbers of subdomains the consideration of a coarse grid will be necessary.

Problems for which the approximate inverse preconditioner here proposed can be applied are being sought.

Acknowledgments. The authors are grateful to M. Benzi, L. Giraud, L. Carvalho, from CERFACS (Toulouse), for helpful suggestions. The International Linear Algebra Year - Iterative Methods workshop organized by CERFACS was very enlightening.

Table 2. Parallelization and speedup - explicit S

	nb. iter.	CPU time (Γ/Ω)	Elapsed time	relative speedup
Grid:32x32				
2 subdom.	15	3/515	529	1
3 subdom.	14	6/239	256	2.1
5 subdom.	14	23/128	246	2.2
Grid:64x64				
2 subdom.	16	21/5173	7200	1
3 subdom.	15	46/4755	4864	1.5
5 subdom.	15	137/1954	3999	1.8

Table 3. Parallelization and speedup - implicit S

	nb. iter.	CPU time (Γ/Ω)	Elapsed time	relative speedup
Grid:32x32				
2 subdom.	15	1/125	136	1
3 subdom.	16	2/100	109	1.2
5 subdom.	21	3/67	92	1.5
Grid:64x64				
2 subdom.	17	4/1326	1402	1
3 subdom.	17	4/761	787	1.8
5 subdom.	22	8/475	607	2.3
Grid:96x96				
3 subdom.	17	8/2137	2246	1
5 subdom.	20	12/1226	1770	1.3
Grid:128x128				
3 subdom.	17	13/5231	5382	1
5 subdom.	21	21/3075	4694	1.2

References

1. M. Ahues, F. d'Almeida, M. Telias: On the defect correction method with applications to iterative refinement techniques. Univ. of Grenoble IMAG Report 324 (1982).
2. F. Dias d' Almeida, F. A. Castro, J. M. L. M. Palma, P. Vasconcelos: Development of a parallel implicit algorithm for CFD calculations. Proceedings of AGARD Symposium on Progress and Challenges in CFD methods and algorithms (1995).
3. R. Barret, et. al.: *Templates*, SIAM, Philadelphia (1994).
4. M. Benzi, C. D. Meyer, M. Tuma: A sparse approximate inverse preconditioner for the conjugate gradient method. SIAM J. Sci. Comput. **17** (1996) (in press).
5. M. Benzi, Mi. Tuma: A Sparse Approximate Inverse Preconditioner for Nonsymmetric Linear Systems. CERFACS Report TR-PA-96-15 (1996).

6. P. Bjorstadt, O. Widlund: Iterative methods for the solution of elliptic problems on systems paritioned into substructures. SIAM J. Sci. Stat. Comp. **10** (**5**) (1986) 103-1061.

7. J. Bourgat, R. Glowinski, P. Le Tallec, M. Vidrascu: Variational formulation and algorithm for trace operators in domain decomposition calculations. In T. Chan, R. Glowinski, J. Periaux, O. Widlund, editors, Proceedings of the 2nd SIAM Int. Conf. on domain decomposition methods. SIAM, Philadelphia (1989).

8. J. Bramble, J. Pasciak, A. Schatz: An iterative method for elliptic problems on systems partitioned on subdomains. Math. Comp. **46** (**173**) (1984) 361-369.

9. F. A. Castro: Método de cálculo acoplado para a resolução das equações de Navier-Stokes e continuidade. Relatório técnico, UP-FEUP-SFC (1993).

10. T., Chan; D. Resasco: A survey of preconditioners for domain decomposition. Yale Univ. Dept. of Comp. Sci. Report YALEU/DSC/RR-414 (1985).

11. T. Chan: Analysis of preconditioners for domain decomposition. SIAM J. Numer. Anal. **24** (**2**) (1987) 382-390.

12. T. Chan, T. Mathew: The interface probing technique in domain decomposition. SIAM J. Matrix Anal. Appl.**13** (**1**) (1992) 212-238.

13. E. Chow, Y. Saad: Approximate Inverse Techniques for Block-Partitioned Matrices. Univ. of Minnesota Supercomputer Institute Research Report UMSI 95/13 (1995).

14. I. S. Duff, A. M. Erisman, J. K. Reid: *Direct Methods for Sparse Matrices*. Oxford Science Publication (1986).

15. L. Fox: *An Introduction to Numerical Linear Algebra*. Oxford University Press, Oxford (1964).

16. D. Keyes, W. Gropp: A comparison of domain decomposition techniques for elliptic partial differential equations and their parallel implementation. SIAM J. Sci. Stat. Comp. **8** (**2**) (1987) 166-202.

17. L. Marini, A. Quarteroni: A relaxation procedure for DD methods using finite elements. Numerisch Math. **56**(1989) 575-548.

18. Y. Saad: ILUT: A dual threshold incomplete LU factorization. Num. Lin. Alg. Appl.**1** (1994) 387-402.

19. Y. Saad, M. Schultz: GMRES: A Generalized Minimum Residual Algorithm for solving nonsymmetric linear systems. SIAM J. Sci. Stat. Comput., **7** (**3**) (1986) 856-869.

20. B. Smith, O. Widlund: A domain decomposition algorithm based on a change of hierarchical basis. Dept. of Comp. Sci. Courant Institute Tech. Report 473 (1989)

21. H. Stetter: The defect correction principle and discretization methods. Numerisch Math. **29** (1978) 425-443.

22. Paulo B. Vasconcelos, Filomena D. d'Almeida: The effects of preconditioning iterative methods in coupled discretization of fluid flow problems. 16th Dundee Biennial Conference on Numerical Analysis, University of Dundee (1995).

23. Paulo B. Vasconcelos, Filomena D. d'Almeida: Preconditioning Iterative Methods in Coupled Discretization of Fluid Flow Problems. CERFACS Report TR-PA-96-04 (1996).

Parallel Implementation of Non Recurrent Neural Networks

T.Calonge*, L.Alonso**, R.Ralha*** & A.L.Sánchez*

e-mail: T.Calonge@infor.uva.es

* Dpto. de Informática. Universidad de Valladolid
Facultad de Ciencias. 47011-Valladolid (Spain)
** Dpto. Informática y Automática
Universidad de Salamanca. 37008-Salamanca (Spain)
*** Dpto. de Matemática. Campus de Gualtar.
Universidade do Minho. 4710 Braga (Portugal)

Abstract. The computational models introduced by neural network theory exhibit a natural parallelism in the sense that the network can be decomposed in several cellular automata working simultaneously. Following this idea, we present in this paper a parallel implementation of the learning process for two of the main non recurrent neural networks: the Multilayer Perceptron (MLP) and the Self-Organising Map of Kohonen (SOM).

The system we propose integrates both neural networks applied to an isolated word recognition task. The implementation was carried out on a transputer based machine following the model given by the CSP (Communicating Sequential Processes) specifications. Several experiments with different numbers of processors were made in order to evaluate the performance of the proposed system. The aspects related to the load balancing and communication overheads are discussed.

Index Terms: Neural Networks, Speech Recognition, Parallel Processing & Transputer

1 Introduction

In general terms, when a neural network is used as a pattern recogniser, a learning phase must be previously executed. Actually, the simulation of the neural network using parallel processing focuses on the learning task since it usually implies a large volume of computation. To achieve this goal, in a real application, several factors must be considered, in order to evaluate different strategies of parallelization. These factors can be divided into three categories depending on aspects related to:

- the specific neural networks being used

- the characteristics of the working platform

- the particularities of the problem being solved

Thereby, although the main objective of this paper is the parallel implementation of MLP and SOM, it is necessary to give an overview of the real application as well as the rest of the previous items.

1.1 Artificial Neural Networks Used For Speech Recognition

The proposed system was designed for static pattern recognition, whose inputs correspond to a set of speech signal features vectors. Each of these vectors derives from a previous processing phase applied to the original speech signal; this produces a vector of 128 components that holds time and frequency information [1][2]. In our case, the speech signal comes from the ten Spanish digits, pronounced by several speakers, male and female. In order to assure the independence of the speaker, the samples are divided into learning and test groups. Both data series correspond to disjoint speakers sets. In summary, the system will try to recognise isolated words, speaker independent, with a restricted vocabulary (see Fig. 1).

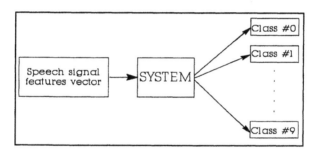

Fig. 1. System functional diagram

There are many classifier systems described in the literature, some of which are based on neural networks paradigms. A good description of these can be found in [3]. An approach that uses recurrent neural networks is needed whenever it is wanted to exploit the time information in an exhaustive way; since this is not the case in our application, non recurrent neural networks are appropriate. Of these, two of the most used in the last decade are the MLP and SOM, since they both have a good behaviour as classifier systems when their inputs consist on a set of static patterns. In previous works [1][2][4], we have tested separately MLP and SOM obtaining success ratios ranging from 90% to 93%. So in order to improve these values, we propose an approach that combines SOM and MLP as illustrated in Fig. 2. Actually, the gain obtained with this hybrid approach is significant since we managed to improve the success ratio to an average of 98%. The explanation for this derives from the fact that in the MLP (or SOM) alone some information produced is considered to be erroneous and discarded. In the proposed system an important part of this information is retrieved successfully.

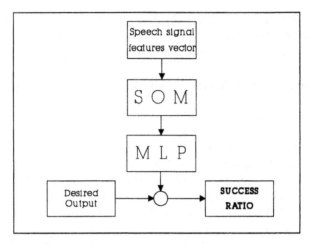

Fig. 2. Functional diagram of the proposed system

Under this configuration, the SOM is the primary pattern recognition system, whose outputs will be the MLP inputs. The SOM is the real feature classifier while the MLP operates as a self organising feature maps recogniser. Hence, this hybrid system is not complex from a conceptual point of view; however, there are many parameters that need to be calculated, specially the weights of neurons. This is known as learning (or training) process and requires a large amount of CPU time.

1.2 Simulation Of Artificial Neural Network

The computational model introduced by a given neural network is enough, in theoretical terms, to construct a particular machine that performs the related tasks. However, it is not affordable to do that, so it is necessary to resort to simulation. Generally speaking, there are two classes of solutions to accelerate the learning process: hardware emulation and hardware implementation, also called neurohardware [5]. The last one is usually a specific purpose hardware that acts as a co-processor of a host computer. There is a number of laboratory prototypes and some of them have been presented as commercial products. Nevertheless, these products are very expensive; yet the main disadvantage is their specific application scope, even within the neural network environment. Such lack of generality is due to the variety of new architectures and computational paradigms that are being introduced everyday.

On the other hand, the hardware emulation is based on a general purpose parallel computer. They are cheaper and more versatile than the neurohardware machines. In some cases, their implementation can even be faster because there is no specific neurohardware for a particular neural network There are five different approaches for parallelism, depending on the level of granularity chosen:

- training-session parallelism (simultaneous execution of different sessions)

- training-example parallelism (simultaneous learning on different training samples)

- layer parallelism (concurrent execution of layers within the network)

- node parallelism (parallel execution of nodes for a single input)

- weight parallelism (simultaneous weighted summation within a node)

Each category refines the level of granularity of the preceding one. This classification was formulated by Nordstrom and Svensson [6], initially for feed-forward networks, but it is valid for back-propagation ones too. The parallel strategy chosen depends on several factors. We may outline the following ones:

- Parallel processing model: shared memory versus distributed local memory.

- Number of available processors.

- The neural network has / has not a layer structure.

- A back-propagation phase is / is not required.

- Number of neurons within each layer.

- Design of experiments.

There are no general rules to select the best solution in advance: a definitive comparison of two different approaches can only be made upon the results of experimentation [7].

In our case, the parallel implementation was done over a T800 transputer network. Of the four physical links each transputer comes with, only two of them are used to build a bidirectional ring topology. Finally, in order to complete the working platform description, it should be necessary to speak about the current programming model. Since we want to optimise the hardware capabilities, we have used OCCAM, which allows us to express the parallelism in a natural way. OCCAM is based on CSP (Communicating Sequential Processes) predicates [8][9]. In summary, the working platform will be a distributed memory system multiprocessor, under a synchronous message passing model [10].

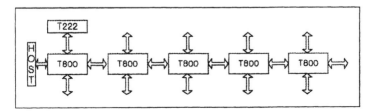

Fig. 3. Hardware working platform

In these circumstances we will show that the "node parallelism", also called "subnet oriented parallelization", may be a successful approach.

2   The Global Task

The proposed system (Fig. 2) can be viewed as a sequence of two components: SOM and MLP, so that each one will be an independent application program. Since the learning process takes a substantial amount of CPU time to complete, it is not reasonable to have it running on line. Therefore, the SOM training process will be the first to be executed, after which MLP will follow. The connection between both applications will be made by a common file where the SOM will store the outputs and from which the MLP will take the inputs.

The learning algorithms for both neural networks are iterative processes, where a distinction is made between:

- "sample" as the input vector for the neural network;

- "iteration" as the set of samples that the system uses for modifying the weights;

- "epoch" as the total number of samples that make up an experiment.

In general terms, these concepts are related as:

$$\{\text{sample}\} \subseteq \{\text{iteration}\} \subseteq \{\text{epoch}\}$$

As we will see later, one of the differences between the two algorithms consists of the size of each iteration. In effect, for the SOM one iteration corresponds to one sample; for the MLP each iteration spawns several samples.

2.1   SOM Training

The architecture of SOM we use can be seen in Fig. 4. It has a bidimensional disposition with a rectangular neighbourhood criterion following the model given in [1][2].

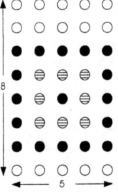

Fig. 4. SOM architecture

As it is well known, the algorithm consists in a non supervised learning process [3], in which the final computation condition can be chosen in several ways. In our case, we set constant the total number of iterations. Briefly, the training process can be described as a set of the following tasks, that have to be executed sequentially for each iteration:

a) Calculate the SOM output.

b) Obtain the neuron winner.

c) Generate the neighbourhood matrix.

d) Update the weights.

At an abstract level, SOM performs a vector quantization, that corresponds to a mapping: $R^M \rightarrow R^N$, where M is the dimension of the input vector and N is the total number of neurons. Since M is greater than N and this transformation must be done with minimal loss of information, then SOM activity can be understood as a process of extraction of features from the original input.

To do that, the weights of each neuron need to be tuned by an algorithm that starts by finding the neuron which best follows the input vector. In this way, we can use the minimum distance criteria $d(\overline{x}^v, \overline{w_i})$, where $\overline{x}^v$ is the vth sample and $\overline{w_i}$ is the ith neuron weights vector. However, if both vectors are normalised and the euclidean distance is used, we simply need to compute the N inner products $(\overline{x}^v \cdot \overline{w_i})$ and find the maximum of them. The previous expression matches the output of the MacCullots-Pits neuron model, so it is also known as SOM output.

In terms of the tasks described above, firstly we have to calculate the SOM outputs, and then, obtain the maximum of them, that is, the winner neuron. Next, the algorithm only updates the weights of the winner neuron and some of its neighbours which are selected by a neighbourhood function. In our case, this function corresponds to a square matrix which contains the neighbour neurons identification. Except for the first two epochs, we use a neigbourhood radius equal to zero which corresponds to a matrix of size one. In the first epoch, such radius is two and the size of the matrix is five; for the second epoch we will have a matrix of size three since the radius is one. We omit the details of the construction of this matrix, although it should be said that the rules followed for this construction are cyclic in the sense that border effects are eliminated.

Finally, to complete an iteration we update the weights of neurons contained in the neighbourhood matrix, following the next expression:

$$\overline{w_i}(t+1) = \frac{\overline{w_i}(t) + \alpha(t)\overline{x}^v}{\left\|\overline{w_i}(t) + \alpha(t)\overline{x}^v\right\|}$$

In the above formula, the variable t represents the iteration number. The parameter α, called training coefficient, is a function of t:

$$\alpha(t) = \frac{25}{1 + t/P}$$

where P represents the size of one epoch. This function follows the initial specification of Kohonen [3], since it accomplishes a fine tuning of the weights for large t

2.2 MLP Training

The architecture of the neural network proposed is shown in Fig. 5, with parameters L, M and N. The size L of the input layer (L) must be 40 according to the SOM output dimension. The output layer has 10 neurons, that fire in response to a given input. Neuron "i" is labelled as corresponding to digit "i".

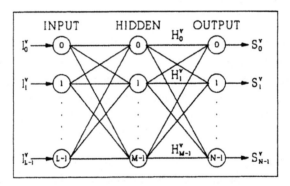

Fig. 5. MLP architecture

The number M of neurons in the hidden layer must be chosen experimentally since there is no general criteria to fix it; we have used M=20.

Numerically, the system works as follows:

$$H_j^v = F(h_j^v); \quad h_j^v = \sum_{i=0}^{L-1} I_i^v J_{ji}^{IH} + J_{jL}^{IH}$$

$$S_k^v = F(s_k^v); \quad s_k^v = \sum_{j=0}^{M-1} H_j^v J_{kj}^{HS} + J_{kM}^{HS}$$

where:

I_i^v is the ith component of the input vector (sample).

h_j^v is the analogic hidden layer of the jth neuron output for the vth sample.

H_j^v is the digital hidden layer of the jth neuron output for the vth sample.

J_{ji}^{IH} is the ith weight for the jth neuron of the hidden layer.

J_{jL}^{IH} is the offset (bias) for the jth neuron of the hidden layer.

s_k^v is the analogic output for the kth neuron of the output layer for the vth sample.

S_k^v is the digital output for the kth neuron of the output layer for the vth sample.

J_{kj}^{HS} is the jth weight of the kth neuron in the output layer.

J_{kM}^{HS} is the offset (bias) of the kth neuron in the output layer.

$F(x)$ is the activation function (usually a sigmoid).

The MLP learning process can be summarised as an iterative algorithm where the weights are modified in each step as follows:

$$w_i(t+1) = w_i(t) + \alpha \frac{\partial E}{\partial w_i} + \beta(w_i(t) - w_i(t-1))$$

The weights' update depends on two terms called training and momentum. The last one is easy to calculate because it is proportional to the difference between the current weight's value and the previous one. However, the training term is more complex since it comes from the application of the gradient algorithm to minimise the quadratic error function:

$$E = \frac{1}{2} \sum_{v=0}^{P-1} \sum_{k=0}^{N-1} (D_k^v - S_k^v)^2$$

where D_k^v is the kth component of the desired output vector for the vth sample.

Therefore, the value E gives the quadratic distance between the desired output and the neural network output for all samples in the experiment. If we apply this algorithm to the output layer weights, we obtain the next expressions:

$$-\frac{\partial E}{\partial J_{mn}^{HS}} = \sum_{v=0}^{P-1} (D_m^v - S_m^V) S_m^v (1 - S_m^v) H_n^v \quad \forall n \neq M$$

$$-\frac{\partial E}{\partial J_{mM}^{HS}} = \sum_{v=0}^{P-1} (D_m^v - S_m^V) S_m^v (1 - S_m^v)$$

And for the hidden layer weights:

$$-\frac{\partial E}{\partial J_{rs}^{IH}} = \sum_{i=0}^{P-1} H_r^v(1 - H_r^v)I_s^v \sum_{k=0}^{N-1}(D_k^v - S_k^v)S_k^v(1 - S_k^v)J_{kr}^{HS} \quad \forall s \neq L$$

$$-\frac{\partial E}{\partial J_{rL}^{IH}} = \sum_{v=0}^{P-1} H_r^v(1 - H_r^v) \sum_{k=0}^{N-1}(D_k^v - S_k^v)S_k^v(1 - S_k^v)J_{kr}^{HS}$$

In these formulae, the following property of the derivative of the sigmoid function has been applied:

$$F'(x) = F(x)(1 - F(x))$$

Finally, the learning algorithm can be written as the following three nested loops:

FOR each epoch IN the experiment.
   FOR each iteration IN one epoch.
   <1>FOR each sample IN one iteration.
      Computation of hidden layer output.
   <2>FOR each sample IN one iteration.
      Computation of output layer output.
   <3> Weights update.
<4>   Computation of hidden layer output with the last weights update.
<5>   Computation of output layer output with the last weights update.
<6>   Calculate maximum error, total error and success ratio. Verify the final training condition.

In the original error back-propagation algorithm [11], each iteration spawns an entire epoch. Nevertheless, in the literature, most simulations are made with one sample iterations because this usually implies a substantial decrease of CPU time on monoprocessor systems. However, in our platform this fact could increase the overhead due to interprocessor communication and decrease the speedup. In practice we have found, on a trial and error basis, that for this particular problem the optimum size of one iteration is equal to 4 samples.

3   Subnet Oriented Parallelization

The same strategy was used for the parallel implementation of the two components (SOM and MLP) of our system. In both cases, the neural network has been decomposed into several subsets of neurons and each one of these subnets assigned to one processor.

Applying this idea to the SOM training, the algorithm is now divided in the following tasks:

<1> Subnet output and winner data
<2> Global winner data
<3> Subnet weights' update

Task <1> corresponds to steps a) and b) in the sequential SOM training (see previous section) but executed for a particular subnet; task <3> relates to steps c) and d). These tasks must be executed simultaneously on each processor and for each iteration. The communication between processors occurs before starting task <2>, since the winner data from the rest of subnets are required to calculate the global winner. Task <2> is global and could be executed in just one processor; in this case, all the other processors would be idle, waiting for the results to be available. Furthermore, the one processor responsible for task <2> would the need to send the results to all the other processors. For these reasons, it is better to replicate task <2> on every processor: the same results will be produced locally and no communication will be required at this point.

Following the horizontal direction in Fig. 5, we will obtain the consequent subnet oriented parallelization for MLP. Therefore each subnet will contain the corresponding part of each layer. With respect to the previous pseudo-code given in *subsection 2.2*, of the six simple tasks in which we have decomposed the algorithm, only the last one <6> should be executed on a single transputer, since it is global. However, as we have already done in the SOM training, we decided to replicate this task on each processor. Finally, we must bear in mind that, as a general rule, before switching to the next task, an *all to all* communication will take place.

4   Communication Overhead

Under this parallelization strategy, different kinds of communication are required and they directly determine the transputer network connection. Three different types of communication will take place:

a) *All to all.*
This kind of transmission is the most frequently required and it implies the largest information volume traffic. In both, SOM and MLP, each processor must send its results to the others before switching tasks. This is the main reason to choose the ring topology as our interconnection network. Although a unidirectional ring could be used, we have found in practice that it is more efficient to use a bidirectional ring. In order to implement the communication algorithms, each processor has an internal variable (*pos*) which is its private identification in the network.

Representing by Q the number of processors in the ring, each processor sends its own packet of data and receives Q-1 packets coming from the other processors. Half of these packets will arrive from "left" and the other half from "right", in two sequences which depends upon the position (pos) the receiving processor, according to the following expressions:

$$right = \left\{(pos + l) \bmod Q\right\}_{l=0}^{(Q-1)/2}$$

$$left = \left\{((Q + pos) - l) \bmod Q\right\}_{l=0}^{Q/2}$$

if Q is even, the "left" sequence has one more packet than the "right" sequence.

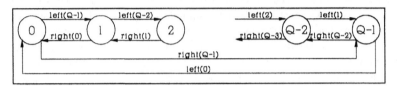

Fig. 6. Bidireccional ring topology

b) *One to all.*

At the beginning the master transputer (pos=0) sends to every processor in the network a copy of the initial data and the initial weights generated by a random function. Since it occurs just once in the algorithm, it does not affect significantly the efficiency. Furthermore, after the master has carried out task <6>, a message for termination (or continuation) must be broadcast.

c) *All to one.*

It takes place only once in the process, when the final training condition has been reached. The results (weights) must be collected by the master transputer and saved in the host computer.

5   Experimental Results

In order to study the effect of communication and granularity in the global performance, several experiments were made with different numbers of processors. The chosen numbers are such that allow a decomposition of the neural network into subnets of the same size (load-balancing is guaranteed in this way). More precisely, for the SOM we use 1, 2, 4 and 5 transputers. The results are shown in the next table, where t_1 is the processing time of task <1>, t_{1c} is the communication time after task <1> execution and $t_{2,3}$ is the processing time of tasks <2> and <3>.

Procs.	1	2	4	5
t_1	0,01490	0,00746	0,00373	0,00299
t_{1c}	0	0,00002	0,00007	0,00009
$t_{2,3}$	0,00092	0,00105	0,00104	0,00103
T_{itera}	0,01580	0,00853	0,00484	0,00411
Speedup	1	1,85	3,27	3,84
Efficiency	100,00%	92,74%	81,69%	76,88%

We do not present the measured times for the first and second epochs for which the neighbourhood radius is 2 and 1, respectively. Although the speedup is better for these epochs, they represent less than 4% of the total number of epochs.

The results obtained for the MLP training are given in the following table, where t_{ip} and t_{ic} are the processing and communication times (in seconds), respectively, of the ith task. T_{epoch} is the time per epoch. Since the global task is an iterative process based on entire epochs, the total time can be obtained multiplying T_{epoch} by the number of epochs required.

Procs.	1	2	5	10
t_{1p}	0,04218	0,02106	0,00845	0,00422
t_{1c}	0	0,00032	0,00045	0,0007
t_{2p}	0,00595	0,00301	0,00128	0,00064
t_{2c}	0	0,00019	0,00025	0,00051
t_{3p}	0,11104	0,06035	0,02394	0,01152
t_{3c}	0	0,00064	0,00083	0,00109
t_{4p}	9,98646	5,09382	2,03668	1,04813
t_{4c}	0	0,04442	0,05389	0,06169
t_{5p}	1,3929	0,71309	0,29994	0,1591
t_{5c}	0	0,01133	0,03277	0,0439
t_{6p}	0,17747	0,17747	0,17747	0,17747
t_{6c}	0	0,00058	0,00237	0,01667
T_{epoch}	43,39	23,15	9,64	5,24
Speedup	1,00	1,87	4,50	8,28
Efficiency	100,00%	93,71%	90,02%	82,81%

6 Conclusions

We developed a parallel implementation of a system consisting of a combination of two neural networks: the Multilayer Perceptron (MLP) and the Self-Organising Map of Kohonen (SOM). The implementation was carried out on a multiprocessor machine with distributed memory where the processors communicate through the exchange of messages.

Different strategies for parallelism are possible and in our experience we found out that node parallelism is the choice that gives the best performance (we are aware of the fact that this is not independent of particular conditions as those related to the architecture of the MLP and SOM and the number of processors used). For this reason, in this paper only the node parallelism approach has been presented, although we have results produced with other parallel implementations. For instance, in previous work [4] we have tested layer parallelism for the MLP network but the resulting system proved to have poor load-balancing properties. In the case of the SOM network, we also evaluated weights parallelism but in this case the communication overheads were the main responsible for the bad performance of the parallel system.

In our current implementation, both MLP and SOM proved to be reasonably well load-balanced systems: except for some exceptions, the workload is evenly distributed among the processors and a good synchronisation in their activity minimises the duration of the idle states in which the processors are waiting for messages to arrive. As far as load-balancing is concerned, the situation is not the ideal since in the MLP one of the tasks (verification of final training condition) must be executed on a single processor but it represents always less than 4% of the total computation time. In the case of the SOM network, two different tasks are carried out by a single processor (the update of the weights and the calculation of "global winner" data). This explains the best performance achieved with the MLP network since the communication overheads are smaller in the SOM system.

Acknowledgements

This work was partially supported by Spanish Government CICYT, and European Union under APDYNN project.

References

1. A.L. Sánchez, L. Alonso & V. Cardeñoso, "A Double Neural Network for Word Recognition", Procc. 10th IASTED International Conference, Innsbruck, February 1992, pp 5-8.
2. Sánchez Lázaro, L. Alonso, C. Alonso, P. de la Fuente, C. Llamas, "Isolated word recognition with a hybrid neural network", International Journal of Mini & Microcomputers, Vol. 16, No. 3, 1994, pp. 134-140.
3. Haykin S, "Neural Networks - A comprehensive foundation", IEEE Press - Macmillan College Publishing Company, Inc. 1994.
4. T. Calonge, A.L. Sanchez & L. Alonso, "A transputer neural network implementation for isolated recognition", Procc. 11th IASTED International Conference, Annecy (France), May 1993, pp 64-66.
5. R. Hetch-Nielsen, "Neurocomputing", Addison-Wesley Publishing Company, 1990.
6. T. Nordstrom and B. Svensson, "Using and Designing Massively Parallel Computer for Artificial Neural Networks", J. Parallel and Distributed Computing, Vol. 14, No. 3, 1992, pp. 260 - 285.
7. Nikola B. Serbedzija, "Simulating Artificial Neural Networks on Parallel Architecture", Computer Magazine, IEEE Computer Society, Vol. 29, No. 3, March 1996, pp. 56-63.
8. C.A.R. Hoare, "Communicating Sequential Processes", Prentice Hall International, 1985.
9. C.A.R. Hoare.- "OCCAM2 Reference Manual". Prentice Hall International, 1988.
10. A. Burns, "Programming in OCCAM2". Addison-Wesley Publishing Company, 1988.
11. B.J.A. Kröse, P.P. van der Smagt, "An Introduction to Neural Networks", University of Amsterdam. Faculty of Mathematics & Computer Science, Fifth edition, 1993.

Parallel Computing of Fragment Vector in Steiner Triple Systems

Erik Urland

Centre Universitaire d' Informatique
Université de Genève, 24 rue Général Dufour
1211 Genève 4, Switzerland
urland@cui.unige.ch

Abstract. In this paper we describe a linear time algorithm using $O(n^2)$ processors for computing the fragment vector in Steiner triple systems. The algorithm is designed for SIMD machine having a grid interconnection network. We show the implementation and some experimental results obtained on the Connection Machine CM-2.

1 Introduction

It has been suggested that combinatorial designs are one of the most active area of research in combinatorial theory. Recently, there has been considerable interest in the study of Steiner triple systems and configurations. One particular aspect of this work is concerned with the decomposition of Steiner systems into configurations with given properties. The articles [5], [6], [9] contain many results, as well as unsolved problems, in this area. Another part investigates avoidance problems: the construction of Steiner systems containing no copies of a particular configuration [1], [7]. At the opposite extreme will be systems which contain the maximum number of given configurations, [13]. The authors of [4] began the study of how the numbers of occurrences of configurations in a Steiner triple system are interrelated. It is not difficult to see that in some ways this is a more fundamental question than decomposition or avoidance problem. In this paper we show an application of parallel processing to the problem of counting the numbers of occurrences of configurations.

A *Steiner triple system* of order n, denoted by STS(n), is a pair (V, B), where V is a set of elements called *points* (or *vertices*) such that $|V| = n$, and B is a collection of 3-subsets of V, called *lines* (*blocks* or *triples*), such that every 2-subset of V is contained in exactly one line. It is well known [10] that STS(n) exists if and only if $n \equiv 1 \ mod \ 6$ or $n \equiv 3 \ mod \ 6$. For example, an STS(7) over $V = \{1, 2, ..., 7\}$ can be formed with the following set of blocks $B = \{[1, 2, 3]; [4, 1, 5]; [1, 6, 7]; [4, 6, 2]; [2, 5, 7]; [3, 4, 7]; [3, 5, 6]\}$. Two STS($n$) (V_1, B_1) and (V_2, B_2) are said to be *isomorphic* if there is a bijection $\phi : V_1 \longmapsto V_2$ such that $[\alpha, \beta, \gamma] \in B_1$ if and only if $[\phi(\alpha), \phi(\beta), \phi(\gamma)] \in B_2$. A $k-line$ *configuration*, $k \geq 1$, is simply any k lines of an STS(n). Consider the above mentioned STS(7). The sets $A_1 = \{[1, 2, 3]; [4, 1, 5]\}$ and $A_2 = \{[2, 5, 7]; [3, 4, 7]\}$ are two different *representations* (or *copies*) of the same configuration C_1 in Fig. 1, while the sets $X_1 = \{[1,2,3]; [4,1,5]; [1,6,7]\}$ and $X_2 = \{[1,2,3]; [4,6,2]; [3,5,6]\}$ represent two different configurations C_2 and C_3 shown in Fig. 1. An *Erdös* configuration of order k is a $k-$line configuration on $k+2$ points which contains no *subconfiguration* of m lines on $m+2$ points for $1 < m < k$. Two $k-$line configu-

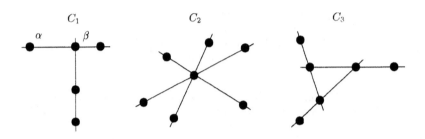

Fig. 1. Configurations having two and three lines

rations C_1 and C_2 are considered to be *isomorphic* if there is a bijection between the vertices of the configurations mapping lines to lines. If $C_1 = C_2$ then the isomorphism is called *automorphism*. By *frequency* (or *number of occurrences*) of the configuration C in a given STS(n) we understand the number of all different representations (copies) of the configuration C in the STS(n). The *degree* of a vertex α is the number of lines in a configuration which contain the vertex α and is denoted by $deg(\alpha)$. Let C be a configuration and let Q be a subset of vertices of C. Then by *partial configuration* P_Q of C we understand a subconfiguration of C which consists only from the lines having at least one point in Q. Let C be a configuration and let α be a vertex of C. Then the *number of symmetries* of C according to vertex α, denoted by $S(C, \alpha)$, is the number of vertices v of C so that there is an automorphism of C mapping α to v. For example, the number of all automorphisms of configuration C_1 in Fig. 1 is eight, but $S(C_1, \alpha) = 4$ and $S(C_1, \beta) = 1$. Let C be a configuration in an STS$(n) = (V, B)$. Let Q be a subset of vertices of C and let P_Q denote the partial configuration of C so that P_Q is formed by a set of lines $L \subseteq B$. Then by $R(C, P_Q)$ we denote the number of all different representations (copies) of C in the STS(n) containing L.

Non-isomorphic Steiner triple systems are frequently used as source data for various kinds of statistical experiments. Similarly, there is no way to determine a linear basis for $k-$line configurations without having a large set of different Steiner systems of a given order, see [4], [8], [14]. The classical approach is to generate randomly STS(n) on a computer by using the *hill-climbing* technique [12] and some enumeration results presented in [11]. This technique appears to be extremely fast but unfortunately cannot guarantee that the Steiner triple systems, constructed in such a way, are non-isomorphic. As there is no known polynomial time algorithm to test isomorphism of STS(n), thus in practice one can use invariants as a proof that two given Steiner systems are non-isomorphic.

In this paper, we concentrate on one invariant called *fragment vector*, originally introduced by Gibbons in [3]. Consider an Erdös configuration of order 4 called *Pasch* configuration (or *quadrilateral*) as shown in Fig. 2. Let α be a point

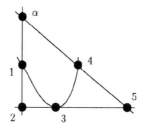

Fig. 2. Pasch configuration

of a Steiner triple system and let $f(\alpha)$ denote the number of Pasch configurations containing α. Then by *fragment vector* of the STS$(n) = (V, B)$ we understand a sequence of integers $f(\alpha_1)$, $f(\alpha_2)$, ..., $f(\alpha_n)$ for $\alpha_i \in V$, $1 \le i \le n$, sorted in non-decreasing order. It is not difficult to see that determining the number of occurrences of the Pasch configuration in a Steiner triple system forms the main part of the algorithm for computing the fragment vector. In fact, the Pasch configuration represents an important combinatorial structure. The significance of such configurations having k lines and $k + 2$ points has also been identified elsewhere [2]. In the next section we present a parallel algorithm for counting the frequency of the Pasch configuration designed for an SIMD machine having a grid interconnection network. We discuss the implementation of the algorithm on the Connection Machine CM-2. In Section 3 we show an optimization of the parallel algorithm together with experimental results obtained on the CM-2. Finally, Section 4 contains some concluding remarks.

2 Parallel Algorithm

Before we describe the algorithm itself we present here the following two lemmas which we make use subsequently in the design of the algorithm.

Lemma 1. *Let α be the vertex of the Pasch configuration as in Fig. 2. Then* $S(Pasch, \alpha) = 6$.

Proof. Let the vertices of the Pasch configuration be denoted as in Fig. 2. The identical mapping is an automorphism of the Pasch configuration which maps α to itself. To see that α can be mapped to the vertices 2 and 5 is trivial. We just simply rotate the triangle and obtain another two automorphisms of the Pasch configuration. Observe that the inner triple $[1, 3, 4]$, not forming the triangle, is always mapped to itself. Thus if α can be mapped to one of vertices 1, 3 or 4 then, by rotating such obtained triangle, we would get the remaining automorphisms. One can check that the required map is $\{\alpha \longmapsto 1, 1 \longmapsto \alpha, 2 \longmapsto 2, 3 \longmapsto 5, 4 \longmapsto 4, 5 \longmapsto 3\}$ and the proof follows. $\qquad\square$

The following corollary is immediate.

Corollary 2. *For any vertex v of the Pasch configuration $S(Pasch, v) = 6$.*

Note that if α is an arbitrary vertex of the Pasch configuration then by Corollary 2, every P_Q, $Q = \{\alpha\}$, is the same partial configuration, up to isomorphism. The following lemma determines the number of different copies of the Pasch configuration, having the same P_Q, in a given $STS(n)$.

Lemma 3. *Let α be an arbitrary vertex of the Pasch configuration so that P_Q is a partial configuration having two lines intersecting in the common point α. Then $R(Pasch, P_Q) \leq 2$.*

Proof. Let $STS(n) = (V, B)$ be a Steiner triple system of order n. Let a and b denote a pair of intersecting blocks in the $STS(n)$ with a common point α. Without loss of generality, let the remaining points of triples a and b be denoted as in Fig. 3. Thus we can assume that a and b form the partial configuration P_Q, $Q = \{\alpha\}$, of the Pasch configuration. We distinguish among two cases:

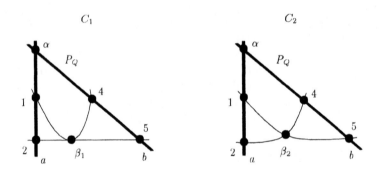

Fig. 3. Two copies of the Pasch configuration

    (*i*) Consider the pair of vertices $\{1, 4\}$ of P_Q. It follows from the definition of $STS(n)$ that there exists a point $\beta_1 \in V$ so that $[1, 4, \beta_1] \in B$. Hence, if $\{2, 5\}$ form a triple with β_1 then we obtain a copy of the Pasch configuration which is in the form of configuration C_1 in Fig. 3.

    (*ii*) The pair of vertices $\{1, 5\}$ form a triple $[1, 5, \beta_2] \in B$. If there is a line in the $STS(n)$ passing through β_2 and containing vertices $2, 4 \in P_Q$ then the resulting configuration is C_2 in Fig. 3 and so we obtain the second copy of the Pasch configuration. $\qquad\square$

Remark. Note that both copies C_1 and C_2 of the Pasch configuration in Fig. 3 can exist simultaneously in a given Steiner triple system.

Concentrate now on a parallel algorithm for computing the frequency of the Pasch configuration in a given STS(n). The algorithm we consider is designed for a massively parallel machine with a set of processors running in SIMD mode and having a grid interconnection network. As one can intuitively see a matrix representation of Steiner triple system is well suited for SIMD-MC^2 parallel machine. According to this representation the proposed algorithm consists of the following three levels.

Algorithm 1.

Level 1. Initialization

- For each point α of STS(n), $0 \le \alpha < n$, let $f(\alpha):=0$. Let $N:=0$, where N denotes the number of occurrences of the Pasch configuration.

- Let $T_{n \times n}$ be a matrix and let $t_{(i,j)}:=-1$ for $0 \le i, j < n$.

- Transform the input list of triples forming STS(n) to matrix $T_{n \times n}$ in such a way that the entry $t_{(i,j)}:=x$, $0 \le i, j < n$, where x is a point of the triple $[i, j, x]$.

- Let $B_{n \times n}$ be an index matrix such that $b_{(i,j)}:=j$ for $0 \le i, j < n$.

Level 2. Precomputation and reduction

- As each triple of the STS(n) is represented six times in T thus we compress the redundant representations. Let $A_{n \times n}$ be a matrix and let $a_{(i,j)}:=t_{(i,j)}$, $0 \le i, j < n$. For all rows i, $0 \le i < n$, and each column j, $0 \le j < n$, if $a_{(i,j)} \ne -1$ then $a_{(i,a_{(i,j)})}:=-1$.

- Delete the entries $a_{(i,j)} < 0$, $0 \le i, j < n$, from the matrix A and the corresponding entries $b_{(i,j)}$ from B. Note that after this reduction both matrices A and B are of the size $n \times \varphi$, where $\varphi = \frac{n-1}{2}$.

- Let $C_{n \times \varphi}$ and $D_{n \times \varphi}$ be two matrices and let $c_{(i,j)}:=a_{(i,j)}$ and $d_{(i,j)}:=b_{(i,j)}$ for $0 \le i < n$ and $0 \le j < \varphi$.

Level 3. Main computation

- Let us shift the entries $c_{(i,j)}$ and $d_{(i,j)}$ of the matrices C and D in such a way that $c_{(i,j)}:=c_{(i,j+1)}$ and $d_{(i,j)}:=d_{(i,j+1)}$ for $0 \le i < n$ and $0 \le j < \varphi - 1$.

- Delete the last column of the matrices A, B, C and D, and set $\varphi := \varphi - 1$. Without loss of generality, assume that a triple $h_1 = [i, b_{(i,j)}, a_{(i,j)}]$ corresponds to line a, in the proof of Lemma 3, for some vertex $\alpha = i$ and for some j-th triple from the list of triples containing α. Similarly for the triple $h_2 = [i, d_{(i,j)}, c_{(i,j)}]$ which corresponds to line b. Thus the pairs of entries $\{a_{(i,0)}, b_{(i,0)}\}$, $\{a_{(i,1)}, b_{(i,1)}\}$, ..., $\{a_{(i,\varphi-1)}, b_{(i,\varphi-1)}\}$ of the matrices A and B, together with some vertex $\alpha = i$, form the triples which represent one line of the partial configuration $P_{\{\alpha\}}$. The second line of $P_{\{\alpha\}}$ is represented in a similar way, by the pairs of corresponding entries of the rows i in C and D. Note that the lines h_1 and h_2, containing the j-th pair of entries of the matrices A, B and C, D, forming $P_{\{\alpha\}}$ for $\alpha = i$ and $0 \leq \alpha < n$, cannot be the same.

- For all $a_{(i,j)}$, $b_{(i,j)}$, $c_{(i,j)}$ and $d_{(i,j)}$, $0 \leq i < n$ and $0 \leq j < \varphi$, check if $t_{(a_{(i,j)}, c_{(i,j)})} = t_{(b_{(i,j)}, d_{(i,j)})}$. If yes, then without loss of generality the two other lines, not forming $P_{\{\alpha\}}$, $\alpha = i$, have a common point β_1. Thus we obtain a representation C_1 of the Pasch configuration as in Fig. 3. In this case let $N := N + 1$ and $f(\pi) := f(\pi) + 1$ for all points π forming the copy of the Pasch configuration.

- Repeat the previous step for all $a_{(i,j)}$, $b_{(i,j)}$, $c_{(i,j)}$ and $d_{(i,j)}$, $0 \leq i < n$ and $0 \leq j < \varphi$, under the condition $t_{(a_{(i,j)}, d_{(i,j)})} = t_{(b_{(i,j)}, c_{(i,j)})}$ which checks the the case (ii) in the proof of Lemma 3.

- Repeat the last four steps until $\varphi = 1$.

- By Lemma 1, let $N := N/6$ and for each point α of the STS(n), $0 \leq \alpha < n$, let $f(\alpha) := f(\alpha)/6$. The computation is performed and the algorithm terminates.

The SIMD-MC^2 computational model can be represented by the Connection Machine CM-2. The physical processors of CM-2 are connected in an n-cube topology but they can be configured as a k-dimensional grid. To represent the set of lines forming an STS(n) we configure the CM-2 as an $N \times N$ grid, where N is n rounded up to the nearest power of 2. The CM-2 performs two different kind of interprocessor communications. One is *general* communication, when a particular processor interacts with an arbitrary processor in the network, and the second is *grid* communication, when the processors communicate along any axis in the grid. Clearly, the operations which involve general communication are computationally much more expensive and we try to avoid them. A Connection Machine can emulate a large number of processors by having each physical processor simulate a number of *virtual* processors. The mapping of virtual processors to the physical processors of the machine is called *geometry*. In our computations we use such a geometry that the virtual processors along two axes in the grid are uniformly mapped to the CM-2 physical processors.

We describe now an implementation of the above presented Algorithm 1 on the Connection Machine CM-2. Assume that each processor in the network contains one element of the parallel variables T, A, B, C and D. Each parallel variable is of the shape $N \times N$ and has the same meaning as in Algorithm 1. We use another two parallel variables Result and Tmp of the shape $N \times N$ to represent the result of the computation and some auxiliary data, respectively. One can see that to implement the first two levels of Algorithm 1 require only a routine transcription to a programming language of CM-2. Note that in Level 2, during precomputation and reduction, we do not delete the negative entries of the matrix A and the related entries of B. Instead of this we rearrange the elements of variable A and the corresponding elements of B in such a way that the first φ elements, in each row of A, have non-negative values, while the rest of the row consists of elements having value -1. This step preserves the one to one correspondence between the elements of variables A and B before and after rearrangement. Thus assume that we have performed the first two levels of Algorithm 1 on a given STS(n). Then one can consider the representation of variables A, B, C and D as in Fig. 4. The dark part of the parallel variables represents the elements having value -1. Here we refer to vertical axis as axis 0 and to horizontal axis as axis 1. We do not present the remaining elements of parallel variables, between n and N along each axis, as these elements are not active. Suppose now that $\varphi = \frac{n-1}{2}$ and all elements of Tmp and Result are

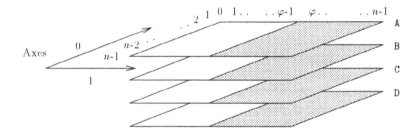

Fig. 4. Representation of parallel variables

initially set to 0. The number of occurrences of the Pasch configuration in the STS(n) is denoted by pn and initially also equals zero. Then an implementation of Level 3 of Algorithm 1 can be described as follows.

Implementation A.

Step 1. Shift the elements of parallel variables C and D by one to the left along axis 1. If we use an axis function with parallel left indexing then this can be done by `[.][.-1]C:=C` and `[.][.-1]D:=D`.

Step 2. Let $\varphi:=\varphi-1$. Set the context to axis $1 < \varphi$. Assign 1 to elements of Tmp satisfying condition `[A][C]T=[B][D]T` and 0 to other elements.

Step 3. Let `Result:=Result+Tmp` and by a global sum reduction assign `pn:=pn+ +=Tmp`. Set `Tmp:=0`.

Step 4. As the context axis $1 < \varphi$ is still set we can assign 1 to elements of `Tmp` satisfying condition `[A][D]T=[B][C]T`, while the remaining elements of `Tmp` have value 0.

Step 5. Perform Step 3.

Step 6. If $\varphi > 1$ then goto Step 1. Else, without any context, perform a grid-spread-add operation along axis 1 on `Result`. Let `pn:=pn/6`. It is not difficult to see that the values $f(\alpha)$, $0 \leq \alpha < n$, are represented by corresponding elements of an arbitrary column of the parallel variable `Result`.

Step 7. End.

Remark. One can see that the Implementation A uses only operations based on grid communication. Moreover, it does not have any part which involves a sequential approach.

Theorem 4. *The Algorithm 1 is correct and using implementation A computes the frequency of the Pasch configuration in a given STS(n) in linear time using $O(n^2)$ processors.*

Proof. Let α be a point of an STS(n). As there are $\varphi = \frac{n-1}{2}$ lines containing α thus we have $\binom{\varphi}{2}$ possibilities for a pair of lines forming P_Q, $Q = \{\alpha\}$. Let $t_0, t_1, \ldots, t_{\varphi-1}$ denote the lines crossing α. Then one can consider the possible cases of P_Q as follows. Assume that the structure in Fig. 5 forms the

$$
\begin{array}{ll}
t_0,\, t_1,\, t_2,\, \ldots & \ldots,\, t_{\varphi-3},\, t_{\varphi-2},\, t_{\varphi-1} \\
t_1,\, t_2,\, t_3,\, \ldots & \ldots,\, t_{\varphi-2},\, t_{\varphi-1} \\
t_2,\, t_3,\, t_4,\, \ldots & \ldots,\, t_{\varphi-1} \\
\vdots & \\
t_{\varphi-3},\, t_{\varphi-2},\, t_{\varphi-1} & \\
t_{\varphi-2},\, t_{\varphi-1} & \\
t_{\varphi-1} &
\end{array}
$$

Fig. 5. Different representations of the partial configuration P_Q

upper left part of a matrix M. In sequential approach, the pairs of lines forming P_Q can be considered column by column, in the following way: $\{t_0, t_1\}$, $\{t_0, t_2\}$, $\ldots$, $\{t_0, t_{\varphi-1}\}$; $\{t_1, t_2\}$, $\{t_1, t_3\}$, $\ldots$, $\{t_1, t_{\varphi-1}\}$ etc. Opposite, in parallel approach we consider the lines forming P_Q as pairs of corresponding triples of two rows of M. The first row, for $\alpha = i$, is $\{[i, b_{(i,0)}, a_{(i,0)}], [i, b_{(i,1)}, a_{(i,1)}],$ ..., $[i, b_{(i,\varphi-1)}, a_{(i,\varphi-1)}]\} = \{t_0, t_1, \ldots, t_{\varphi-1}\}$. By Step 1 of Implemetation A we obtain the other row r of M, $1 \le r < \frac{n-1}{2}$. Initially $r = 1$, thus the pairs of lines $\{t_0, t_1\}, \{t_1, t_2\}, \ldots, \{t_{\varphi-2}, t_{\varphi-1}\}$, represented by $\{[i, b_{(i,0)}, a_{(i,0)}],$ $[i, d_{(i,0)}, c_{(i,0)}]\}$, $\{[i, b_{(i,1)}, a_{(i,1)}], [i, d_{(i,1)}, c_{(i,1)}]\}$, ..., $\{[i, b_{(i,\varphi-2)}, a_{(i,\varphi-2)}],$ $[i, d_{(i,\varphi-2)}, c_{(i,\varphi-2)}]\}$, form P_Q. Let $\varphi := \varphi - 1$ and let the context be set to axis $1 < \varphi$. Then in Step 2 we check the representation C_1 of the Pasch configuration, see Fig. 3, for all φ cases of P_Q at once. Moreover, we can do this for all vertices $\alpha = i$, $0 \le i < n$, at the same time. Similarly, in Step 4 we check the second possible representation C_2 of the Pasch configuration. In Steps 3 and 5 we count the obtained copies of the Pasch configuration, variable **pn**, while the elements of parallel variable **Result** denote the numbers of copies of the Pasch configuration found for a particular P_Q and vertex α. The loop condition $\varphi > 1$ in Step 6 of Implementation A guarantees that in $\frac{n-1}{2} - 1$ cycles each row r of M is considered. Thus we have examined $\binom{\varphi}{2}$, $\varphi = \frac{n-1}{2}$, different pairs of lines forming P_Q for each point α, $0 \le \alpha < n$. As an STS(n) contains precisely $\frac{1}{8}n(n-1)(n-3)$ pairs of intersecting lines, it follows that there is no other P_Q which can lead to a new copy of the Pasch configuration. Further assume, that each element of a parallel variable is handled by a processor. Then the Implementation A requires initially $n\varphi$, $\varphi = \frac{n-1}{2}$, processors to shift the elements of **C** and **D**. There are $\varphi - 1$ cycles. As in each cycle only a less number of processors, determined by the context, is used thus the computation run in $O(n)$ time using $O(n^2)$ processors. $\qquad\qquad\square$

Remark. Note that in Step 6 of Implementation A only the variable **pn** is divided by 6. The sum of each row i of the parallel variable **Result**, $0 \le i < n$, represents the number $f(\alpha)$, $\alpha = i$.

3 Optimization and Experimental Results

It follows from the proof of Theorem 4 that Implementation A initially requires $n\varphi$ processors, $\varphi = \frac{n-1}{2}$. During the computation this number decreases by n in each cycle, and finally only n processors are used. One can see that this tail effect reduces the possibility of *parallelism*, where by parallelism we understand here the average number of active processors in a computation cycle of implementation of Algorithm 1, with respect to $n\varphi$ available processors. We consider this number in percentage. In order to avoid this shortcoming of Algorithm 1 we propose the following optimization. Consider the matrix structure in Fig. 5. The last computation cycle of Implementation A checks the pair of triples $\{t_0, t_{\varphi-1}\}$. The previous cycle checks the pairs of triples $\{t_0, t_{\varphi-2}\}$ and $\{t_1, t_{\varphi-1}\}$ etc. One can see that these pairs of triples can be reconfigured as it is shown in Fig. 6. So assume, that there are *wrap-around* connections along axis 1 between the elements of **C** and **D**, respectively. Hence, in Step 1 of Implementation A the elements $[i][\varphi-1]$**C** receive the values from $[i][0]$**C** for $0 \le i < n$, see Fig. 6. Similarly for the elements of **D**. It is clear that in this case the context is fixed,

$$t_0, \ t_1, \ t_2, \ \ldots \qquad\qquad \ldots, \ t_{\varphi-3}, \ t_{\varphi-2}, \ t_{\varphi-1}$$
$$\longleftarrow t_1, \ t_2, \ t_3, \ \ldots \qquad\qquad \ldots, \ t_{\varphi-2}, \ t_{\varphi-1}, \ t_0 \qquad \longleftarrow$$
$$\longleftarrow t_2, \ t_3, \ t_4, \ \ldots \qquad\qquad \ldots, \ t_{\varphi-1}, \ t_0, \qquad t_1 \qquad \longleftarrow$$
$$\vdots$$
$$\longleftarrow t_{[\frac{\varphi-1}{2}]}, \ t_{[\frac{\varphi-1}{2}]+1}, \ \cdots \qquad \ldots, \ t_{\varphi-1}, \ t_0, \ t_1, \ \ldots, \ t_{[\frac{\varphi-1}{2}]-1} \qquad \longleftarrow$$

$$t_{[\frac{\varphi-1}{2}]+1}, \ t_{[\frac{\varphi-1}{2}]+2}, \ \ldots, \ t_{\varphi-1}$$

Fig. 6. Representations of P_Q using wrap-around connections

axis $1 < \frac{n-1}{2}$. As the pairs of lines forming P_Q, $Q = \{\alpha\}$ where $\alpha \in STS(n)$, are not ordered thus using this modification we do not miss any representation of the Pasch configuration. Note that if $\varphi = \frac{n-1}{2}$ is even, then the last step of computation is performed under the context axis $1 < \frac{n-1}{4}$, which corresponds to the last row of the structure in Fig. 6. Otherwise we would get a multiple of the result in this cycle. In both cases (φ odd, even) the number of occurrences of the Pasch configuration is obtained after $[\frac{n-1}{4}]$ steps. Unfortunately, this does not change anything on the fact that the computation run in linear time.

For purposes of comparison we present in Table 1 some experimental results obtained on 8K CM-2 Connection Machine using modified size random instances. Note that there is an $O(n^3)$ sequential algorithm for computing the fragment

n	S-Alg	Imp-A	Par-A	Imp-B	Par-B
49	0.51	1.88/0.86	52.08 %	1.63/0.73	95.83 %
99	7.26	3.43/2.22	51.02 %	3.03/1.87	100.00 %
151	42.52	5.16/4.08	50.66 %	4.87/3.38	100.00 %
199	127.14	6.91/5.77	50.50 %	5.85/4.79	100.00 %
249	312.17	9.90/8.47	50.40 %	8.26/7.04	99.19 %

Table 1. Evaluation of experimental results

vector of an $STS(n)$. In Table 1 the column S-Alg presents the running time of such algorithm obtained on a Sun SPARCstation 4 computer using Solaris 2.4 system. The order of a Steiner triple system is denoted by n. The columns Imp-A and Imp-B show the total and CM time for Implementation A and its optimized version, respectively. By *total time* we understand here the whole computation time of the front-end machine together with the processor elements of CM, while the *CM time* denote only the time consumed by the processor elements of CM. In all columns the time is given in seconds. In the light of definition of parallelism, in the beginning of this section, we present in columns Par-A and Par-B the average number of active processors of these two implementations during the computation. One can observe that as n is increasing, the speedup of the parallel approach is evident. The optimized version is only slightly faster than the original Implementation A, but the parallelism is significantly better explored. Finally we notice, that the total and CM time are nearly the same, what corresponds to the fact that both implementations do not contain a sequential part.

4 Concluding Remarks

We have presented an application of parallel processing in the area of combinatorial designs. A parallel algorithm for computing the fragment vector in Steiner triple systems has been described. The algorithm is designed for SIMD machine having a grid interconnection network. It has a linear time complexity using $O(n^2)$ processors. We have implemented and tested the algorithm on a Connection Machine CM-2 which represents an SIMD-MC^2 computational model.

There are many papers concerning parallel algorithms and graph theory. Unfortunately, this is not the case of combinatorial designs. Ilustrating our approach, we would like to show a new area of applications of parallel processing in combinatorial theory. For example, restricting attention to designs having block size 3, especially Steiner triple systems, one can ask for parallel algorithms computing configurations with different properties and structures. This question is of great importance in the case of determining a linear basis for configurations or in various statistical experiments. There are also many other problems concerning combinatorial designs where the application of parallel processing would be of great interest.

References

1. Brouwer, A. E.: Steiner triple systems without forbidden subconfigurations. Math. Centrum Amsterdam, ZW 104/77.
2. Brown, W. G., Erdős, P., Sos, V. T.: Some extremal problems on r-graphs. New directions in the theory of graphs, Acad. Press NY (1973) 53–63
3. Gibbons, P. D.: Computing techniques for the construction and analysis of block designs. Ph.D. Thesis, University of Toronto (1976)
4. Grannell, M. J., Griggs, T. S., Mendelsohn, E.: A small basis for four-line configurations in Steiner triple systems. Journal of Combinatorial Designs, Vol. 3, **1** (1995) 51–59
5. Griggs, T. S., Mendelsohn, E., Rosa, A.: Simultaneous decomposition of Steiner triple systems. Ars Combinatoria **37** (1994) 157–173
6. Griggs, T. S., de Resmini, M. J., Rosa, A.: Decomposing Steiner triple systems into four-line configurations. Annals of Discrete Math. **52** (1992) 215–226
7. Griggs, T. S., Murphy, J. P., Phelan, J. S.: Anti-Pasch Steiner triple systems. J. Combin. Inform. System Sci. **15** (1990) 79–84
8. Horak, P., Phillips, N., Wallis, W. D., Yucas, J.: Counting frequencies of configurations in Steiner triple systems. Journal of Combinatorial Designs (to appear)
9. Horak, P., Rosa, A.: Decomposing Steiner triple systems into small configurations. Ars Combinatoria **26** (1988) 91–105

10. Kirkman, T. P.: On a problem in combinations. Cambridge and Dublin Mathematical Journal (1847) 191-204
11. Mathon, R. A., Phelps, K. T., Rosa, A.: Small Steiner systems and their properties. Ars Combinatoria **15** (1983) 3–110
12. Stinson, D. R.: Hill-climbing algorithms for the construction of combinatorial designs. Annals of Discrete Math. **26** (1985), 321–334
13. Stinson, D. R., Wei, Y. J.: Some results on quadrilaterals in Steiner triple systems. Discrete Math. **105** (1992) 207–219
14. Urland, E.: A linear basis for the 7-line configurations. Journal of Comb. Math. and Comb. Computing (to appear)

Stabilizing Large Control Linear Systems on Multicomputers

Vicente Hernández[1] and Enrique S. Quintana-Ortí[2]

[1] Depto. de Sistemas Informáticos y Computación, Universidad Politécnica de Valencia, Aptdo. 22.012, 46.071 - Valencia, Spain
[2] Depto. de Informática, Universidad Jaime I de Castellón, Aptdo. 242, 12.071 - Castellón, Spain.

Abstract. In this paper we present several parallel algorithms for solving the stabilization problem of control linear systems. The first stabilizing algorithm, based on Bass' method, consists of matrix computations which result difficult to parallelize. A different two-stage approach, based on highly parallel spectral division techniques, is then described and used to develop parallel algorithms for the stabilization of large linear systems. The new approach consists of two well-defined stages. First, an efficient spectral division technique is used to identify the stable part of the linear system. Then, the unstable part of the system is stabilized by means of Bass' algorithm. The experimental results on a multicomputer show considerable performance improvements of these two-stage approaches over Bass' algorithm.

1 The stabilization problem

Consider a continuous time-invariant control linear system (CLS)

$$\dot{x}(t) = Ax(t) + Bu(t), \quad x(0) = x_0, \tag{1}$$

where $A \in \mathbb{R}^{n \times n}$ is the state matrix and $B \in \mathbb{R}^{n \times m}$ is the input (or control) matrix. System (1) is stable if the eigenvalues of the state matrix have negative real part, i.e., $\text{Re}(\lambda) < 0$, $\forall \lambda \in \Lambda(A)$. Stability is an important property of CLS since it ensures that a bounded input $u(t)$ produces a bounded output

$$y(t) = Cx(t)$$

where $C \in \mathbb{R}^{p \times n}$ is the output matrix.

One of the most efficient methods to stabilize a CLS consists of applying an state feedback control. In this approach an input $u(t) = -Kx(t) + \bar{u}(t)$, $K \in \mathbb{R}^{m \times n}$, is computed such that the state matrix of the closed-loop system

$$\dot{x}(t) = (A - BK)x(t) + B\bar{u}(t), \quad x(0) = x_0, \tag{2}$$

is stable. In this case, the CLS (or the pair (A, B)) is said to be stabilizable [44].

A different important property of CLS is controllability. CLS (1) is controllable if for any given initial state x_0 and final state x_f there exist a finite time

t_f and an input $u(t)$, $0 \leq t \leq t_f$, such that if $x(0) = x_0$ then $x(t_f) = x_f$. Controllability can be characterized as a numerical rank problem, e.g., (A, B) is controllable iff rank$([B, AB, \ldots, A^{n-1}B]) = n$. More interesting for our purposes, controllability is also characterized as a pole assignment problem; thus, (A, B) is controllable iff for any set of n complex numbers Λ_s, closed under complex conjugation, $\bar{\Lambda}_s = \Lambda_s$, there exists a matrix F such that $\Lambda(A + BF) = \Lambda_s$. Therefore, controllability guarantees stabilizability.

In this paper we present parallel algorithms for solving the stabilization problem on multicomputers. The first approach is based on a parallel implementation of Bass' algorithm [3]. This method basically requires the solution of a linear matrix equation (LME) and an algebraic linear system. Parallel distributed solvers for algebraic linear systems have been extensively studied in the last years (see references in [25]). The results are well-known, basic data layouts (1-D column/row block wrap distributions) achieve high efficiencies. On the other hand, numerically stable algorithms for solving LME consist of three stages: reduction of the coefficient matrices to real Schur form, solution of a reduced LME, and computation of the solution of the initial LME [9, 24, 26]. The parallelism of the LME solvers is analyzed in [30]. The computation of the real Schur form, by means of the iterative QR algorithm, is the most-weighting stage of this process and the results about its parallelization are somewhat dissapointing. Parallel implementations of the iterative QR algorithm obtain poor performances on basic distributions [11, 51]. Advanced data layouts, as block Hankel-wrap distribution [29], improve the performance and scalability of parallel algorithms. However, it is difficult to combine block Hankel-wrap data layouts with basic data layouts to produce a global efficient parallel algorithm for the stabilization problem. 2-D data layouts are generally more efficient for large size problems. Moreover, these data layouts usually offer better scalability properties. However, the matrices that arise in the stabilization problem are dense and of moderate size (a few hundreds of rows/columns). Therefore, scalability is not such an important restriction for our parallel algorithms.

A different approach is proposed in this paper. First, some of the most efficient techniques for spectral division [4, 6] are employed to identify the stable part of the CLS. Then, an state feedback control is computed to stabilize the unstable part of the CLS. The advantage of this approach is the high degree of parallelism of the spectral division techniques: the matrix sign function and the generalized inverse free spectral divide and conquer (SDC) algorithms.

Furthermore, as a direct consequence of the described two-stage process, the stable part of the state matrix of the closed-loop system does not change which, in some problems, may be a desirable property. In this way, for example, the required stabilizing effort K may be smaller.

This two-stage approach was also noted in [27], where it was also proposed as a stabilizing algorithm for large scale systems. In that paper the stabilization problem was solved by means of a quadratic matrix equation, the algebraic Riccati equation [1].

The paper is structured as follows. In the next section we describe Bass' algorithm. In section 3, first the two-stage generic algorithm is introduced and then the spectral division techniques are analyzed. Section 4 describes the interface of a parallel distributed library for matrix algebra kernels. In section 5 the parallel algorithms for the stabilization problem are presented. Finally, in sections 6 and 7 the experimental results are analyzed and the conclusions are outlined.

2 Stabilizing algorithms based on the real Schur form

Bass' algorithm, described in [3], is a stabilizing method with favorable numerical properties. However, Bass' algorithm consists of high cost matrix computations which, besides, are poorly efficient on high performance architectures [2, 4, 29]. The high cost of this algorithm is particularly due to the fact that it does not take into account those eigenvalues of the CLS which are already stable and therefore it carries out an useless reallocation of some eigenvalues of the CLS.

Bass' algorithm for stabilizing a controllable CLS defined by the matrix pair (A, B) is composed of the following stages.

Algorithm 1 { *Let (A, B) define a controllable CLS. The following algorithm computes an input feedback matrix K such that $A - BK$ is stable.* }

1. *Select $\beta > \max_i\{|\lambda_i|: \lambda_i \in \Lambda(A)\}$.*
2. *Compute U orthogonal such that $\bar{A} = U^T A U$ is in real Schur form.*
3. *Compute $\bar{B} = U^T B$ and $\bar{Q} = 2\bar{B}\bar{B}^T$.*
4. *Obtain the solution $\bar{Z}$ of the quasi-triangular Lyapunov LME*

$$-(\bar{A} + \beta I_n)\bar{Z} + \bar{Z}\left(-(\bar{A} + \beta I_n)\right)^T = -\bar{Q}.$$

5. *Compute the feedback matrix*

$$K = \bar{B}^T \bar{Z}^{-1} U^T.$$

In the first stage the value for the parameter β may be chosen as $\beta > \|A\|$. In stage 2 the real Schur form of A is computed. In stage 3 the transformation is applied to B and $\bar{Q}$ is computed. This is a necessary process prior to solving the LME in stage 4. Finally, in stage 5 the feedback matrix K is computed from a special algebraic linear system.

An extension of Bass' algorithm can be applied to uncontrollable systems. In such case $\bar{Z}$ is a symmetric positive semidefinite matrix and the inverse in stage 5 is indeed a generalized More-Penrose inverse [25].

A similar approach was developed in [52] and is also based on the real Schur form of the state matrix.

3 Two-stage stabilizing algorithms based on the spectral decomposition

Consider an orthogonal matrix $U \in \mathbb{R}^{n \times n}$,

$$U = \begin{bmatrix} U_1 & U_2 \end{bmatrix}, \\ {\scriptstyle s \quad n-s}$$

such that

$$U^T A U = \bar{A} = \begin{bmatrix} \bar{A}_{11} & \bar{A}_{12} \\ 0 & \bar{A}_{22} \end{bmatrix} \begin{matrix} s \\ n-s \end{matrix}, \quad U^T B = \bar{B} = \begin{bmatrix} \bar{B}_1 \\ \bar{B}_2 \end{bmatrix} \begin{matrix} s \\ n-s \end{matrix}.$$
$$\phantom{U^T A U = \bar{A} =}{\scriptstyle s \quad n-s}$$

The partition in $\bar{A}$ characterizes the first s columns of U as a basis of an invariant subspace of A. Furthermore, if the eigenvalues of $\bar{A}_{11}$ are those of A with negative real part then U_1 is a basis of the stable invariant subspace of A. Therefore, if the feedback matrix $\bar{K}_2 \in \mathbb{R}^{m \times (n-s)}$ stabilizes the subsystem $(\bar{A}_{22}, \bar{B}_2)$ then the feedback matrix

$$\bar{K} = \begin{bmatrix} 0 & \bar{K}_2 \end{bmatrix} \\ \phantom{\bar{K} =}{\scriptstyle s \quad n-s}$$

stabilizes $(\bar{A}, \bar{B})$. Finally, $K = \bar{K} U^T$ stabilizes the initial CLS (A, B) since

$$\begin{aligned} \Lambda(A - BK) &= \Lambda(U^T(A - BK)U) \\ &= \Lambda(U^T A U - U^T BKU) \quad = \Lambda \left(\begin{bmatrix} \bar{A}_{11} & \bar{A}_{12} \\ 0 & \bar{A}_{22} \end{bmatrix} - \begin{bmatrix} \bar{B}_1 \\ \bar{B}_2 \end{bmatrix} \begin{bmatrix} 0 & \bar{K}_2 \end{bmatrix} \right) \\ &= \Lambda \left(\begin{bmatrix} \bar{A}_{11} & \bar{A}_{12} - \bar{B}_1 \bar{K}_2 \\ 0 & \bar{A}_{22} - \bar{B}_2 \bar{K}_2 \end{bmatrix} \right) = \Lambda(\bar{A}_{11}) \bigcup \Lambda(\bar{A}_{22} - \bar{B}_2 \bar{K}_2) \end{aligned}$$

where $\bar{A}_{11}$ is stable and $\bar{K}_2$ is designed (for example, by means Bass' algorithm) to stabilize $(\bar{A}_{22}, \bar{B}_2)$.

The described procedure defines a two-stage stabilizing algorithm. First, a basis of the stable invariant subspace of the state matrix must be computed to identify the stable part of the system. Then Bass' algorithm is applied to stabilize only the unstable part. This two-stage approach presents the advantage over Bass' algorithm of requiring a smaller effort (control feedback) to stabilize the system. Moreover, it presents a higher degree of parallelism, as will be shown later.

There exist several spectral division techniques that allow the computation of an invariant (deflating) subspace of a matrix (pair of matrices). When combined with Möbius transformations these techniques produce a basis for an invariant (deflating) subspace of a matrix, related with almost any region of the complex plane containing a part of the spectrum of such matrix. Two of the most efficient spectral division techniques are the matrix sign function [50] and the generalized inverse free SDC algorithms [6].

3.1 The matrix sign function and the spectral division

The matrix sign function was introduced in [50] as a reliable method for solving algebraic Riccati equations [1]. In the last years this matrix function has experimented an increasing research activity due to its numerical and parallel properties [4, 5, 21, 43, 45].

As shown next, the matrix sign function may be defined as a generalization of the scalar sign function.

Definition 1. Consider matrix $A \in \mathbb{R}^{n \times n}$ and let

$$Y^{-1}AY = D + N$$

be its Jordan canonical form where Y is the matrix of eigenvectors and principal vectors of A, $D = \operatorname{diag}(\lambda_1, \lambda_2, \ldots, \lambda_n)$ is a diagonal matrix which consists of the eigenvalues of A, and N is a nilpotent matrix which commutes with D.

The matrix sign function of A is then defined as

$$\operatorname{sign}(A) = Y \operatorname{diag}(\alpha_1, \alpha_2, \ldots, \alpha_n) Y^{-1} \qquad (3)$$

where

$$\alpha_i = \begin{cases} +1 & \text{if } \operatorname{Re}(\lambda_i) > 0, \\ -1 & \text{if } \operatorname{Re}(\lambda_i) < 0. \end{cases}$$

If A has an eigenvalue with zero real part then the matrix sign function is not defined.

The following property relates the matrix sign function and the computation of the required basis for the stable invariant subspace.

Proposition 2. *Consider $A \in \mathbb{R}^{n \times n}$ and the spectral projector*

$$P_- = \frac{1}{2}(I_n - \operatorname{sign}(A)).$$

Let $P_- = QR\Pi$ be a rank-revealing QR decomposition of P_-, and $\operatorname{rank}(P_-) = s$, then

$$Q^T AQ = \bar{A} = \begin{bmatrix} \bar{A}_{11} & \bar{A}_{12} \\ 0 & \bar{A}_{22} \end{bmatrix} \begin{matrix} s \\ n-s \end{matrix}$$
$$\quad\;\; s \quad\; n-s$$

where the eigenvalues of $\bar{A}_{11}$ are those of A with negative real part.

The following theorem (see, for example, [50]) describes a numerical method for computing the matrix sign function.

Theorem 3. *Applying Newton's method to the matrix equation $\operatorname{sign}(A)^2 = I_n$, the following iteration is obtained*

$$W_{k+1} = \tfrac{1}{2}(W_k + W_k^{-1}), \quad W_0 = A, \qquad (4)$$

which converges quadratically to $\operatorname{sign}(A)$.

Although the convergence of the iterative scheme (4) is quadratic, in the first iterations it may be a slow process. In [8] several acceleration techniques are described. Most of these techniques consists of a scaling like

$$W_{k+1} = \tfrac{1}{2}(\gamma_k W_k + (\gamma_k W_k)^{-1}), \quad W_0 = A. \tag{5}$$

Recently it has been shown that it is possible to apply an optimal scaling only if the spectrum of W_k is known [31]. However, there exist several quasi-optimal scaling techniques which besides are easy to use. The most well-known acceleration technique is the determinantal scaling, introduced in [8, 14]. The combination of this scaling strategy and iteration (5) is known as the *classic iterative scheme for the matrix sign function*

$$W_{k+1} = \tfrac{1}{2}(\gamma_k W_k + (\gamma_k W_k)^{-1}), \quad W_0 = A,$$

$$\gamma_k = |\det(W_k)|^{-1/n}. \tag{6}$$

An appropriate stopping criterion for the previous iterative scheme is

$$\|W_{k+1} - W_k\|/\|W_k\| < \tau.$$

A different approach for the computation of the matrix sign function is proposed in [43]. It is based on a wide theory, developed in [32], about rational expressions for the matrix sign function such as

$$W_{k+1} = W_k P_r(W_k^2) Q_m^{-1}(W_k^2) \tag{7}$$

where P_r y Q_m are polynomials of degree r and m, respectively.

From the rational expressions a *rational iterative scheme for the matrix sign function* is obtained

$$W_{k+1} = W_k \sum_{i=1}^{m} \tfrac{1}{m x_i} \left(W_k^2 + \alpha_i^2 I_n\right)^{-1}, \quad W_0 = A. \tag{8}$$

In this expression, m is the order of the Padé approximant and

$$x_i = \tfrac{1}{2}\left(1 + \cos\left(\tfrac{\Pi(2i-1)}{2m}\right)\right) \quad y \quad \alpha_i^2 = \tfrac{1}{x_i} - 1 > 0.$$

It is possible to show that $\lim_{k\to\infty} W_k = \text{sign}(A)$ and an appropriate stopping criterion for (8) is

$$\|W_{k+1}^2 - I_n\| < \tau.$$

The previous iterative schemes are globally convergent. On the other hand, locally convergent iterative schemes only guarantee the convergence if some initial conditions are satisfied. However, these locally convergent iterations may offer a lower cost per iteration, faster convergence, etc. [32, 33, 35, 38].

If the initial matrix $W_0 = A$ (or the current approximation W_k to $\text{sign}(A)$) is close to the matrix sign function (specifically, if $\|W_k^2 - I_n\| < 1$) the *Newton-Schulz iterative scheme for the matrix sign function*

$$W_{k+1} = \tfrac{1}{2}W_k(3I_n - W_k^2), \quad W_0 = A, \tag{9}$$

quadratically converges to $\text{sign}(A)$ [32]. Although this iteration presents a higher computational cost per iteration, it may be more efficient in most of the current high performance architectures since it is composed of matrix products.

The application of this iterative scheme is specially appropriate after the rational iterative scheme. The rational iteration can be stopped when

$$\|W_k^2 - I_n\| < \tau$$

and, if $\tau = 1$, when it has converged, the convergence of the Newton-Schulz iterative scheme is guaranteed. Moreover, an appropriate stopping criterion for Newton-Schulz iterative scheme is

$$\|W_{k+1} - W_k\|/\|W_k\| < \tau.$$

The combination of the rational iteration and Newton-Schulz iteration produces the *mixed iterative scheme for the matrix sign function*.

3.2 The generalized inverse free SDC algorithms and the spectral division

These algorithms, which have been recently developed, allow the division of the spectrum of a matrix pencil so that a smaller size problem, with a specified spectrum, is obtained [6]. For this purpose, these algorithms compute an appropriate deflating subspace.

Although there are several other spectral division algorithms [4, 12, 23, 41] the most important advantage of the algorithms presented in [6] is that no matrix inversion is required.

The generalized problem of spectral division may be defined as follows.

Definition 4. Consider a regular pencil $A - \lambda B$ of size $n \times n$ and a region of the complex plane, $\mathcal{D} \subset \mathbb{C}$. We have to find a couple of unitary matrices $Q_L = [Q_{L1}, Q_{L2}]$ and $Q_R = [Q_{R1}, Q_{R2}]$, such that

$$Q_L^H A Q_R = \begin{bmatrix} A_{11} & A_{12} \\ 0 & A_{22} \end{bmatrix} \quad \text{and} \quad Q_L^H B Q_R = \begin{bmatrix} B_{11} & B_{12} \\ 0 & B_{22} \end{bmatrix}, \quad (10)$$

and the eigenvalues of $A_{11} - \lambda B_{11}$ are those of $A - \lambda B$ in $\mathcal{D}$. The subspace $\mathcal{L} = \text{span}(Q_{L1})$ ($\mathcal{R} = \text{span}(Q_{R1})$) is a left (right) deflating subspace of $A - \lambda B$.

The method was initally designed to divide the spectrum of the matrix pencil in the interior or exterior of the unit circle. Using Möbius transforms this method allows the division in other types of regions. For example, in [7] the algorithm is used to compute a basis of the stable invariant subspace of a real Hamiltonian matrix and then to solve an algebraic Riccati equation.

The same algorithm can be used to compute a basis of the stable invariant subspace of a real matrix with no special structure.

Algorithm 2 { *Let A be a real $n \times n$ matrix with no eigenvalues on the imaginary axis. The following algorithm computes an orthonormal basis $\bar{Q}$ of the stable invariant subspace of A.* }

1. Let $A_0 = I_n - A$, $B_0 = I_n + A$ and τ a tolerance parameter that plays the role of a convergence criterion.
2. For $k = 0, 1, \ldots$ until convergence or $k >$ maxit
   (a) Compute the QR decomposition

$$\begin{bmatrix} B_k \\ -A_k \end{bmatrix} = \begin{bmatrix} Q_{11} & Q_{12} \\ Q_{21} & Q_{22} \end{bmatrix} \begin{bmatrix} R_k \\ 0 \end{bmatrix}.$$

   (b) $A_{k+1} = Q_{12}^T A_k$.
   (c) $B_{k+1} = Q_{22}^T B_k$.
   (d) If $\|R_k - R_{k-1}\|_1 \leq \tau \|R_{k-1}\|_1$ then $p = k + 1$, end loop.
   End for.
3. Compute the rank-revealing QR decomposition

$$(A_p + B_p)^{-1} A_p = \bar{Q} R \Pi.$$

The explicit inversion of $(A_p + B_p)$ can be avoided by computing a rank-revealing QR decomposition $A_p = Q_1 R_1 \Pi$ and then an RQ decomposition, $Q_1^T(A_p + B_p) = R_2 \bar{Q}^T$. Thus, $(A_p + B_p)^{-1} A_p = \bar{Q}(R_2^{-1} R_1) \Pi$ and $\bar{Q}$ is the basis of the required invariant subspace. Several new rank-revealing algorithms are presented in [10, 49] and its parallelization is studied in [48].

4 Parallel matrix algebra kernels

4.1 Interface of parallel matrix kernels

We have developed a library of parallel subroutines for solving several matrix algebra problems based on a cyclic column block layout of the matrices. We will only define the interface of each subroutine since the complete description of these subroutines would be extense. Thus, the description of more complex parallel algorithms for the computation of the matrix sign function, stabilization of linear systems, etc. will be considerably simplified.

The parallel algorithms which have been developed can be grouped as follows. In the notation we have used similar criteria to those used in LAPACK [2] with the prefix "par_" to point out that these are parallel subroutines.

Some of the parameters have the same meaning in all the subroutines. Thus, nb is the block size of the distribution, p is the number of processors and μ is the identifier of the local processor.

Matrix product:

– Matrix product.
Subroutine par_gemm(*TrasA, TrasB, m, n, l,* α*, A, B,* β*, C, nb,* μ*, p*)
This subroutine computes and stores in $C \in \mathbb{R}^{m \times n}$ the matrix product
$\alpha C + \beta op(A) \times op(B)$ where $\alpha, \beta \in \mathbb{R}$, $op(A) \in \mathbb{R}^{n \times l}$, $op(B) \in \mathbb{R}^{l \times n}$, and
$op(M) = M$ if *TrasM*='N' or $op(M) = M^T$ if *TrasM*='T'.

Matrix decompositions based on non-orthogonal transformations:

– Gaussian elimination with partial pivoting.
Subroutine par_getrf(*m, n, A, piv, nb,* μ*, p*)
This subroutine computes the LU decomposition with column pivoting of
matrix $A \in \mathbb{R}^{m \times n}$ and stores the resulting triangular factors L and U on A.
The sequence of pivots is stored in vector *piv*.

Triangular linear systems:

– Solution of triangular linear systems.
Subroutine par_trsm(*UpLo, Side, Unit, m, n, B, T, nb,* μ*, p*)
This subroutine solves a triangular linear system with coefficient matrix T
and independent term matrix $B \in \mathbb{R}^{m \times n}$. The parameters *UpLo, Side* and
Unit define the form of the matrix T (upper or lower), the side (left or right)
where this matrix appears in the system and whether the triangular matrix
has unit diagonal elements or not (unit or non-unit).
– Inversion of triangular matrices.
Subroutine par_trtri(*UpLo, Unit, n, T, nb,* μ*, p*)
This subroutine computes the inverse of a triangular matrix $T \in \mathbb{R}^{n \times n}$.
The parameters *UpLo* and *Unit* have the same meaning than in the previous
subroutine.

Matrix decompositions based on orthogonal matrices:

– QR decomposition.
Subroutine par_geqrf(*m, n, A,* β*, nb,* μ*, p*)
This subroutine computes the decomposition $A = QR$ where $A \in \mathbb{R}^{m \times n}$. The
orthogonal matrix $Q \in \mathbb{R}^{m \times m}$ (in compacted form) and the upper triangular
matrix $R \in \mathbb{R}^{m \times n}$ overwrite A. β stores the parameters of the Householder
transformations.
– QR decomposition with column pivoting.
Subroutine par_geqpf(*m, n, A, piv,* β*, r, nb,* μ*, p*)
This subroutine computes the decomposition $A\Pi = QR$ where $A \in \mathbb{R}^{m \times n}$.
The orthogonal matrix $Q \in \mathbb{R}^{m \times m}$ (in compacted form) and the upper triangular matrix $R \in \mathbb{R}^{m \times n}$ overwrite A. Π is a permutation matrix that is
stored in vector *piv*. β stores the parameters of the Householder transformations.
– Construction of orthogonal matrices.
Subroutine par_orgqr(*TrasQ, n, Q,* β*, nb,* μ*, p*)
This subroutine constructs the matrix $Q \in \mathbb{R}^{n \times n}$ or Q^T that performs the
reduction of A to triangular form. On entry the matrix is stored in compacted
form on β and the strictly lower triangular part of Q.

- Application of orthogonal matrices.
  Subroutine par_ormqr(*TrasQ*, *Side*, m, n, U, β, C, nb, μ, p)
  This subroutine computes the product of Q and $C \in \mathbb{R}^{m \times n}$. Matrix Q is stored in compacted form on β and the strictly lower triangular part of Q. The parameters *TrasQ* and *Side* have the same meaning described in previous subroutines.
- *RQ* decomposition.
  Subroutine par_gerqf(m, n, A, β, nb, μ, p)
  This subroutine computes the decomposition $A = RQ$ where $A \in \mathbb{R}^{m \times n}$. The matrices R and Q (in compacted form) of the decomposition overwrite A. β stores the parameters of the Householder transformations.

Condensed forms:

- Reduction to Hessenberg form.
  Subroutine par_gehrd(n, A, β, nb, μ, p)
  This subroutine computes $U \in \mathbb{R}^{n \times n}$ orthogonal and $H \in \mathbb{R}^{n \times n}$ upper Hessenberg such that $H = U^T A U$. Matrices H and U (in compacted form) overwrite A. β stores the parameters of the Householder transformations.
- Construction of orthogonal matrices.
  Subroutine par_orghr(*TrasQ*, n, Q, β, nb, μ, p)
  This subroutine constructs the matrix $Q \in \mathbb{R}^{n \times n}$ or Q^T that performs the reduction of A to Hessenberg form. On entry the orthogonal matrix Q is stored in compacted form on the last $n - 2$ subdiagonals of Q and β stores the parameters of the Householder transformations.
- Reduction to the real Schur form.
  Subroutine par_lahrd(n, H, V, nb, μ, p)
  This subroutine computes $U \in \mathbb{R}^{n \times n}$ orthogonal and $S \in \mathbb{R}^{n \times n}$ in real Schur form such that $S = U^T H U$. Matrix S overwrites the Hessenberg matrix H. On exit, V stores the product $U^T V$.

Linear matrix equations:

- Solution of a Sylvester matrix equation.
  Subroutine par_trsyl(*TrasA*, *TrasB*, m, n, A, B, C, nb, μ, p)
  This subroutine solves the equation $op(A)X + X op(B) = C$ where $C \in \mathbb{R}^{m \times n}$ and $op(A) \in \mathbb{R}^{m \times m}$, $op(B) \in \mathbb{R}^{n \times n}$ are upper quasi-triangular matrices. The solution X overwrites C.

4.2 Parallelization of the matrix sign function

The parallel algorithms for computing the matrix sign function [50] are obtained from the classic (6), rational (8) and Newton-Schulz (9) iterative schemes. We only describe the parallel algorithm for the classic iterative scheme. In the parallel algorithm, matrix $A \in \mathbb{R}^{n \times n}$ is cyclically distributed by blocks of columns of size nb, τ is the convergence threshold, and $maxit$ is the maximum number of iterations allowed. At the end of the computations, $\text{sign}(A)$ is stored on A with the same pattern.

The parallel algorithm is described next.

Algorithm par_gesgnd(n, A, τ, $maxit$, nb, μ, p)
for $k = 1:maxit$
    $W \leftarrow A$

    /* 1. $W_{k+1} = \frac{1}{2}(\gamma_k W_k + (\gamma_k W_k)^{-1})$, $\gamma_k = |\det(W_k)|^{-1/n}$.   */
    par_getrf(n, n, A, piv, nb, μ, p)
    $\gamma \leftarrow \prod_{i=1}^{n} |A(i,i)|^{-1/n}$
    par_trtri($Upper$, n, A, nb, μ, p)
    $B \leftarrow$ upper_triangular_part(A)
    par_trsm($Lower$, $Right$, $Unit$, n, n, B, A, nb, μ, p)
    $B \leftarrow B \cdot piv^T$
    $A \leftarrow (\gamma W + \gamma^{-1} B)/2$

    /* 2. If $\|W_{k+1} - W_k\|/\|W_k\| < \tau$ then end loop.   */
    if ($\|A - W\|/\|W\| < \tau$)
        end loop
    else if ($k = maxit$)
        error
    end if
end for
end algorithm

In stage _1_, a new matrix of the sequence W_{k+1} is computed in each iteration of loop k. For this purpose, the inverse of W_k is computed in three steps. First, we obtain the LU decomposition with column pivoting of A (subroutine par_getrf). Then, the triangular factors L and U will stored on A, and the scaling factor γ is obtained from the diagonal elements of U. Next, the inverse of the triangular matrix U is computed by means of subroutine par_trtri. Finally, the inverse is obtained by solving a lower triangular system and applying the column pivoting. From the stability point of view, this is an appropriate process for computing the explicit inverse of a matrix [20]. The column pivoting may be carried out locally if a special fan-out triangular linear system solver [28] is used to solve $XL = B$. In this algorithm the solution of the triangular linear system is computed by blocks of rows and replicated in all the processors, without any communication overhead. The pivoting is then performed locally. The size of the blocks of rows depends on the size of the workspace available.

In stage _2_ the convergence of the algorithm is checked.

We have developed similar parallel algorithms for the computation of the matrix sign function based on the rational iterative scheme (par_gesgnr) or the Newton-Schulz iterative scheme (par_gesgnn):

  - **Algorithm** par_gesgnr(n, A, τ, $maxit$, m, nb, μ, p).
  - **Algorithm** par_gesgnn(n, A, τ, $maxit$, nb, μ, p).

We have also combined both iterative schemes to produce a mixed iterative scheme, where the rational iteration is applied until $\|W_k^2 - I_n\|_F < 1$, and then the Newton-Schulz iteration is applied:

Algorithm par_gesgnm(n, A, τ, $maxit$, m, nb, μ, p).

The parameter m in algorithms par_gesgnr and par_gesgnm stands for the order of the Padé approximant.

4.3 Parallelization of the generalized inverse free SDC algorithms

The parallel algorithm which computes a basis of the stable invariant subspace is obtained from the algorithm described in subsection 3.2. Consider the matrix $A \in \mathbb{R}^{n \times n}$, cyclically distributed by blocks of columns of size nb, and let τ be the convergence threshold and $maxit$ the maximum number of iterations. The following algorithms computes the number of stable eigenvalues, s, in A and an orthonormal basis of the stable invariant subspace of A.

Algorithm par_gegesd(n, A, Q_T, s, τ, $maxit$, nb, μ, p)

```
/* 1.                                                              */
R ← 0
B ← A − I_n
A ← A + I_n

/* 2.                                                              */
for k = 1: maxit

        /* (a) Compute the QR decomposition of [B^T, A^T]^T.       */
        W ← [B^T, A^T]^T
        par_geqrf( 2n, n, W, β, nb, μ, p )
        Q̄ ← W
        par_orgqr( 'N', 2n, Q̄, β, nb, μ, p )
        W ← upper_triangular_part(W)

        /* (b) A_{k+1} = Q_{12}^T A_k.                             */
        par_gemm( 'T', 'N', n, n, n, 1, Q̄(1:n, n+1, :2n), A, 0, A, nb, μ, p )

        /* (c) B_{k+1} = Q_{22}^T B_k.                             */
        par_gemm( 'T', 'N', n, n, n, 1, Q̄(n+1:2n, n+1, :2n), B, 0, B ,
                  nb, μ, p )

        /* (d) If ‖R_k − R_{k−1}‖_1 ≤ τ‖R_{k−1}‖_1 then end loop.  */
        if (‖W − R‖/‖R‖ < τ)
                end loop
        else if (k = maxit)
                error
```

```
      end if
        R ← W
      end for
```

```
/* 3. Compute the rank-revealing QR decomposition              */
/*      (A_p + B_p)^{-1} A_p = Q_R R_R Π_R.                     */
      B ← A + B
      par_geqpf( n, n, A, piv, β, s, nb, μ, p )
      par_ormqr( 'T', Left, n, n, A, β, B, nb, μ, p )
      par_gerqf( n, n, B, β, nb, μ, p )
      Q_T ← B
      par_orgqr( 'T', n, Q_T, β, nb, μ, p )
      end algorithm
```

In stage *1* matrices R, B and A are initialized.

Inside loop k, in sub-stage *(a)* the QR decomposition of $[B^T, A^T]^T$ is computed by means of subroutine par_geqrf. The upper triangular matrix R is then stored on W and the orthogonal matrix Q is constructed explicitly on $\bar{Q}$ with subroutine par_orgqr. Next, in sub-stages *(b)* and *(c)* A and B are multiplied by the appropriate blocks of the orthogonal matrix Q. In the last sub-stage the convergence of the iteration is tested.

In stage *3* the basis of the invariant subspace of A is computed by means of a rank-revealing QR decomposition which, is basically a QR decomposition with column pivoting.

5 Parallel stabilizing algorithms

In the parallel algorithms that we will describe next, matrices $A \in \mathbb{R}^{n \times n}$ and $B \in \mathbb{R}^{n \times m}$ are cyclically distributed by blocks of size nb. The parameter τ is a convergence threshold for the iterative algorithms and $maxit$ is the maximum number of iterations of these algorithms. On exit, $K \in \mathbb{R}^{m \times n}$ stores the stabilizing feedback matrix, distributed with the same pattern. Matrices A and B are overwritten by intermediate results.

5.1 Parallelization of Bass' algorithm

The parallel stabilizing algorithm based on Bass' method is described next.

Algorithm par_gestly(m, n, A, B, K, nb, μ, p)

```
/* 1. Let β = ||A||_F > max_i{|λ_i| : λ_i ∈ Λ(A)}.             */
      β ← ||A||_F
```

```
/* 2. Find the similarity transformation defined by an orthogonal   */
/*      matrix U that reduces A to the real Schur form, Ā = U^T A U. */
```

```
par_gehrd( n, A, β, nb, μ, p )
U_T ← A
par_orghr( 'T', n, U_T, β, nb, μ, p )
par_lahrd( n, A, U_T, nb, μ, p )
```

/* 3. Compute $\bar{B} = U^T B$ y $\bar{Q} = 2\bar{B}\bar{B}^T$.   */
```
par_gemm( 'N', 'N', n, m, n, 1, U_T, B, 0, B, nb, μ, p )
par_gemm( 'N', 'T', n, n, m, 2, B, B, 0, Z, nb, μ, p )
```

/* 4. Obtain the solution $\bar{Z}$ of the quasi-triangular continuous-time   */
/* Lyapunov equation   */
/* $\left(-(\bar{A} + \beta I_n)\right) \bar{Z} + \bar{Z} \left(-(\bar{A} + \beta I_n)\right)^T = -\bar{Q}$.   */
```
A ← A + βI_n
par_trsyl( 'N', 'T', n, n, A, A, Z, nb, μ, p )
```

/* 5. Compute the feedback matrix $K = \bar{B}^T Z^{-1} U^T$.   */
```
par_getrf( n, n, Z, piv, nb, μ, p )
U_T ← piv^T · U_T
par_trsm( Lower, Left, Unit, n, n, U_T, Z, nb, μ, p )
par_trsm( Upper, Left, Nounit, m, n, U_T, Z, nb, μ, p )
par_gemm( 'T', 'N', m, n, n, 1, B, U_T, 0, K, nb, μ, p )
```
end algorithm

In stage *1* of the algorithm β is chosen as the Frobenius norm of matrix A.

In stage *2*, A is overwritten by its real Schur form. The reduction is computed in three steps. First, A is reduced to Hessenberg form by means of subroutine par_gehrd. Therefore, matrix A is overwritten by its Hessenberg form and the orthogonal matrix U that performs the reduction overwrites β and the last $n-2$ subdiagonals of A. Next, matrix U^T is explicitly formed on U_T (subroutine par_orghr). Finally, in subroutine par_lahrd, the Hessenberg matrix stored in A is reduced to real Schur form and the required transformations are accumulated on U_T.

In stage *3* the independent term matrix, $2U^T BB^T U$, of the continuous-time Lyapunov matrix equation is computed. It is worth noticing that U_T stores U^T.

In the fourth stage, a continuous-time Lyapunov LME is solved. The coefficient matrix $A + \beta I_n$ of this LME is in real Schur form. Although for this purpose we use a Sylvester LME solver, par_trsyl, the computational overhead is small when compared to the global cost of the algorithm.

In the last stage, the feedback matrix is obtained by solving for Y an algebraic linear system $ZY = U^T$. First, we compute the LU decomposition with column pivoting of Z (subroutine par_getrf). Then, the triangular factors L and U are stored in the strict lower triangular part and the upper triangular part of Z, respectively. Next, the row pivoting is applied on U^T and the triangular linear systems are solved (subroutine par_trsm). The application of row pivoting can be carried out locally in the processors. Finally, $K = BY$ is computed by means of subroutine par_gemm.

5.2 Parallelization of the stabilizing algorithm based on the matrix sign function

The parallel algorithm that computes the stabilizing feedback matrix by means of the matrix sign function is obtained from the method presented in subsection 3.1 and the parallel subroutines described in subsections 4.1 and 4.2.

The parallel algorithm is described next.

Algorithm par_gestsd(m, n, A, B, K, τ, $maxit$, nb, μ, p)

```
/* 1. Compute the matrix sign function of A.                          */
P ← A
par_gesgnd( n, P, τ, maxit, nb, μ, p )

/* 2. Compute a rank-revealing QR decomposition                      */
/*    of P_ = (I_n - sign(A))/2.                                      */
P ← (I_n - P)/2
par_geqpf( n, n, P, piv, β, r, nb, μ, p )

/* 3. Apply the transformation to the matrix pair (A, B).            */
par_ormqr( 'T', Left, n, n, P, β, A, nb, μ, p )
par_ormqr( 'N', Right, n, n, P, β, A, nb, μ, p )
par_ormqr( 'T', Left, n, n, P, β, B, nb, μ, p )

/* 4. Compute the stabilizing feedback matrix of the unstable subsystem.*/
K ← 0_{m×n}
par_gestly( m, n - r, A(r + 1:n,r + 1:n), B(r + 1:n,:), K(:,r + 1:n),
            nb, μ, p)
/* 5. The feedback matrix that stabilizes (A,B)                      */
/*    is then obtained from K = [0, K_2] Q^T.                         */
par_ormqr( 'T', Right, n, n, P, β, K, nb, μ, p )
end algorithm
```

First, the algorithm computes the matrix sign function of A by means of the parallel algorithm par_gesgnd, based on the classic iterative scheme with determinantal scaling.

Then, in stage 2 the spectral projector P_-, and an orthonormal basis, are computed from a QR decomposition with column pivoting, par_geqpf. After this subroutine, matrix P is overwritten by the triangular matrix R and the Householder reflectors are stored in β and the strictly lower triangular part of P. The number of stable eigenvalues of A is $r = \text{rank}(P_-)$.

In stage 3 the transformation defined by Q is applied to the system by means of subroutine par_ormqr. Therefore, the transformation is applied from the left and the right to A and only from the left to B.

In the fourth stage we compute the feedback matrix K_2 which stabilizes the unstable part of the linear system (Bass' algorithm, par_gestly).

353

In the fifth and last stage the stabilizing feedback matrix of the original system is obtained by applying transformation Q^T from the right to K_2 (subroutine par_ormqr).

We have also implemented other parallel stabilizing algorithms based on different iterative schemes for the matrix sign function:

- **Algorithm** par_gestsr(m, n, A, B, K, $\bar{m}$, τ, $maxit$, nb, μ, p).
- **Algorithm** par_gestsm(m, n, A, B, K, $\bar{m}$, τ, $maxit$, nb, μ, p).

Algorithm par_gestsr is based on the rational iterative scheme for the matrix sign function. Algorithm par_gestsm is based on a mixed iterative scheme for the matrix sign function, i.e., the rational iterative scheme is applied until the convergence of the Newton-Schulz iterative scheme is guaranteed and then, the Newton-Schulz iteration is applied. In both algorithms the parameter $\bar{m}$ stands for the order of the Padé approximant.

5.3 Parallelization of the stabilizing algorithms based on the generalized inverse free SDC approach

The parallel stabilizing algorithm based on the generalized inverse free SDC algorithms is obtained from the algorithm presented in subsection 3.2 and the parallel subroutines described in subsections 4.1 and 4.3.

The code of the parallel algorithm is presented next.

Algorithm par_gestgs(m, n, A, B, K, τ, $maxit$, nb, μ, p)

```
/* 1. Compute the basis of the stable invariant subspace.          */
par_gegesd( n, A, Q_T, r, τ, maxit, nb, μ, p )

/* 2. Apply the transformation to the matrix pair (A, B).          */
par_gemm( 'N', 'N', n, n, n, 1, Q_T, A, 0, A, nb, μ, p )
par_gemm( 'N', 'T', n, n, n, 1, A, Q_T, 0, A, nb, μ, p )
par_gemm( 'N', 'N', n, n, n, 1, Q_T, B, 0, B, nb, μ, p )

/* 3. Compute the stabilizing feedback matrix of the unstable subsystem.*/
K ← 0_{m×n}
par_gestly( m, n − r, A(r + 1:n,r + 1:n), B(r + 1:n,:), K(:,r + 1:n),
            nb, μ, p )

/* 4. The feedback matrix that stabilizes (A, B)                   */
/*     is then obtained from K = [ 0, K_2 ] Q^T.                   */
par_gemm( 'N', 'T', n, n, n, 1, K, Q, 0, K, nb, μ, p )
end algorithm
```

First, a basis Q_T of the stable invariant subspace of A is computed by means of algorithm par_gegesd.

Then, in stage 2 the transformation defined by Q_T is applied to the system (A, B) by means of subroutine par_gemm. For this purpose, the transformation Q^T is applied from the left to A and B and only from the right to A.

In the third stage a feedback matrix K_2 that stabilizes the unstable part of the system is computed (Bass' algorithm, par_gestly).

In the fourth stage we compute the feedback matrix which stabilizes the initial linear system just by applying transformation Q^T from the right to K_2 (subroutine par_gemm).

6 Experimental results

The performances of the sequential and parallel stabilizing algorithms have been analyzed for two problems. The first one is a random problem where the matrices are uniformly generated. The second one is a modified version of the circulant systems which arise in certain linear-quadratic optimal control problems [36].

The state and input matrices of the linear system, A and B respectively, of the first case study are randomly generated with a uniform distribution between -0.5 and 0.5. This problem will be denoted as Random. With this distribution, approximately half of the eigenvalues of A are stable and the other half unstable. The results are obtained from the arithmetic average of 5 executions on different matrices.

The second study case is a modified version of the problem of circulant matrices (MCM) where the state matrix is defined by

$$
A = \begin{bmatrix}
-2 & 1 & 0 & \dots & 0 & 0 & 0 \\
1 & -2 & 1 & \dots & 0 & 0 & 0 \\
0 & 1 & -2 & \dots & 0 & 0 & 0 \\
\vdots & \vdots & \vdots & \ddots & \vdots & \vdots & \vdots \\
0 & 0 & 0 & \dots & -2 & 1 & 0 \\
0 & 0 & 0 & \dots & 1 & -2 & 1 \\
0 & 0 & 0 & \dots & 0 & 1 & -2
\end{bmatrix},
$$

and the input matrix is $B = I_n$. In this case A is almost an stability matrix (only a few of its eigenvalues are unstable) and therefore the simplest feedback stabilizing matrix is $\bar{K} = 0$ (but for a few columns). This problem allows to analyze the degree of parallelism of the algorithms based on spectral division techniques since these algorithms should detect that the linear system is almost already stable and therefore Bass' algorithm should be applied to a very small size problem.

The size of the matrices ranges between $n, m = 32$ and $n, m = 768$ for the MCM problem. In the Random problem it was not possible to test such large matrices due to the increasingly bad condition number of the problem. Furthermore, the generalized inverse free approach requires workspace matrices much larger than Bass' algorithm and the algorithms based on the matrix sign function. Therefore, some of the result for this approach were obtained by extrapolation.

In the parallel algorithms the block sizes is $nb = 4, 8, 16$ and 32, but only the results corresponding to the optimum block size are shown. In these problems we analyze the CPU time of the sequential and parallel algorithms as a function of two parameters, the problem size and the number of processors (4, 8 or 16).

The precision of the results is evaluated by means of the parameter

$$d = \max\{\text{Re}(\lambda) \ : \ \lambda \in \Lambda(A - BK)\}.$$

Thus, we measure the distance to the imaginary axis of the eigenvalues of the closed-loop state matrix $A - BK$. This measure is not definitive since, in some cases, a small perturbation in matrices A or B may unstabilize $A - BK$. The solution to this problem is given by the so-called robust stabilizing algorithms [44], which are out of the scope of this work.

Three different architectures were used for the experimental evaluation of algorithms. Single computational elements of the shared memory multiprocessors Alliant FX/80 and Silicon Graphics PowerChallenge were used to test the performances of the sequential algorithms. Each computational element of the Alliant FX/80 is a vector processor with a peak performance of 23 Mflops. On the other hand, the Silicon Graphics PowerChallenge has of superscalar RISC processors which perform up to 360 Mflops. The parallel algorithms were evaluated on a Cray T3D multicomputer. This architecture is composed of 256 superscalar processors connected through a 2-D torus topology. The computational elements of this architecture perform up to 150 Mflops.

The precision results only include those obtained by the sequential algorithms on the Silicon Graphics PowerChallenge. The *machine precision* of this architecture is $\epsilon = 2, 2 \times 10^{-16}$. The results of the sequential and parallel algorithms in the Alliant FX/80 and Cray T3D do not differ significantly.

6.1   Performance results

In the comparative study, first we compare the CPU time of the sequential algorithms:

- dgestly: Bass' algorithm.
- dgestsg: Matrix sign function (results corresponding to the best of classic, rational or mixed iterative scheme) + Bass' algorithm.
- dgestgs: Generalized inverse free SDC algorithm + Bass' algorithm.

Figures 1 and 2 show the CPU time of the sequential algorithms. In these figures, the results of the sequential algorithm dgestgs with $n = 256$ are obtained by extrapolating the available data.

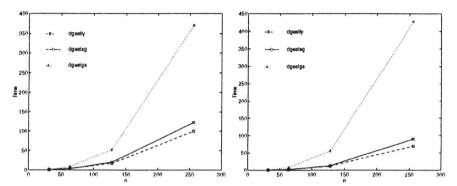

Figure 1. CPU time (sec.) of the sequential stabilizing algorithms on the Alliant FX/80 and the problems Random (left) and MCM (right).

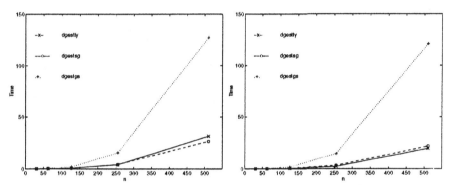

Figure 2. CPU time (sec.) of the sequential stabilizing algorithms on the Alliant FX/80 and the problems Random (left) and MCM (right).

In both problems, the method based on the matrix sign function offers the most efficient results. In the Silicon Graphics, the CPU time of Bass' algorithm is closely similar (even better in the MCM problem) due to the high performance in this architecture of the matrix algebra computations that form this method (basically, the iterative QR algorithm). The CPU time of the generalized inverse free SDC approach is rather high due to the computational overhead that this algorithm requires.

Next, we compare the CPU time of the parallel algorithms:

- par_dgestly: Bass' algorithm.
- par_dgestsg: Matrix sign function (results corresponding to the best of classic, rational or mixed iterative scheme) + Bass' algorithm.
- par_dgestgs: Generalized inverse free SDC algorithm + Bass' algorithm.

Figures 3 through 5 show the CPU time of the parallel algorithms on 4, 8 and 16 processors of the Cray T3D. In the figures, the CPU time of the parallel algorithm par_dgestgs with $n = 512$ and 768 was obtained by extrapolating the available data.

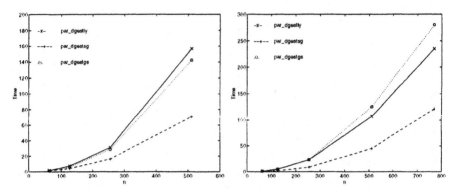

Figure 3. CPU time (sec.) of the parallel stabilizing algorithms on 4 processors of the Cray T3D and the problems Random (left) and MCM (right).

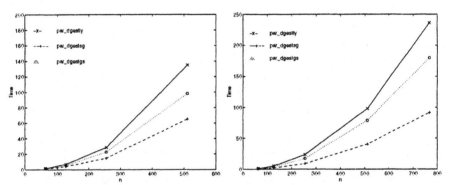

Figure 4. CPU time (sec.) of the parallel stabilizing algorithms on 8 processors of the Cray T3D and the problems Random (left) and MCM (right).

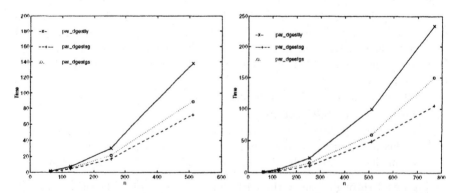

Figure 5. CPU time (sec.) of the parallel stabilizing algorithms on 16 processors of the Cray T3D and the problems Random (left) and MCM (right).

In a first analysis of the figures we relate the performances of the stabilizing algorithms and the type of the problem.

The solution of the MCM problem requires a lower CPU time in all the algorithms (note that the sizes of the largest Random and MCM problems are different). In Bass' algorithm the lower cost of the MCM problem is due to the structure of the state matrix. This matrix is already in Hessenberg form, which reduces the computational cost of the algorithm.

The reason in the two-stage algorithms is quite different. The matrix computations of the spectral division techniques present a higher degree of parallelism (see figures 7 and 8) than those of Bass' algorithm (see figure 6). Moreover, the spectrum of the state matrix of the MCM problem is almost stable. Therefore, the size of the subsystem that must be stabilized by means of Bass' algorithm in this problem (i.e., the "serial" part of the process) is smaller than in the Random problem.

Next, we compare the performances of the different stabilizing algorithms.

The method based on the matrix sign function (par_dgestsg) offers the most efficient results in all the cases. The performances of this method on 4 processors are much higher (2.5 times faster) than those of Bass' algorithm (par_dgestly). In this case, the generalized inverse free SDC approach (par_dgestgs) obtains results very similar to Bass' algorithm due to the high computational cost of this algorithm.

The difference between the CPU time of algorithms par_dgestsg and par_dgestly is even larger when 8 processors are employed (a factor of 3). Furthermore, the increase in the number of processors also produces an improvement on the performances of par_dgestgs, which are now closer to those of the matrix sign function approach.

For 16 processors the results are in the same line. The algorithms based on the matrix sign function are more efficient than Bass' algorithms and the CPU time of the generalized inverse free SDC approach keeps on improving.

Finally, we analyze the behavior of the stabilizing algorithms with the number of processors.

Let us focus the study on the MCM problem. The CPU time of the algorithms experiment a reduction when the number of processors is increased from 4 to 8. However, the reduction is smaller than expected and, in some algorithms (Bass' algorithm and methods based on the matrix sign function) the CPU time is larger if 16 processors are employed. This is due to the high efficiency of the parallel algorithms on 4 processors (figures 6-8). For example, the efficiency of the parallel algorithm based on the matrix sign function is around 0.9 for 4 processors. This value is almost twice the efficiency of the parallel algorithm for 8 processors. Therefore, when the number of processors is multiplied by two, the CPU time only experiments a slight improvement.

However, the behavior of the efficiency of the two-stage algorithms allow us to consider that the gap between the efficiency of these algorithms is smaller for problems of larger size than those presented here. In this way, the two-stage algorithms present a higher efficiency and better scalability properties than Bass' algorithm.

A similar analysis is derived from the Random problem if we consider that the degree of parallelism of this problem is more reduced.

Figures 6 through 8 show the efficiency of the parallel algorithms.

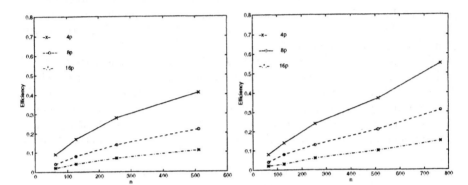

Figure 6. Efficiency of the parallel algorithm par_dgestly on the Cray T3D and the problems Random (left) and MCM (right).

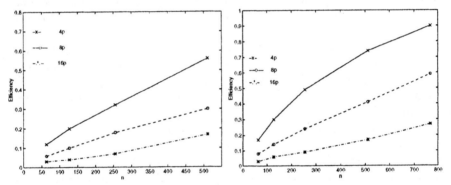

Figure 7. Efficiency of the parallel algorithm par_dgestsg on the Cray T3D and the problems Random (left) and MCM (right).

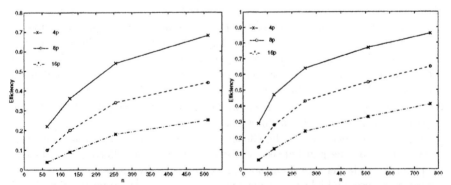

Figure 8. Efficiency of the parallel algorithm par_dgestgs on the Cray T3D and the problems Random (left) and MCM (right).

6.2 Precision results

Finally, we compare the precision of the sequential stabilizing algorithms

- dgestly: Bass' algorithm.
- dgestsd: Classic iterative scheme for the matrix sign function + Bass' algorithm.
- dgestsr: Rational iterative scheme for the matrix sign function + Bass' algorithm.
- dgestsm: Mixed iterative scheme for the matrix sign function + Bass' algorithm.
- dgestgs: Generalized inverse free SDC algorithm + Bass' algorithm.

Table 1 shows the distance

$$d = \max\{\mathrm{Re}(\lambda) \ : \ \lambda \in \Lambda(A - BK)\}$$

obtained by means of these algorithms. We also present the number of iterations that the two-stage algorithms require to compute a basis of the stable invariant subspace (two different types of iterations in the mixed iterative scheme corresponding to the rational and Newton-Schulz iterations). Note that the results of the algorithm based on the generalized inverse free SDC approach for large problems are not known.

n	dgestly	dgestsd	dgestsr	dgestsm	dgestgs
			Random		
32	$-0,18 \times 10^2$	$-0,14$ (10)	$-0,14$ (5)	$-0,14$ (3+3)	$-0,14$ (10)
64	$-0,36 \times 10^2$	$-0,12$ (16)	$-0,12$ (8)	$-0,12$ (6+4)	$-0,12$ (16)
128	$-0,73 \times 10^2$	$-0,14$ (11)	$-0,13$ (6)	$-0,13$ (4+3)	$-0,13$ (11)
256	$-0,14 \times 10^3$	$-0,10$ (14)	$-0,28$ (7)	$-0,28$ (5+4)	$-0,28$ (14)
512	$-0,29 \times 10^3$	$-0,13$ (13)	$-0,28$ (7)	$-0,28$ (6+2)	
			MCM		
32	$-0,13 \times 10^2$	$-0,57$ (8)	$-0,57$ (4)	$-0,57$ (2+5)	$-0,57$ (9)
64	$-0,19 \times 10^2$	$-0,13$ (10)	$-0,13$ (6)	$-0,13$ (3+6)	$-0,13$ (12)
128	$-0,27 \times 10^2$	$-0,11$ (10)	$-0,11$ (6)	$-0,11$ (3+6)	$-0,11$ (11)
256	$-0,39 \times 10^2$	$-0,10$ (13)	$-0,10$ (7)	$-0,10$ (5+4)	$-0,10$ (14)
512	$-0,55 \times 10^2$	$-0,10$ (13)	$-0,10$ (7)	$-0,10$ (5+4)	
768	$-0,67 \times 10^2$	$-0,05$ (15)	$-0,05$ (8)	$-0,05$ (6+4)	

Table 1. Distance d and number of iterations of the stabilizing algorithms on the Silicon Graphics PowerChallenge.

Table 1 shows that all the spectral division algorithms obtain the same results. However, these results differ significantly of those obtained by Bass' algorithm. The reason is that Bass' algorithm computes a stabilizing feedback matrix that reallocates both the stable and unstable eigenvalues of the initial system. On the other hand, the algorithms based on the spectral division respect the stable eigenvalues of the initial system.

In the Random problem we have used Möbius transforms to divide the spectrum along the line $\lambda = t$, where $t < 0$ is close to 0. In this way we avoid numerical difficulties which appear in large-scale problems, where there may exist eigenvalues very close to 0.

7   Conclusions

In this paper we have presented several parallel implementations of stabilizing algorithms. The first of these algorithms, based on Bass' method, shows poor parallel performances due to the difficulties of parallelizing the most weighting stage, the computation of the real Schur form by means of the iterative QR algorithm.

Therefore, two different two-stage approaches, based on highly parallel spectral division techniques, have been described and used to develop parallel stabilizing algorithms for large scale systems. The first stage of these approaches is a spectral division technique that allows the identification of the stable part of the CLS. In the second stage, Bass' algorithm is applied to the unstable part of the CLS, which requires the stabilization of a problem of smaller size.

The matrix sign function is an efficient spectral division technique due to its low cost and high parallel efficiency. The generalized inverse free SDC algorithms, though highly parallel and with better numerical properties, present a higher computational cost. Furthermore, the generalized inverse free SDC algorithms require a considerable large amount of workspace.

Both spectral division algorithms allow the development of two-stage algorithms which reduce the seriality of the approach based on Bass' algorithm.

References

1. B. D. O. Anderson, J. B. Moore. *Linear Optimal Control*. (Prentice-Hall Int., Englewood Cliffs, USA, 1971).
2. E. Anderson et al. *Lapack user's guide*. SIAM, 1992.
3. E. S. Armstrong. *An extension of Bass' algorithm for stabilizing linear continuous systems*. IEEE Trans. on Automatic Control, AC-20, pp. 153-154, 1975.
4. Z. Bai, J. Demmel. *Design of a parallel nonsymmetric eigenroutine toolbox, part 1*. Computer Science Division Report UCB//CSD-92-718, University of California at Berkeley, 1992.
5. Z. Bai, J. Demmel, J. Dongarra, A. Petitet, H. Robinson. *The spectral decomposition of nonsymmetric matrices on distributed memory parallel machines*. Computer Science Division Report, Work in progress, University of California at Berkeley, 1994.

6. Z. Bai, J. Demmel, M. Gu. *Inverse free parallel spectral divide and conquer algorithms for nonsymmetric eigenproblems.* Computer Science Division Report, UCB//CSD-94-793, University of California at Berkeley, 1994.

7. Z. Bai, Q. Qian. *Inverse free parallel method for the numerical solution of algebraic Riccati equations.* Proceedings of the Fifth SIAM Conference on Applied Linear Algebra (Ed. J. G. Lewis), SIAM, pp. 167-171, 1994.

8. L. A. Balzer. *Accelerated convergence of the matrix sign function.* Int. Journal of Control, 32, pp. 1057-1078, 1980.

9. R. Bartels, G. W. Stewart. *Algorithm 432: The solution of the matrix equation AX-XB=C.* Comm. of the ACM, 15, pp. 820-826, 1972.

10. C. H. Bischof, G. Quintana. *Computing rank-revealing QR factorizations of dense matrices.* Tech. Report MCS-P559-0196, Mathematics and Computer Science Division, Argonne National Laboratory, 1996.

11. D. Boley, R. Maier. *A parallel QR algorithm for the non-symmetric eigenvalue problem.* Technical Report TR-88-12, Dept. of Computer Science, Univ. of Minnesota, 1988.

12. A. Ya. Bulgakov, S. K. Godunov. *Circular dichotomy of the spectrum of a matrix.* Siberian Math. J., 29, pp. 734-744, 1988.

13. P. A. Businger, G. H. Golub. *Linear least squares solutions by Householder transformations.* Numerische Mathematik, 7, pp. 269-276, 1965.

14. R. Byers. *Solving the algebraic Riccati equation with the matrix sign function.* Lin. Algebra & Its Appl., 85, pp. 267-279, 1987.

15. J. Choi, J. Dongarra, D. Walker. *Level 3 BLAS for distributed memory concurrent computers.* Proceedings of the CNRS-NSF Workshop, 1992.

16. J. M. Claver, V. Hernández, E. S. Quintana. *Solving discrete-time Lyapunov equations for the Cholesky factor on a shared memory multiprocessor.* Parallel Processing Letters, (to be published) 1996

17. J. J. Dongarra et al. *A set of level 3 BLAS basic linear algebra subprograms.* Tech. Report ANL-MCS-TM-88, Dept. of Computer Science, Argonne National Laboratory, 1988.

18. J. J. Dongarra, S. Hammarling, D. C. Sorensen. *Block reduction of matrices to condensed form for eigenvalue computations.* Tech. Report ANL-MCS-TM88, Dept. of Computer Science, Argonne National Laboratory, 1987.

19. J. J. Dongarra et al. *An extended set of Fortran basic linear algebra subprograms.* ACM Trans. on Mathematical Software, 14, pp. 1-17, 1988.

20. J. Du Croz, N. J. Higham. *Stability of methods for matrix inversion.* IMA J. of Numerical Analysis, 12, pp. 1-19, 1992.

21. J. D. Gardiner. *A stabilized matrix sign function algorithm for solving algebraic Riccati equations.* SIAM J. Scientific & Statistical Computing, (to be published), 1996.

22. G. A. Geist et al. *PVM: Parallel Virtual Machine. A users' guide and tutorial for networked parallel computing.* (The MIT Press, Cambridge, USA, 1994).

23. S. K. Godunov. *Problem of the dichotomy of the spectrum of a matrix.* Siberian Math. J., 27, pp. 649-660, 1986.

24. G. H. Golub, S. Nash, C. F. Van Loan. *A Hessenberg-Schur method for the problem $AX + XB = C$.* IEEE Trans. on Automatic Control, AC-24, pp. 909-913, 1979.

25. G. H. Golub, C. F. Van Loan. *Matrix computations.* (The Johns Hopkins University Press, Baltimore, USA, 1989).

26. S. J. Hammarling. *Numerical solution of the stable, non-negative definite Lyapunov equation*. IMA Journal of Numerical Analysis, 2, pp. 303-323, 1982.

27. C. He, V. Mehrmann. *Stabilization of large linear systems*. Proceedings of the IEEE Workshop (Eds. M. Karny and K. Warwick), Praga, Eslovaquia, 1994.

28. M. T. Heath, C. H. Romine. *Parallel solution of triangular systems on distributed-memory multiprocessors*. SIAM J. Scientific & Statistical Computing, 9, pp. 558-588, 1988.

29. G. Henry, R. van de Geijn. *Parallelizing the QR algorithm for the unsymmetric algebraic eigenvalue problem: myths and reality*. Lapack Working note #79, 1994.

30. V. Hernández, E. S. Quintana, M. Marqués. *Solving linear matrix equations in control problems on distributed memory multiprocessors*. Proceedings of the 33rd Conference on Decision and Control, pp. 449-454, Lake Buena Vista, USA, 1994.

31. C. Kenney, A. J. Laub. *On scaling Newton's method for polar decomposition and the matrix sign function*. Report No. SCL 89-11, Dept. of Electrical and Computer Engineering, University of California, 1989.

32. C. Kenney, A. J. Laub. *Rational iterative methods for the matrix sign function*. SIAM J. Matrix Analysis & Appl., 12, pp. 273-291, 1991.

33. C. Kenney, A. J. Laub, P. M. Papadopoulos. *A Newton-squaring algorithm for computing the negative invariant subspace of a matrix*. IEEE Trans. on Automatic Control, AC-38, pp. 1284-1289, 1993.

34. D. L. Kleinmann. *An easy way to stabilize a linear constant system*. IEEE Trans. on Automatic Control, AC-15, pp. 692, 1970.

35. Z. Kovarik. *Some iterative methods for improving orthonormality*. SIAM J. Numerical Analysis, 7, pp. 386-389, 1970.

36. A. J. Laub. *A Schur method for solving algebraic Riccati equations*. IEEE Trans. on Automatic Control, AC-24, pp. 913-921, 1979.

37. C. L. Lawson et al. *Basic linear algebra subprograms for Fortran usage*. ACM Trans. on Mathematical Software, 5, pp. 308-323, 1979.

38. R. B. Leipnik. *Rapidly convergent recursive solution of quadratic operator equations*. Numerische Mathematik, 17, pp. 1-16, 1971.

39. G. Li, T. Coleman. *A Parallel triangular solver for a distributed-memory multiprocessor*. SIAM J. Scientific & Statistical Computing, 9, pp. 485-502, 1989.

40. G. Li, T. Coleman. *A new method for solving triangular systems on distributed-memory message-passing multiprocessor*. SIAM J. Scientific & Statistical Computing, 10, pp. 382-396, 1989.

41. A. N. Malyshev. *Parallel algorithm for solving some spectral problems of linear algebra*. Linear Algebra & Its Appl., 188-189, pp. 489-520, 1993.

42. V. Mehrmann. *The autonomous linear quadratic control problem*. (Springer-Verlag Berlin, Alemania, 1991).

43. P. Pandey, C. Kenney, A. J. Laub. *A parallel algorithm for the matrix sign function*. Int. J. of High Speed Computing, 2, pp. 181-191, 1990.

44. P. Hr. Petkov, N. D. Christov, M. M. Konstantinov. *Computational methods for linear control systems*. (Prentice-Hall International Ltd., United Kingdom, 1991).

45. E. S. Quintana, V. Hernández. *Parallel algorithms for solving the algebraic Riccati equation via the matrix sign function*. Proceedings of the 3rd IFAC Workshop on Algorithms and Architectures for Real-Time Control - AARTC'95, pp. 533-542, Ostend, Belgium, 1995.

46. E. S. Quintana, V. Hernández. *Parallel solvers based on the Schur vectors for the algebraic Riccati equation*. Conference on Mathematical Theory of Networks and Systems, St. Louis, USA, (to be published) 1996.

47. E. S. Quintana, V. Hernández. *Algoritmos por bloques y paralelos para la estabilización de sistemas dinámicos lineales.* Tech. report DSIC-II/5/96, Dpto. de Sistemas Informáticos y Computación, Universidad Politécnica de Valencia, 1996.
48. G. Quintana, E. S. Quintana. *Parallel bidimensional algorithms for computing rank-revealing QR factorizations.* Conference on High Performance Computing and Local Area Gigabyte Networks, Essen, Germany, (to be published), 1996.
49. G. Quintana, X. Sun, C. H. Bischof. *A BLAS-3 version of the QR factorization with column pivoting.* Tech. Report ANL-MCS-P551-1295, Dept. of Computer Science, Argonne National Laboratory, 1996.
50. J. Roberts. *Linear model reduction and solution of the algebraic Riccati equation by the use of the sign function.* Int. Journal of Control, 32, pp. 677-687, 1980.
51. G. W. Stewart. *A parallel implementation of the QR algorithm.* Parallel Computing, 5, pp. 187-196, 1987.
52. V. Sima. *An efficient Schur method to solve the stabilization problem.* IEEE Trans. on Automatic Control, AC-26, pp. 724-725, 1981.
53. R. A. van de Geijn, D. G. Hudson. *An efficient parallel implementation of the non-symmetric QR algorithm.* Proceedings of the 4th Conference on Hypercube Concurrent Computers and Appl., 1989.

An Interface Based on Transputers to Simulate the Dynamic Equation of Robot Manipulators Using Parallel Computing

J.C. Fernández

Dpto. Informática. Universidad Jaume I, 12071-Castellón, Spain
e-mail : jfernand@inf.uji.es

L. Peñalver

Dpto. Ing. Stmas., Comp. y Automática, Univ. Politécnica, 46071-Valencia, Spain
e-mail : lourdes@disca.upv.es

M. Sobejano, J. Martínez

Dpto. Stmas. Inform. y Comp., Univ. Politécnica, 46071-Valencia, Spain

V. Hernández

Dpto. Stmas. Inform. y Comp., Univ. Politécnica, 46071-Valencia, Spain
e-mail : vhernand@dsic.upv.es

J. Tornero

Dpto. Ing. Stmas., Comp. y Automática, Univ. Politécnica, 46071-Valencia, Spain
e-mail : jtornero@disca.upv.es

Abstract. The inverse dynamics control of robot manipulators is based on the application of a nonlinear feedback control law. The implementation of this control requires the computation of all the terms of the dynamic equation at each sample instant. In order to solve this problem, distributed memory parallel algorithms using the Lagrange-Euler formulation are presented. This formulation permits us to establish matrix structures that are distributed among the processors by rows with a good computational load balance. A new Windows interface, called WinServer, based on transputers, which permits us to modify the robot parameters and to simulate the dynamic equation is also presented. This interface can be used as a monitoring tool that allows the user to know the situation of the different processes.

1 Introduction.

The dynamic equation of a rigid manipulator with n arms is given by

$$\tau = D(q)\ddot{q} + h(q, \dot{q}) + c(q), \tag{1}$$

where τ is the $n \times 1$ vector of nominal driving torques, q is the $n \times 1$ vector of nominal generalized coordinates, and $\dot{q}$ and $\ddot{q}$ are the $n \times 1$ vectors of the first and second derivatives of the vector q, respectively.

The idea of inverse dynamics is to seek a nonlinear feedback control law

$$\tau = f(q, \dot{q}), \tag{2}$$

which, in the ideal case, exactly linearizes the nonlinear system, resulting in a linear closed loop system

$$D(q)\ddot{q} + h(q, \dot{q}) + c(q) = f(q, \dot{q}). \tag{3}$$

Many advanced manipulator control schemes are based on inverse dynamics calculation [7, 18, 19]. There are also different approaches based on adaptative inverse dynamics [1, 3, 4, 16].

The implementation of the control law (2) requires the on-line computation of the dynamic equation terms at each sample instant, that is the inertia matrix, $D(q)$, the vector of centrifugal and Coriolis forces, $h(q, \dot{q})$, and the vector of gravitational forces, $c(q)$.

The main problem here is the need to implement algorithms to obtain the dynamic equation in real-time. One way to solve this problem is to use a parallel computing approach.

Various robot formulations have been proposed during the past two decades. The Lagrange-Euler formulation has a very well structured and systematic representation that accommodates different control applications. The recursive Lagrangian formulation has a better computation time but destroys the structure of the equation. The Newton-Euler formulation is based on a set of highly recursive equations which makes it very difficult to use in control applications. In this paper the Lagrange-Euler formulation is employed. This formulation incurs in a heavy computational cost. Many attempts had been made to reduce the order of computations of the dynamics [2, 13, 14]. We introduce an optimization of the matrix products to obtain the dynamic equation. Moreover distributed memory parallel algorithms for computing all the terms of the dynamic equation in the Lagrange-Euler formulation are described.

A friendly user interface in Windows, called WinServer, is proposed in order to develop parallel algorithms introducing the robot parameters and obtaining the simulation results. The main goal is to have a monitoring tool which permits us to apply directly the results obtained for the dynamic simulation to a robot manipulator.

The paper is organized as follows. Section 2 describes the dynamic equation of robot manipulators and introduces different data structures. In section 3 the parallel algorithms developed are introduced. The interface based in transputers for the simulation of the dynamic equation is described in section 4. Section 5 shows the results obtained. Finally, conclusions and remarks are given in section 6.

2 Dynamic Equations of Rigid Manipulators.

The dynamic equation (1) of a rigid manipulator with n links can be expressed in explicit form as

$$\begin{bmatrix} \tau_1 \\ \vdots \\ \tau_n \end{bmatrix} = \begin{bmatrix} d_{11} & \cdots & d_{1n} \\ \vdots & \ddots & \vdots \\ d_{n1} & \cdots & d_{nn} \end{bmatrix} \begin{bmatrix} \ddot{q}_1 \\ \vdots \\ \ddot{q}_n \end{bmatrix} + \begin{bmatrix} h_1(q, \dot{q}) \\ \vdots \\ h_n(q, \dot{q}) \end{bmatrix} + \begin{bmatrix} c_1(q) \\ \vdots \\ c_n(q) \end{bmatrix}. \tag{4}$$

The elements of the $n \times n$ symmetric positive definite inertia matrix $D(q)$ are given by

$$d_{ij} = \sum_{k=max\{i,j\}}^{n} tr(U_{jk} J_k U_{ik}^T).$$ (5)

for $i, j = 1 : n$. In this expression $tr(\cdot)$ represents the trace of a matrix (the sum of its diagonal elements) and J_k, $k = 1 : n$, is the inertia tensor related to the link k.

To obtain D, it is necessary to compute U_{jk}, $j, k = 1 : n$, given by

$$U_{jk} = \frac{\partial^0 A_k}{\partial q_j} = \begin{cases} {}^0A_{j-1} Q_j {}^{j-1}A_k, \ j \leq k \\ 0, \ j > k \end{cases}$$ (6)

where Q_j is the well-known constant matrix that allows us to calculate the partial derivative of ${}^{j-1}A_k$ related to the link j. For open-chaine robot arms, a block upper triangular matrix can be defined, where each block is one of the matrices U_{jk}

$$U = \begin{bmatrix} U_{11} \ U_{12} \ \cdots \ U_{1n} \\ U_{22} \ \cdots \ U_{2n} \\ \ddots \ \vdots \\ U_{nn} \end{bmatrix}.$$ (7)

To obtain matrix U it is necessary to compute matrices ${}^j A_i$. From the Denavit-Hartenberg convention [6], ${}^j A_i$ is an $n \times n$ matrix that transforms the coordinates of a point from frame i to frame j. This matrix is defined in a recursive way as

$${}^j A_i = {}^j A_{i-1} {}^{i-1}A_i.$$ (8)

where

$${}^{i-1}A_i = \begin{bmatrix} cos\theta_i & -cos\alpha_i sen\theta_i & sen\alpha_i sen\theta_i & a_i cos\theta_i \\ sen\theta_i & cos\alpha_i cos\theta_i & -sen\alpha_i cos\theta_i & a_i sen\theta_i \\ 0 & sen\alpha_i & cos\alpha_i & d_i \\ 0 & 0 & 0 & 1 \end{bmatrix}$$ (9)

in the case of a revolution joint, or

$${}^{i-1}A_i = \begin{bmatrix} cos\theta_i & -cos\alpha_i sen\theta_i & sen\alpha_i sen\theta_i & 0 \\ sen\theta_i & cos\alpha_i cos\theta_i & -sen\alpha_i cos\theta_i & 0 \\ 0 & sen\alpha_i & cos\alpha_i & d_i \\ 0 & 0 & 0 & 1 \end{bmatrix}$$ (10)

for the case of a prismatic joint. Taking this expression into account, the following block upper triangular matrix can be defined

$$A = \begin{bmatrix} {}^0A_1 \ {}^0A_2 \ \cdots \ {}^0A_n \\ {}^1A_2 \ \cdots \ {}^1A_n \\ \ddots \ \vdots \\ {}^{n-1}A_n \end{bmatrix}.$$ (11)

The components of the $n \times 1$ vector of centrifugal and Coriolis forces, $h(q, \dot{q})$, could be expressed as $h_i(q, \dot{q}) = \dot{q}^T H_i \dot{q}$, $i = 1 : n$, where $H_i = [h_{ijk}]$ is an $n \times n$ matrix given by

$$h_{ijk} = \sum_{l=max\{i,j,k\}}^{n} tr(U_{kjl} J_l U_{il}^T) \tag{12}$$

for $j, k = 1 : n$. These matrices define the block column matrix

$$H = \begin{bmatrix} H_1 \\ H_2 \\ \vdots \\ H_n \end{bmatrix}. \tag{13}$$

To obtain H, it is necessary to compute matrices U_{kjl}, which represent the second derivative terms given by

$$U_{kjl} \equiv \frac{\partial U_{jl}}{\partial q_k} = \begin{cases} {}^0 A_{j-1} Q_j \; {}^{j-1} A_{k-1} Q_k \; {}^{k-1} A_l, & j \leq k \leq l \\ {}^0 A_{k-1} Q_k \; {}^{k-1} A_{j-1} Q_j \; {}^{j-1} A_l, & k \leq j \leq l \\ 0, & j > l \text{ or } k > l. \end{cases} \tag{14}$$

Note that $U_{kjl} = U_{jkl}$. As a result, matrices H_i are symmetric. From matrices U_{kjl}, the following block column matrix can be defined

$$DU = \begin{bmatrix} DU_1 \\ DU_2 \\ \vdots \\ DU_n \end{bmatrix}, \tag{15}$$

where

$$DU_i = \begin{bmatrix} 0 \cdots & 0 & U_{i1i} & U_{i1i+1} & \cdots & U_{i1n} \\ 0 \cdots & 0 & U_{i2i} & U_{i2i+1} & \cdots & U_{i2n} \\ \vdots \; \vdots & \vdots & \vdots & \vdots & \vdots & \vdots \\ 0 \cdots & 0 & U_{iii} & U_{iii+1} & \cdots & U_{iin} \\ 0 \cdots & 0 & 0 & U_{ii+1i+1} & \cdots & U_{ii+1n} \\ \vdots \; \vdots & \vdots & \vdots & 0 & \ddots & \vdots \\ 0 \cdots & \cdots & \cdots & \cdots & \cdots & U_{inn} \end{bmatrix}. \tag{16}$$

Finally, the components of the $n \times 1$ vector of gravitational forces are given by

$$c_i = \sum_{j=i}^{n} (-m_j g^T U_{ij} \; {}^j r_j) \tag{17}$$

for $i = 1 : n$. In this expression, ${}^j r_j$ is the position of the center of mass of link j with respect to the origin of the coordinates of the link j.

3 Parallel Algorithms for the Dynamic Equation.

The parallel algorithms to compute the terms of the dynamic equation have been developed, for the case of a rigid manipulator with p arms, using a multicomputer with p processors, $P_0, P_1, \ldots, P_{p-1}$, and a bidirectional ring topology. In order to obtain a low computational cost special attention has been paid to the following aspects:

- It is possible to optimize the matrix products involved in the problem due to the characteristics of the matrices (9), (10) and the matrices Q_j. These matrices have enough null elements that permit us to reduce the number of arithmetic operations.
- It is possible to use different properties of the matrices, such as symmetry, repeated rows, etc., in order to reduce the computational cost.

Now the calculation of the different terms of the dynamic equation by using this approach is presented.

3.1 Compute matrix A

The diagonal blocks of matrix A are given by (9) and (10). The rest of blocks of A are obtained from (8). Taking these expressions into account, matrices ${}^j A_i$, for $j = 0 : n - 2$ and $i = j + 2 : n$, have the following structure

$$
{}^j A_i = \begin{bmatrix} a_{11} & a_{12} & a_{13} & a_{14} \\ a_{21} & a_{22} & a_{23} & a_{24} \\ a_{31} & a_{32} & a_{33} & a_{34} \\ 0 & 0 & 0 & 1 \end{bmatrix}. \tag{18}
$$

This structure permits us to reduce the computational cost of the matrix product in expression (8). In the parallel approach it is possible to compute matrix A by rows or by columns because the number of matrix products involved is the same. In this paper the computation of matrix A is distributed among the processors by rows, row 1 in processor P_0, row 2 in processor P_1 and so on. The diagonal blocks ${}^j A_{j+1}$ can be computed in parallel because they are independently obtained from the parameters of the links. To obtain the rest of the blocks it is necessary to apply expression (8).

3.2 Compute matrix U

From expression (6), matrix U has the following structure:

- First row of matrix U, U_{1j}, for $j = 1 : n$
  - If Q_1 is a prismatic joint then

$$
U_{1j} = \begin{bmatrix} 0 & 0 & 0 & 0 \\ 0 & 0 & 0 & 0 \\ 0 & 0 & 0 & u_{34} \\ 0 & 0 & 0 & 0 \end{bmatrix}. \tag{19}
$$

- If Q_1 is a revolution joint then

$$U_{1j} = \begin{bmatrix} u_{11} & u_{12} & u_{13} & u_{14} \\ u_{21} & u_{22} & u_{23} & u_{24} \\ 0 & 0 & 0 & 0 \\ 0 & 0 & 0 & 0 \end{bmatrix}. \tag{20}$$

- Rest of the rows of matrix U, U_{ij}, for $i = 2 : n$ and $j = i : n$
  - If Q_i is a prismatic joint then

$$U_{ij} = \begin{bmatrix} 0 & 0 & 0 & u_{14} \\ 0 & 0 & 0 & u_{24} \\ 0 & 0 & 0 & u_{34} \\ 0 & 0 & 0 & 0 \end{bmatrix}. \tag{21}$$

  - If Q_i is a revolution joint then

$$U_{ij} = \begin{bmatrix} u_{11} & u_{12} & u_{13} & u_{14} \\ u_{21} & u_{22} & u_{23} & u_{24} \\ u_{31} & u_{32} & u_{33} & u_{34} \\ 0 & 0 & 0 & 0 \end{bmatrix}. \tag{22}$$

Using these structures, it is possible to reduce the computational cost. In the parallel algorithm, if matrix U is computed by block of columns it is necessary to calculate more matrix products than in the case oriented by block of rows. In a distributed computation by block of rows, it is possible to use the blocks that have been previously obtained.

3.3 Compute matrix DU

Using expression (14), matrix DU has the following structure:
- Matrix DU_1, $U_{1ij} = Q_1 U_{ij}$, for $i = 1 : n$ and $j = i : n$
  - If Q_1 is a revolution joint and Q_i is a prismatic joint then

$$U_{1ij} = \begin{bmatrix} 0 & 0 & 0 & du_{14} \\ 0 & 0 & 0 & du_{24} \\ 0 & 0 & 0 & 0 \\ 0 & 0 & 0 & 0 \end{bmatrix}. \tag{23}$$

  - If Q_1 and Q_i are revolution joints then

$$U_{1ij} = \begin{bmatrix} du_{11} & du_{12} & du_{13} & du_{14} \\ du_{21} & du_{22} & du_{23} & du_{24} \\ 0 & 0 & 0 & 0 \\ 0 & 0 & 0 & 0 \end{bmatrix}. \tag{24}$$

– Matrices DU_i, for $i = 2 : n$

- Row i of DU_i, $U_{iij} =^0 A_{i-1}Q_i^2{}^{i-1}A_j$, for $j = i : n$. If Q_i is a revolution joint then

$$
U_{iij} = \begin{bmatrix} du_{11} & du_{12} & du_{13} & du_{14} \\ du_{21} & du_{22} & du_{23} & du_{24} \\ du_{31} & du_{32} & du_{33} & du_{34} \\ 0 & 0 & 0 & 0 \end{bmatrix}. \tag{25}
$$

- Row k, $i < k \le n$, $U_{ikj} = U_{ik-1}Q_k{}^{k-1}A_j$, for $j = k : n$
  * If Q_k is a prismatic joint and Q_i is a revolution joint then

$$
U_{ikj} = \begin{bmatrix} 0 & 0 & 0 & du_{14} \\ 0 & 0 & 0 & du_{24} \\ 0 & 0 & 0 & du_{34} \\ 0 & 0 & 0 & 0 \end{bmatrix}. \tag{26}
$$

  * If Q_k and Q_i are revolution joints then

$$
U_{ikj} = \begin{bmatrix} du_{11} & du_{12} & du_{13} & du_{14} \\ du_{21} & du_{22} & du_{23} & du_{24} \\ du_{31} & du_{32} & du_{33} & du_{34} \\ 0 & 0 & 0 & 0 \end{bmatrix}. \tag{27}
$$

Low computational cost algorithms are obtained by using these data structures. In the parallel approach, the computation of each matrix $DU_i(i : p, i : p)$, $i = 1 : p$ is distributed among the processors by rows. Matrix DU_1 can be computed in parallel without previous communications, because each processor has the data blocks needed.

3.4 Compute matrix D

Matrix D is symmetric, so it is only necessary to compute its upper triangular part. J_k is the inertia tensor related to link k. In the parallel algorithm these matrices are replicated in all the processors. The elements d_{ij}, for $i, j = 1 : n$, of the same row are computed in the same processor. To compute matrix D, it is necessary to use blocks U_{jk} that have been previously obtained. In this case

$$
d_{ij} = \sum_{k=max\{i,j\}}^{n} tr(U_{jk}J_kU_{ik}^T) = \sum_{k=max\{i,j\}}^{n} tr(U_{jk}Y_{ik}). \tag{28}
$$

Using this expression, it is possible to simplify the necessary operations to obtain the trace of the matrix and then the matrix D. Matrix Y permits us to reduce the number of operations. This matrix is necessary in order to compute matrix H. Matrix Y has the following structure:

- First row of Y, $Y_{1k} = J_k U_{1k}^T$, for $k = 1 : n$
  - If Q_1 is a prismatic joint then

$$Y_{1k} = \begin{bmatrix} 0 & 0 & y_{13} & 0 \\ 0 & 0 & y_{23} & 0 \\ 0 & 0 & y_{33} & 0 \\ 0 & 0 & y_{43} & 0 \end{bmatrix}. \tag{29}$$

  - If Q_1 is a revolution joint then

$$Y_{1k} = \begin{bmatrix} y_{11} & y_{12} & 0 & 0 \\ y_{21} & y_{22} & 0 & 0 \\ y_{31} & y_{32} & 0 & 0 \\ y_{41} & y_{42} & 0 & 0 \end{bmatrix}. \tag{30}$$

- Rest of the rows of matrix Y, $Y_{ik} = J_k U_{ik}^T$, for $i = 2 : n$ and $k = i : n$. In this case

$$Y_{ik} = \begin{bmatrix} y_{11} & y_{12} & y_{13} & 0 \\ y_{21} & y_{22} & y_{23} & 0 \\ y_{31} & y_{32} & y_{33} & 0 \\ y_{41} & y_{42} & y_{43} & 0 \end{bmatrix}. \tag{31}$$

With these considerations, it is possible to optimize the calculation of matrix D. To calculate the trace of a product matrix it is only necessary to obtain the diagonal elements of that matrix.

3.5 Compute matrix H

Taking into account that matrix H_i is symmetric, it is only necessary to compute its lower triangular part. In the parallel algorithm, the computation of each matrix H_i is also distributed among the processors by rows. In this case

$$h_{ijk} = \sum_{l=max\{i,j,k\}}^{n} tr(U_{kjl} J_l U_{il}^T) = \sum_{l=max\{i,j,k\}}^{n} tr(U_{kjl} Y_{il}). \tag{32}$$

Using this expression, it is possible to simplify the necessary operations to obtain the trace of the matrix and then the matrix H. In this case the structures of the matrices Y and DU are used.

3.6 Compute vector c

The computation of vector c is completely distributed because each processor has the data blocks that it needs to compute a component of c.

4   An Interface Based on Transputers for the Simulation of the Dynamic Equation of Robot Manipulators.

4.1   Introduction

The algorithms described in Section 3 have been implemented using a multiprocessor based on a PC486 plus a board with 8 T805 transputers. The different transputers were connected by using a bidirectional ring topology . The processor P_j is connected to the left processor P_{j-1} and the right processor P_{j+1}. C Language and communication routines were used for the implementations of the distributed algorithms.

Although the transputer is very useful for real-time control problems, it has some disadvantages when several transputers are connected. In general, one chip has insufficient memory, and this is why the TRAM motherboard was developed. The IMS B008 is a TRAnsputer Module (TRAM) motherboard designed to be plugged into a PC bus. The board has ten TRAM slots, an interface to PC bus and an IMS C004 link switched to allow networks of TRAMs to be set up under software control. INTEL and other manufacturers provide many different TRAMs; in this case the IMS B404 and the IMS B426 have been used. The TRAMs used have T805 transputers. The INMOS C004 is a programmable crossbar network that allows the user to establish the connections between the transputers.

INMOS provides software tools which permit us to do a lot of operations with the transputers and the board which contains these transputers. These tools are:

- Iserver: A server program that facilitates the communication between the transputer board installed in the PC and the PC. This program loads the executable program in the transputers and carries out the requirements of the program. The general operations of Iserver are the following:
  - It analyses the command introduced and identifies the program that is going to be loaded. If there is not any error the program can continue. It also modifies internal variables that will be used later.
  - It tries to access the root transputer across one of its links to observe whether the transputer is prepared and active. If any error occurs the program is aborted.
  - Different actions are carried out according to the requirements of the program. Iserver is in an infinite loop where the possible errors of the executable program in the transputers are tested. It attends the requirements of the transputer using the existing links. This loop is repeated until an error is produced or the executable program is finished.
  - When the link is free Iserver finishes the execution.
- Worm: It can be used to perform different operations with the transputers and the IMS B008 board. The most important functions are:
  - Check: It checks the transputers and the board to obtain the topology of the physical links of the transputers, the number of transputers, transmission speed, etc. This program has important information that must

be used for the other utilities of the Worm package. It provides information about the processor and the links. This information shows how the links are connected using the scheme

$$transputer : link$$

When the link is equal (...) then there are no connections. The general operation of Check is the following. It tests the first transputer of the network. Then it continues with the rest of them in the order fixed by the used links. The information about the transputer is kept by Check in a data structure. This information is showed according to the order that the network is gone across.

- C004: It allows the establishment of the transputer network. It is a utility that permits us to modify the topology of the network by the creation or destruction of the existing connections. It allows one-directional or bi-directional connection between two transputers. C004 needs the information that Check obtains from the current state of the network and it needs the server kernel that Check puts in the transputers to do the configuration of the crossbar network IMS C004. This information goes to the utility C004 by a pipe of MS-DOS which analyses this information to establish the network.
- Other utilities are:
  * Mtest: It tests the memory of the transputers.
  * Ftest: It tests the transputers of the network.
  * Kong: It compares the actual topology of the transputer network with the necessary topology to execute a program.
  * Load: It loads the program in the transputer.

The utilities Check and C004 have been integrated in the Windows interface, WinServer.

4.2 WinServer Utility

To obtain the new interface, it is necessary to take into account the following properties:
- The interface should be user friendly.
- It should permit data graphics to be obtained.
- Iserver, Check and C004 should be integrated.

The application has been developed using the Windows operating system. Different modifications in Iserver, Check and C004 have been developed.

Iserver modifications. Two different modifications have been carried out:
- New calls have been joined to the original Iserver.
  - A call to obtain the direction of the parameters of the robot and their size. This call is necessary because the program in the transputers must receive different input data (robot parameters, etc.). These data are in

the memory of the PC, then it is very important to obtain the directions of these variables.

- A new call to do new simulations without loading the program on the transputers. Using this call it is possible to indicate to Iserver that the simulation has finished. Then Iserver can serve to the user requirements from the PC. In this moment, the user can carry out a new simulation.
- A call to view the results easily. It is possible to view the simulation results and the run-time of the sequential or parallel algorithms.

– Integration in Windows. The original Iserver has been developed in C language and MS-DOS operating system. Different modifications have been done to integrate Iserver in Windows:

- The command line used in MS-DOS has been removed. Using menus and dialog boxes, the user can choose different options of Iserver, such as how to execute the program on the transputer, error checking, the possibility to do a program trace, etc.

Fig. 1. Iserver options.

Morever the user can load the executable program using a command button.

- It is possible to modify the functions for the presentation of the results to integrate them in the new interface.

Check and C004 modifications. The Check and C004 utilities have been modified to obtain:

– An improvement of the output of Check. It is interesting to note that this utility does not modify the order of the transputers in the network. With this modification it is possible to obtain a simple and clear presentation of the transputer network. The operation consists of a double run of the list of transputers. In the first run the link 2 of each transputer is used. This information is printed. In the second, the complete list of transputers is run to print the information of the transputers that were not run using the link 2. The following figures (Fig. 2, Fig. 3 and Fig. 4) show the information printed by the original Check and the Windows Check for the ring topology.

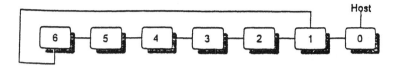

Fig. 2. Ring topology.

```
check 1.5
# Part rate  Mb Bt  [ Link0      Link1      Link2    Link3    ]
0 T805d-25   0.27 0  [ HOST       1:1        2:1      ...      ]
1 T2    -20 1.75 1  [    ...      0:1        ...      C004     ]
2 T805d-25   1.75 1  [    3:3      0:2        3:1      ...      ]
3 T805d-25   1.35 3  [    ...      2:2        5:1      2:0      ]
4 T805d-25   1.75 1  [           6:2         7:1      ...      ]
5 T805d-25   1.75 2  [    ...      3:2        8:1      ...      ]
6 T805d-25   1.75 1  [    ...      8:2        4:1      ...      ]
7 T805d-25   1.75 1  [    ...      4:2        9:1      ...      ]
8 T805d-25   1.75 2  [    ...      5:2        6:1      ...      ]
9 T805d-25   1.75 1  [    ...      7:2        ...      ...      ]
```

Fig. 3. Information printed for the original Check.

- The improvement of the internal communication between Check and C004 using shared variables instead of pipes.
- The integration of these utilities into Windows. This integration is based on different modifications in the Check and the C004 functions, such as:
  - To remove unnecessary program code.
  - To modify the output information routines and error checking.
  - To convert the original program code in C language into C++ language.
  - To make a friendly user interface.

The objetive of this interface is to allow a configuration of the transputer network. It is possible to see the links used and the connections established.

```
Check for Windows 1.5
# Part rate   Mb Bt  [ Link0      Link1     Link2    Link3    ]
0 T805d-25    0.21 0  [ HOST    9:1        1:1      ...      ]
1 T805d-25    1.75 1  [    6:3      0:2        2:1      ...      ]
2 T805d-25    1.75 1  [    ...      1:2        3:1      ...      ]
3 T805d-25    1.77 1  [    ...      2:2        4:1      ...      ]
4 T805d-25    1.75 1  [    ...      3:2        5:1      ...      ]
5 T805d-25    1.75 2  [    ...      4:2        6:1      ...      ]
6 T805d-25    1.33 0  [    ...      5:2        7:1      1:0      ]
7 T805d-25    1.75 1  [    ...      6:2        8:1      ...      ]
8 T805d-25    1.77 1  [    ..       7:2        ...      ...      ]
9 T2    -20  1.75 1  [    ..       0:1        ...      C004     ]
```

Fig. 4. Information printed for the WinServer Check.

The most important characteristics of this interface are:

- To fix the number of transputers automatically. It permits us to establish a board IMS B800 with 10 transputers (9 plus a T222).
- To fix the existing configuration automatically. It is possible to obtain a new configuration by adding or removing connections.

The following figure shows the interface to see the links used. This figure

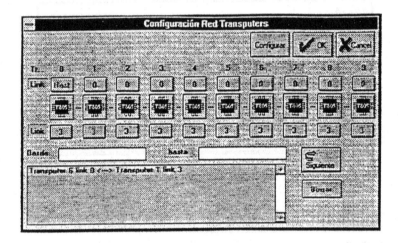

Fig. 5. Interface to see the links used.

shows the existing transputers in the board and the free links that each transputer has. The transputers are numerated from 0. The transputer 0 is connected to the Host, that is the PC. The user can use the links 0 and 3 to establish the connection with other transputers. The links 1 and 2 of the transputers are physically connected with the neighbour transputers. This is the reason why the user can not use them.

5 Experimental Results.

The following tables show the experimental results. The number of processors p is equal to the number of links. In Table 1 the execution time (in seconds) of the sequential algorithm without and with optimization of the involved matrix products are described. This table is given to show the reduction of the cost of computing the dynamic equation, following the approach presented in this paper, in comparison with the initial results given in [15]. The second table shows the speed-ups, efficiencies and the results of the execution times (in seconds) of the parallel algorithms with the optimization of the matrix products.

A detailed description of these parallel algorithms, without the optimization of the matrix products, is also described in [15].

Table 1. Sequential results (p = number of links).

	Seq. without opt.	Seq. with opt.
p=2	0.0196	0.00307
p=3	0.048	0.00774
p=4	0.102	0.0164
p=5	0.197	0.0302
p=6	0.345	0.0508

Table 2. Parallel results (p = number of links = number of processors).

	Sequential	Parallel	Speed-up	Efficiency
p=2	0.00307	0.00224	1.34	67.14%
p=3	0.00774	0.00454	1.70	56.8%
p=4	0.0164	0.00742	2.20	55.17%
p=5	0.0302	0.0120	2.48	49.7%
p=6	0.0508	0.0173	2.93	48.83%

6 Conclusions.

The Lagrange-Euler formulation permits us to compute the dynamic equation of robot manipulators by means of a useful matrix approach that can be implemented in parallel. Some matrices involved in the problem have important structural properties, such as symmetry (matrix D) and repeated rows (matrices DU_i), that can be used to obtain efficient parallel algorithms. The characteristics of the matrices A and Q allow us to optimize the matrix products involved because these matrices have enough null elements. Using these characteristics and the parallel approach, the computational cost of the dynamic equation has been reduced. Thus, it is possible to solve the problem of inverse dynamics control of robot manipulators in real time by using distributed parallel computing.

The new Windows interface is a useful tool to carry out simulations. It is a friendly simulator of the dynamic equation because it allows us to modify the robot parameters, to view the results and to obtain the execution time of the implemented algorithms. With the simulator, it is possible to run any program developed for transputers and to establish their topology.

7 Acknowledgements.

This work was partially supported by the CICYT TAP95-0883-C03-02/03 Projects.

References

1. Balestrino, A. D., A. S. L. Zinober. *Nonlinear Adaptative Model- following Control.*, Automatica, vol. 20. no. 5. pp. 559-568 (1984).
2. Bejzcy, A. K. *Robot arm dynamics and control.*, NASA-JPL Technical Memorandum, 33-669 (1974).
3. Berghuis, H., R. Ortega, H. Nijmeijer. *A robust adaptative controrller for robot manipulators.*, IEEE Trans. on Robotics and Automation, vol. 10. no. 6. pp. 825-830 (1993).
4. Craig, J., *Adaptative Control of Mechanical Manipulators.*, Int. Conf. on Robotics and Automation (San Francisco, 1986),
5. Fijancy, A. and A. K. Bejczy. *A class of parallel algorithms for computation of the manipulator inertia matrix.*, Proc. Int. conf. Robotics and Automation (Scottsdale, Ariz., 1989), pp. 1818-1826.
6. Fu, K. S., R.C. González and C.S.G. Lee. *Robótica, Control, Detección, Visión e Inteligencia.* Ed. MCGraw-Hill.
7. Good, M. C., L.M. Sweet, K. L. Strobel. *Dynamic models for control system design of integrated robot and drive systems.*, ASME J. dynamic System Measurement Control, vol. 107. pp. 53-59 (1985).
8. Graham, I. and T. King. *The Transputer Handbook.*, Pretince Hall International Ltd., U.K., (1990).
9. INMOS Ltd. INMOS B008 User guide and Reference Manual.
10. Kasahara, H. and S. Narita. *Parallel Procesing of Robot-Arm Control Computaion on a Multimicroprocessor System.*, IEEE J. of Robotics and Automatica, vol. RA-1. no. 2. pp. (1985).
11. Khatib, O. and J. Burdick. *Motion and force control of robots manipulators.*, Proc. IEEE Int. Conf. on Robotics 1986), pp. 1381-1386.
12. Lathrop, R. H. and P.R. Chang. *Parallelism in manipulator dynamics.*, Int. J. Robot., pp. 80-102 (1985).
13. Lewis, R. A. *Autonomous manipulation on a robot: Summary of manipulator software functions.*, Tec. Memorandum 33-679, Jet Propulsion Lanboratory, Pasadena, California (1974).
14. Neuman, C. P. and J. J., Murray. *Customized computational robot dynamics.*, Int. Robotics Systems, vol. 4, no. 4, pp. 503-526 (1987).
15. Peñalver, L., J. C. Fernández, V. Hernández and J. Tornero *Distributed Parallel Algorithms for the Inverse Dynamics Control of Robot Manipulators.*, Proc. 3rd IFAC/IFIP Workshop on Algorithms and Architectures for Real-Time Control AARTC95 (Ostende, Belgium, 1995), pp. 59-68.
16. Slotine, J.J. and W. Li. *On Adaptative Control of Robot Manipulators.*, Int. J. Robotics Research,, vol. 6. no. 3. pp. 49-59 (1987).
17. Spong, M. W. and M. Vidyasagar. *Robot Dynamics And Control.*, (John Wiley & Sons, 1989).
18. Tarn, T., A.K. Bejczy, A. Isidori and Y.L. Chen. *Nonlinear feedback in robot arm control.*, Proc. of the 23rd IEEE Conf. on Decision and Control. 1984),
19. Verdier, M., M. Rouff and J.G. Fontaine. *Nonlinear control robot: A phenomenological approach to linearization by static feedback.*, Robotica, vol. 7. pp. 315-321 (1989).

Large Scale Traffic Simulations

Kai Nagel[1,2], Marcus Rickert[1,3], and Christopher L. Barrett[1,2],

[1] Los Alamos National Laboratory, TSA-DO/SA MS M997, Los Alamos NM 87545, U.S.A., kai@lanl.gov, barrett@tsasa.lanl.gov, rickert@tsasa.lanl.gov
[2] Santa Fe Institute, 1399 Hyde Park Rd, Santa Fe NM 87501, U.S.A.
[3] Zentrum für Paralleles Rechnen ZPR, Universität zu Köln, 50923 Köln, Germany

Abstract. Large scale microscopic (i.e. vehicle-based) traffic simulations pose high demands on computational speed in at least two application areas: (i) real-time traffic forecasting, and (ii) long-term planning applications (where repeated "looping" between the microsimulation and the simulated planning of individual person's behavior is necessary). As a rough number, a real-time simulation of an area such as Los Angeles (ca. 1 million travellers) will need a computational speed of much higher than 1 million "particle" (= vehicle) updates per second. This paper reviews how this problem is approached in different projects and how these approaches are dependent both on the specific questions and on the prospective user community. The approaches reach from highly parallel and vectorizable, single-bit implementations on parallel supercomputers for Statistical Physics questions, via more realistic implementations on coupled workstations, to more complicated driving dynamics implemented again on parallel supercomputers.

1 Introduction

Nobody likes traffic jams. Yet, they are only the most visible feature among a variety of related problems: Subways which fail to go where or when you need them; pollution of inner cities; etc. Many of these are results of poorly designed transportation systems. However, what is a good design? In our complex world, such a question is not easy to answer. Addition of new streets may *increase* congestion by concentrating formerly spread-out traffic onto one through route [1]; introduction of a transit system may *increase* pollution by making the car which previously was taken to work now available for short trips all with a cold engine [2]; a transportation infrastructure investment payed for by a certain group may actually turn out to benefit a completely different sub-population (winner/looser analysis); a new major arterial meant to relieve congestion may attract new developments along this new arterial, making congestion worse in the long run (induced demand).

There is more and more agreement between transportation professionals that a useful planning tool for such situations is a transportation microsimulation. In such a microsimulation, each traveler is represented as an individual object in the simulation. That makes it straightforward to separate out winners and losers; to "look" for vehicles with cold engines causing excessive pollution; etc. Yet,

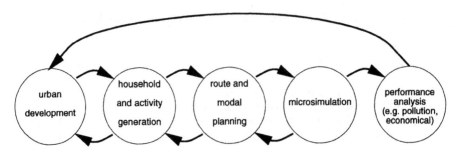

Fig. 1. The TRANSIMS design.

a problem is how do you "drive" such a microsimulation, i.e., how do travelers decide their next move at intersections or at transfer points (train stations etc.)?

The traditional answer to this questions has been "turn counts", i.e. numbers at each intersection which tell you which percentage of vehicles makes left or right turns, respectively. It is fairly clear that this will not work as soon as the infrastructure changes affects people's behavior; e.g. a left turn which has been used heavily before the change is now no longer used much, rendering the turn counts for this particular intersection irrelevant.

It is also fairly obvious that random selection (annealed randomness) often favored by (statistical) physicists has a good chance of not being very helpful in the very non-homogeneous structure of transportation systems.

2   TRANSIMS

It thus seems that the only answer is to drive the traveler objects in the simulation by something which emulates real-world behavior, i.e. by intentions. This, and its realization into a practical computer code, is the core of the TRANSIMS project [3, 4, 5, 6, 7] (see also [8]).

With intentions, one is very soon faced with a consistency problem. Real people presumably plan their trip before they leave their current location (meaning one needs simulated planning), but are open to deviate from the plans for example when the conditions they encounter are much different from what they expected (meaning one needs on-trip planning). Furthermore, the intentions somehow have to be generated in the computer in the first place.

The TRANSIMS approach to this problem is to parcel out the different parts of this process (see Fig. 1):

- Population and activities generation: Stochastically generate a population of individuals for a given geographic area such that the demographics of this generated population matches demographic data. Then, for each individual, generate activities such as work, shopping, social activities, which that individual wants to perform under some scheduling restrictions (e.g. go to work at 9 am; go shopping once a week; etc.).

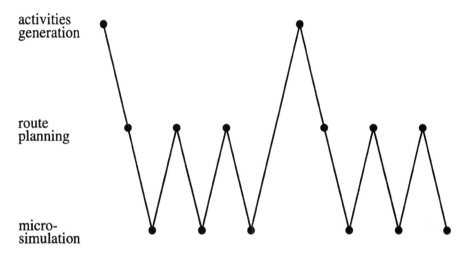

activities
generation

route
planning

micro-
simulation

Fig. 2. Visualization of feedback cycles between activity generator, planner, and microsimulation.

- Trip chaining, modal choice and route planning: The activities are combined into trip chains, and the trip is routed on the transportation system, including modal choice
- Microsimulation: The trip plans are executed in a detailed microsimulation of the transportation system
- Analysis: The output of the traffic microsimulation is input into analysis modules, such as air quality, or measures of efficiency analysis

Despite the separation of the modules, it is clear that there are backward causalities. For example, if the microsimulation displays congestion, people will react by choosing different routes. If that does not help, they will re-schedule their activities. If nothing helps, they will maybe relocate to a more convenient location.

This backwards causality is what causes one of the computational challenges. In order to sort out the causal interdependencies, it is necessary to run the microsimulation which results in a new cost function (travel-time) on the infrastructure, then to re-plan according this new cost function, then to run the microsimulation again, etc., until some relaxation criterion is fulfilled. After that, one probably has to re-run the activities generation, adapting activities to what can actually be achieved in the given transportation system. This triggers in return a completely new relaxation cycle between route planner and microsimulation, etc. (see Fig. 2). In consequence, for certain problems the microsimulation may have to be run hundreds of times, thus demanding ultra-high computing speeds.

3   Other computational challenges

There are at least two other computational challenges in the transportation simulation area: systematic analysis, and real time forecasting.

3.1   Simulation as a dynamical system

Reasonable traffic simulations will be Monte Carlo simulations, meaning that different random seeds will produce different system trajectories.[4] That means that only ensembles of multiple simulations actually produce meaningful results. The extent of robustness of the results has to be analyzed; if simulations turn out to be consistently non-robust in certain aspects, then only the distributions of outcomes may be seen as the result. That is, many simulation cycles will have to be run both to find robust and non-robust elements of the simulation and to generate the distributions for the non-robust outcomes.

3.2   Real time forecasting

Above, transportation simulation has been described in the context of a long-term planning problem. Having a twenty years planning horizon, it may be inconvenient but not impossible if the generation of meaningful simulation results for, say, a representative day takes several weeks. Yet, there are other problems such as traveler information (sometimes called ATIS – Advanced Traveler Information Systems) or traffic management (ATMS – Advanced Traffic Management Systems) where at least some of the questions involve real time forecasting. That means that the forecasting procedure has to run faster than real time, i.e. the forecast better be available from the computer before the situation is there in reality.

4   The TRANSIMS microsimulation

Transportation simulation is, as outlined above, an intricate process involving a multitude of different scales, ranging from the representation of the movements of individual travelers to the representation of planning decisions on a weekly scale or more. It is clear from the above remarks that the transportation microsimulation poses one of the major challenges here, both because of the enormous scale of the problem (large regional areas often have 10 million or more travelers simultaneously on route) and because it is in the innermost loop of the relaxation as described above. Yet, also the other modules pose considerable computational loads — for example, generating the 10 million trips which are necessary for a 24 hour run of the Dallas/Fort Worth area takes about 24 hours

---

[4] Arguably, a deterministically chaotic simulation would do the same when started from slightly varied initial conditions.

on a workstation, and produces several Gigabytes of output. However, reasonably systematic results exist so far only for vehicular traffic microsimulations, which we will concentrate on for the remainder of this paper.

A useful model for traffic microsimulation has to fulfill several fairly obvious criteria: (i) It has to resolve at least individual cars, so that individual plan following can be implemented. (ii) It has to generate reasonable traffic dynamics. (iii) It has to be computationally fast.

TRANSIMS uses a cellular automaton (CA) approach for the traffic microsimulation. The approach is microscopic, it generates realistic traffic dynamics including the characteristic variance structure of certain traffic measurements, and it is computationally fast.

For the CA approach, the road is separated into cells which are either empty, or occupied by exactly one car. The size of the cell, l, is given by the inverse density of a traffic jam: $l = 1/\rho_{jam}$. A good approximation is $l = 7.5$ m. Cars move by jumping from one cell to another. Before they move, they adjust their velocity according to three simple rules: (a) Accelerate if you can up to some maximum velocity v_{max}. (b) Decelerate if you must (i.e. if another vehicle is too close ahead of you). (c) With a certain probability, be slower than that (randomization). See, e.g., [9, 10, 11] for more information. Lane changing and other features are implemented in the same simple, straightforward way, see, e.g., [12, 13, 14].

An update of the system consists of several completely parallel sub-steps (lane changing, intersection dynamics, velocity update and movement). This makes a parallel implementation straightforward: As long as boundary information is exchanged after each sub-step, all sub-steps can be done concurrently.

As a result, the CA approach is computationally very efficient: Due to the simplified dynamics, it already fast on desktop computers, and due to the parallel update, it can be implemented efficiently on parallel computers, yielding linear speed-ups for large enough system sizes.

This approach has been extensively tested on many computer architectures. A simple single-lane circle was implemented using a wide variety of computer architectures (coupled workstations, parallel and/or vectorizing supercomputers) using different algorithms (list-based, bit-coded, etc.). After settling down on a certain implementation technique targeted for certain computer architectures, large realistic road networks were implemented or are under implementation.

5 Implementations of the CA traffic model

For the practical coding, we considered three different approaches: site oriented, particle (= vehicle) oriented, and an intermediate scheme. *Site oriented* directly implements the CA: A street is represented by an array v_j, $j = 1 \ldots L$ (L = system size) of integers with values between -1 and v_{max}. -1 means that there is no vehicle at this site, whereas the other values denote a vehicle and its velocity. In contrast, *vehicle oriented* means that two lists $(x_i)_{i=1,\ldots,N}$ and $(v_i)_{i=1,\ldots,N}$ contain position x_i and velocity v_i of each vehicle i ($i = 1, \ldots, N$).

This is similar to a molecular dynamics algorithm [15], except that vehicles are constrained to integer positions and velocities.

Obviously, the vehicle-oriented approach will always be faster than the site-oriented one for sufficiently low vehicle densities. Yet, for the site-oriented approach single-bit coding (see, e.g., [16]) is possible. This means that the model is formulated in logical variables, which may be stored bitwise into computer words. Logical operations on computer words treat all bits of the word simultaneously, giving a theoretical speedup of b, where b is the number of bits per word (usually 32 or 64). However, the *practical* gain for traffic simulations on a workstation is much lower because the bit-oriented approach cannot take advantage of the fact that only a fraction of all sites is occupied by a vehicle. Nevertheless, we found that, on a workstation, the single-bit algorithm is faster than the vehicle-oriented one for densities above 0.05 (for $v_{max} = 5$). In addition, the single-bit code runs very efficiently on a Thinking Machines CM-5 using data parallel CM-Fortran and on a NEC-SX/3 traditional vector computer.

Once passing of vehicles is allowed (multi-lane traffic), single-bit coding becomes tiresome. Moreover, for realistic simulations, one needs access to more vehicle information, making a pointer to further vehicle data necessary. Since a pointer typically is a 32-bit number which would have to be moved along in memory with the vehicle in the space-based single-bit approach, single-bit coding is no longer efficient.

With respect to the vehicle oriented approach, the problem is the same as in parallel Molecular Dynamics approaches: How to find the neighbors for interaction, for example for intersection dynamics, lane changing, or car following. Solutions to this are possible but elaborate, and the expected computational speed gain (not more than a factor of about four, see Table 1) did not warrant the additional programming complexity at the current state of the project.

These observations led to a third, *intermediate* approach. As in the site oriented approach, each site is in one of ($v_{max} + 2$) states, but for the update only the relevant sites are considered. It turns out (see below) that on parallel but not vectorizing computers this algorithm is about as fast as the single-bit version.

In all cases, the parallelization was done geometrically, i.e. dividing the one-dimensional system of size L into p pieces of length $l = L/p$, where p is the number of CPNs (computational nodes). For single-bit coding this is a bit tricky: The standard trick of having the parallel direction(s) different from the bit-coding direction does not work any longer in one dimension. More details can be found in [17].

6 Computational speeds on different supercomputers

Table 1 gives an overview of the computational speeds on selected computers. When comparing performance data, it is necessary to give the size of the simulated system. This becomes imperative for parallel computers, since too small systems perform poorly due to the communication overhead. All values of Table 1 have been obtained by simulations of systems of size $L = 10,000$ single-lane km $(1,333,333$ *sites*$)$ with an average traffic density of 13.4 vehicles/km (0.1 veh/site, 134,000 vehicles in the whole system). This is a system size which is relevant for applications. Moreover, it is a system size small enough to still fit into memory of our single node machines, but which is at the same time large enough to run relatively efficient on our parallel machines. Quantitatively, this means that both the GCel and the CM-5 were operating at 40% efficiency. A larger system size would be even more efficient.

References in the literature often give a "real time limit" as measure of their model's performance, which then is the *extrapolated* system size (or number of vehicles) where simulation is as fast as reality. We found these values practically useless in the area of parallel computing, except when given in conjunction with the system size which has actually been simulated.

In consequence and in order to avoid confusion, our primary table entries are the CPU times we needed on the different machines in order to simulate the system as defined above. For convenience, we also calculated the real time limits in km and in vehicle sec/sec from these values. But it should be kept in mind that, if one really simulates system sizes near 1 million km on the parallel machines, one will find much better real time limits for these system sizes (e.g. 2 million km instead of 900,000 km on the GCel).

Noteworthy features of the table are: (i) The bit-coded CA-algorithm is far superior over the "intermediate" one on the vectorizing machines (NEC SX-3/11 and CM-5), slightly faster on the workstation-based architectures, and slightly slower on the massively parallel Parsytec GCel-3. (ii) All algorithms can take good advantage of the parallelism. (iii) Already on a relatively modest machine such as an Intel Paragon with 64 nodes, our real time limit including network handling overhead (intersections etc.) is 280 000 single lane kilometers. For comparison, the freeway network of Germany is about 60 000 single lane kilometers long (12 000 km × 2 directions × 2.5 lanes). We are therefore confident to reach, for a realistic network setup, real time limits of 1,700,000 single lane kilometers $(23,000,000$ veh sec/sec) on 512 nodes of a CM-5, even without using the vector nodes.

Although traffic dynamics is quite different, our computational speeds are comparable to those of Ising models: On the NEC-SX3/11, 0.0025 sec for 10 000 sites correspond to 533 MUPS (Mega-Updates Per Second), which may be compared to 1050 MUPS of a very fast implementation of the (two-dimensional) Ising model on the same computer [18].

7 Different system sizes

The above observations were made for one fixed system size (10 000 km). If one wants to make predictions for different system or computer sizes, one needs more results. This section shows results of systematic measurements.

Fig. 3 gives, for different computers, the best computational speed as a function of system size in terms of the "real to simulation time ratio", which is the factor the computer model runs faster than the simulated reality. It is obvious

	s.bit (F77)	particle (F77)	intermed. (C)	netw. (C)
Sparc10	0.33 sec	0.15	0.71 sec	1.14 sec
	30 000 km	66 000 km	14 000 km	8 800 km
	0.4 e 6 veh	0.89 e 6 veh	0.19 e 6 veh	0.12 e 6 veh
PVM	0.07 sec		0.15 sec	
(5× Sp10)	140 000 km		65 000 km	
	1.9 e 6 veh		0.87 e 6 veh	
SX-3/11[1]	0.0025 sec		0.48 sec	
1 CPN	4 000 000 km		21 000 km	
	53 e 6 veh		0.28 e 6 veh	
GCel-3	0.013 sec	0.0065 sec	0.011 sec	
1024 CPNs	750 000 km	1 550 000 km	900 000 km	
	10 e 6 veh	20.7 e 6 veh	12 e 6 veh	
iPSC	0.016 sec		0.038 sec	
32 CPNs	630 000 km		260 000 km	
	8 e 6 veh		3.5 e 6 veh	
Paragon				0.034 sec
64 CPNs				**290 000 km**
				3.9 e 6 veh
CM-5[1]	0.0077 sec[2]		0.045 sec[3]	
32 CPNs	1 300 300 km		220 000 km	
	17 e 6 veh		2.9 e 6 veh	
CM-5[1]				
1024 CPNs				[> 1.7 e 6 km]
				[> 23 e 6 veh]

[1] CPN(s) has/have vector units (SIMD instruction set)
[2] using data parallel Fortran (CMF)
[3] using message passing (CMMD)

Table 1. Computing speed of different algorithms on different computer architectures. "s(ingle) bit", "particle", "intermed(iate)", and "netw(ork)" mean the corresponding algorithms described in the text. For each machine and algorithm, the first table entry gives the time each computer needed to simulate a system of size 10,000 km. From this figure, we derive the other two entries: the real time limits in km and in vehicle sec/sec.

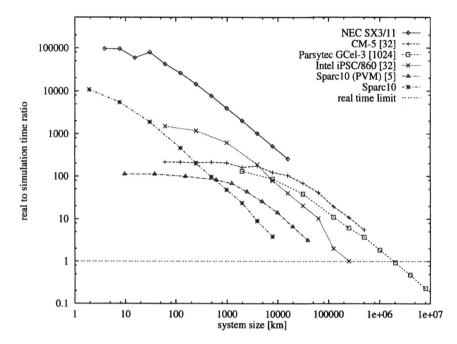

Fig. 3. Comparison between real time and computer time for different system sizes, i.e. the factor r the computer model runs faster than reality. The system sizes where the curves cross $r = 1$ are the so-called real time limits.

from this figure that for fast simulations of small systems single CPN machines are far better than parallel machines. For example, a system of roads of 100 km could be simulated in 1/40000 of the real time on the NEC vector computer. For slower simulations of large systems, parallel machines become equivalent or even better due to their larger overall memory. Our largest system (on the Parsytec GC-el/3 with 1024 CPNs) had a length of nearly 10 million km single-lane road, which would correspond to 1.6 million km of freeway with six lanes (three in each direction). In view of practical applications, this result indicates that memory consumption is not a critical issue for microscopic traffic simulations with our approach.

An efficient system size is reached when doubling of the system size results in an approximate doubling of the computer time, which is visible as a slope of -1 in the double-logarithmic plot of Fig. 3. For some computer architectures one needs quite large systems in order to make efficient use of the computer. In these cases, adding more processing CPNs to the parallel machine only allows processing of larger systems in the same time (scale-up), but not the same system size in shorter time (i.e., no speed-up).

The performance gain of the single-bit algorithm against the intermediate one can be seen in Fig. 4. The gain is relatively low (between a factor of 1 and 4) on workstation-like CPNs (SUN Sparc 10, Intel iPSC/860) and is completely

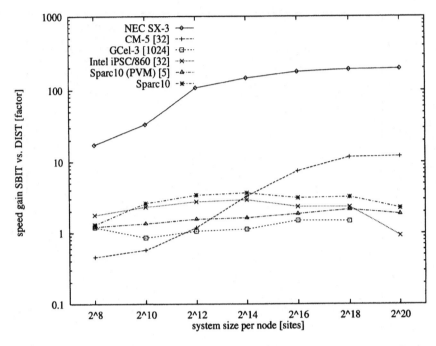

Fig. 4. Computing speed ratio (SBIT divided by DIST) between the two main coding schemes used for our traffic model as a function of system size. Note the low gain on the massively parallel GCel–3 and the huge gains on the vectorizing computers NEC–SX3/11 and the CM–5 (the latter only for large system sizes).

lost (between 0.4 and 2) on the massively parallel Parsytec GCel–3 with 1024 CPNs. On the other hand, this gain is considerable on the vectorizing machines: We found a factor of about 200 on the traditional vector computer NEC–SX3/11 and still a factor of more than 10 on the CM–5 (the latter only for large system sizes).

In order to make reliable predictions on scaling behavior, some theory is helpful. Results on a model for the parallel time complexity of the intermediate algorithm can be found in [17].

8 General observations for the implementation of road networks on parallel computers

After the tests with the single-lane model in simple geometries, the CA was enhanced to include multi-lane traffic and to handle complex networks. As mentioned in previous sections, traffic simulations have to be both computationally fast (e.g. at least real-time) and cover a large street network. Due to the limitations in performance of a single workstation, large-scale simulations have to be moved to parallel computers, as soon as the street network size exceeds a certain limit. The straight-forward approach to parallelization is to perform a domain decomposition of the street-network, assigning each sub-network to a computational node (CPN) of the parallel machine. The three most important issues are then: (1) how to cut the network, (2) what state information to exchange along the CPN boundaries, and (3) how to control parallel execution, in order to guarantee a consistent CA update. But, before we address these issues, let us summarize in what aspects the network traffic simulation differs from the original simple geometry.

Network Elements Looking at a graph as the representation of a street network, it is fairly easy to associate edges (streets) as multi-lane CA segment and nodes (intersections) as points where vehicles are transferred from incoming to outgoing segments in an orderly fashion. The actual implementation turns out to be more tedious because on the one hand the effort to administer a network of edges and nodes with individual characteristics is enormous compared to the simple core implementation of the CA. On the other hand, each intersection represents a disruption of the CA grid which is usually regarded as either infinite (periodic boundary conditions) or at least large compared to the boundary length. The algorithms connecting CA segments have to be designed carefully to deliver satisfactory traffic behavior without producing artifacts.

Complex Vehicles and Route-Plans In contrast to earlier investigations of circular traffic in which vehicles were regarded as completely equivalent, each vehicle in the TRANSIMS traffic simulation has an individual route-plan guiding it from its source to its destination, making it unique from any other vehicle in the system. Moreover, due to the requirements regarding the statistical properties which have to be retrieved from the simulation, vehicles also may have to have local memory to store history information of their trips (travel time, e.g. engine temperature, fuel consumption). Therefore, the data which is associated with a vehicle entity is increased many-fold with respect to the few bits (usually coded in on integer) of the original model.

Input/Output A typical setup of CA traffic in a circle looked like this: Generate vehicles with a certain density homogeneously distributed in the system. Start the simulation and let the transients die out. Then, gather aggregated statistics about average velocity and average flux until the run terminates. It should be obvious to the reader that the amount of data produced by the simulation is negligible (maybe a few hundred values) and the input is restricted to a few parameters such as overall density.

The situation looks considerably different in a realistic network simulation: During the preparation phase, each intersection and each segment of the network has to retrieve certain characteristics (e.g. signal phasing, turning prohibitions, length, speed limit, number of lanes) from a data base. After the simulation has started, each route-plan has to be transferred to the source of the route resulting in the instantiating of a vehicle exactly at the departure time-step. As soon as a vehicle reaches its destination, trip data may have to be stored, before the vehicle instance is destroyed. For the Dallas/Fort-

Worth area the number of trips per day can be as high as 3,000,000 resulting in route-plan input files of about 3 Gigabyte.

Most of the questions concerning efficient input/output of the traffic simulation have not been answered yet. The task becomes all the more difficult, since the simulation is only one part in a series of applications (router, traffic simulation, environmental simulation) exchanging large amounts of data. A generalized concept of parallelization has to integrate all these steps of data processing, instead of regarding them stand-alone applications, unnecessarily multiplying data volumes. Insights into these issues will be published in future papers.

8.1 Domain Decomposition

In our opinion, domain decomposition is the most natural approach to parallelization due to the underlying (almost) planar street network. As long as no re-planning (rerouting) is done on-line, the interaction-range of vehicles is restricted to a couple of hundred meters in direction of their travel and even less in the reverse direction. Therefore, for a consistent CA update, only state information of immediate neighborhood is necessary, reducing the boundary area to narrow strip along the borders between neighboring CPN. It should be obvious that the domain decomposition of the network has to be done in such way as to generate the smallest possible boundary area. An easy to implement recursive orthogonal bisection has proven to be (a) computationally efficient, (b) capable to produce satisfactory results, and (c) accessible to a analytical heuristic description, which can be used to predict the parallel efficiency. In our case, the criterion for bisection is defined by load-estimates of the traffic network nodes (e.g. intersection), each of which is computed as the sum Euclidean lengths of all incident edges. This is based upon the assumption that the execution time of the traffic CA strongly depends on the grid size and not so much on the actual vehicle density.

Due the complexity of the intersections, splitting the network at the center of street segments (with one half of the segment updated by either CPN) turned out to be easier than splitting at intersections between nodes and incident segments. The boundary size is then proportional to the total number of segments that had to be split to form the sub-networks. See Figure 5 for an example using 7 CPNs.

Figure 6 depicts the number of inter-CPN segments returned by the recursive orthogonal bisection method and a heuristic formula [19, 20] for three different maps. The similarity between plots and the data points is striking, especially if one considers the different nature of the street networks: (a) the arterials of the Dallas / Fort Worth area, (b) an excerpt of the former, also including local streets, and (c) the Autobahn network of the Federal Republic of Germany.

8.2 Simulation Timing and Boundary Exchange

Due to definition of the CA rule-set, boundary information has to be exchanged between neighboring CPNs before the CA rules can be applied, resulting in

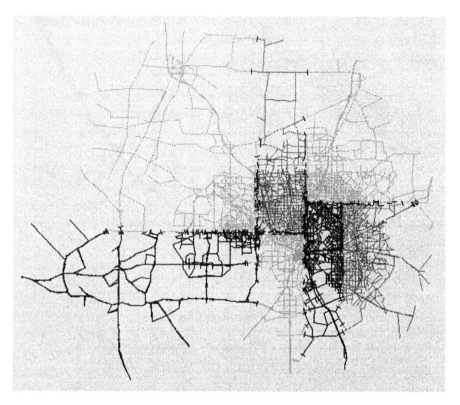

Fig. 5. Domain decomposition example for 7 workstations. On the right is Dallas; for Fort Worth on the left only the major arterials are included. The corresponding microsimulation runs faster than real time.

a basically time-step driven simulation. Depending on the specific coding of the CA rules, more than one sub-time-step may be necessary, each requiring an exchange of boundaries. The basic parallel update scheme works as follows: Each CPN scans its inter-CPN segments and transfers boundary information to its neighbors. Then, it enters a loop waiting for incoming boundaries from its neighbors. As soon as it has received a complete set, it executes a sub-time-step and the cycle is repeated. Note that there is no necessity to trigger a time-step through a master CPN, since each CPN only depends on its neighbors for consistent execution.

As for the length of boundaries, an optimization can be made by taking advantage of specific characteristics of the CA rule set. Usually the boundary that has to be transferred is as large as the *interaction range*[5] of the CA rules,

---

[5] Actually, the real value that defines the boundary length is the maximum of both *interaction range* and *maximum velocity*, but in a consistent, collision-free CA update the first is always at least as large as the second. In our current rule-set they happen to be equal.

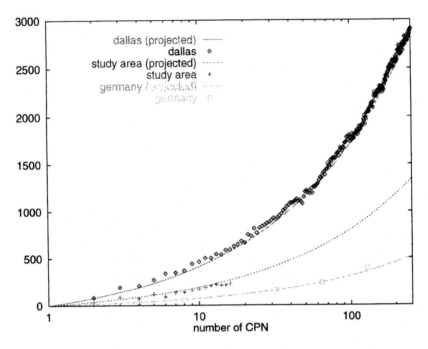

Fig. 6. Number of inter-CPN segments versus number of CPNs.

which is currently v_{max}. This would result in encoding and decoding of all vehicle data that are located within a range of v_{max} sites from a boundary. If the local density is high, that is, the boundary is located within a traffic jam, there may be more than one vehicle per lane. The CA rules, however, only refer to the *immediate* predecessor or successor on each lane, reducing the maximum number of vehicles in a boundary to one per lane. A more detailed description of aspects related boundary exchange can be found in [21].

8.3 Dynamic Load-balancing

Since the simulation is time-step driven, it is the goal to equalize the update times on all CPN. Unfortunately, there may disparities among the CPN, mainly for two reasons: The initial domain decomposition which was based upon a simple estimate, (a) did not include the computational load of the intersection functionality, and (b) did not cover inhomogeneous distribution of traffic during rush hours.

Optional dynamic load balancing can be performed to decrease the load disparities. In one traffic micro simulation [19], the implemented method corresponds to a *local decision, local migration* ($LDLM_S$, see [22]) strategy applied to the network nodes. Incident edges are transferred or split accordingly. When a part of a local network has to be off-loaded, nodes are sequentially transferred along the boundaries with the node furthest away from the center of the sub-network being selected first. As an optional restriction only those nodes can be selected that maintain one connected component on the CPN.

The flexible data structures which are required to perform dynamic load balancing prove to have another advantage: it is possible to remove or add CPN during the run-time of the simulation by (a) systematically reducing the sub-network on a CPN to zero before removing it, or (b) transferring a single seed node on a newly inserted CPN. In the latter case, the on-going dynamic load-balancing will transfer more and more load the new CPN until it is indistinguishable from the other ones.

8.4 Performance Estimates

We have made a first attempt to deduce an upper bound for the efficiency $e(p)$ of a large scale traffic simulation running on p CPN. It is based upon a set of parameters which can be retrieved from simple measurements. We only cite results here. A more detailed version (including an additional estimate for a two-dimensional communication topology) can be found in [20].

Assuming a time of $T(1)$ required for one time-step on a single-node machine, the necessary input parameters turn out to be:

- the size of the street network (number of edges, number of nodes) resulting in estimates for the number of neighbors $N_n(p)$ (with an average number of $n_n(p) = N_n(p)/p$ per CPN) and the number of boundaries $B(p)$ (with an average number of $b(p)$ per CPN),
- the number of sub-timesteps n_{sub} per time-step causing a relative performance loss for administration overhead $f_{adm}(n_{sub})$ and additional communication volume,
- the average boundary length b_{size} and the boundary message header size b_{header}, the application-level boundary transmission time t_{c1} and transmission latency t_{cl} both given as fractions of $T(1)$,
- the low-level communication bandwidth C_{net} (measured in byte per $T(1)$) of the computer network, and
- the relative load gradient $f_{grad}(p)$ generated by the granularity of the street network.

Using these parameters we define four major contributions to the time spent on one timestep:

- The *raw simulation* fraction mainly represents the traffic simulation itself, although it includes the administrative overhead for multiple sub-time-steps. It is equivalent with the efficiency $e(p)$ of the simulation.
- The *load-gradient* fraction represents the loss of execution time due to the load gradient which builds up throughout the CPN network.
- The *application-level (a-l) communication* fraction represents the time spent on retrieving, coding, transferring, decoding, and storing boundary data.

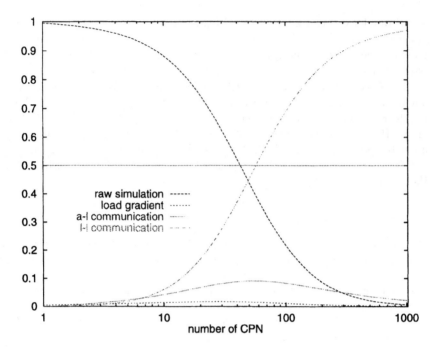

Fig. 7. Sparc-5 Cluster with Ethernet (Bus-Topology)

– Finally, the *low-level (l-l) communication* represents the additional time spent on low-level communication due to the saturation of the underlying communication network.

Figure 7 depicts the estimate for a cluster of Sparc 5 workstations connected by Ethernet. We assumed a relative exponential gradient of 0.01 per layer. Efficiency quickly drops below 0.5 for 30 CPN due to the network saturation by *low-level communication*. Estimates like these can be used to determine the optimal number of CPN to use for a given network size and additional time restrictions.

9   Summary and implications of the computational results

The purely site-based approach is very fast when it uses single-bit coding and runs on machines with vectorizing CPUs. However, in the context of the TRAN-SIMS project, a simulated vehicle needs additional information associated with it besides position and velocity. That makes a bit-coded approach problematic. In addition, comparable computing speeds can be reached on parallel but non-vectorizing supercomputers.[6]

---

[6] The single-bit circle implementation on the CM-5 is really fast. However, one will not reach the same computing speeds for realistic networks, which are composed of relatively short links.

For single-lane traffic, the purely vehicle-based approach is very fast for low traffic densities and still a factor of two to three faster than the intermediate approach for realistic traffic densities. For multi-lane traffic, many approaches turn out to be slower than the intermediate approach [23, 24]. And more sophisticated approaches are problematic at the current state of the project since all additional elements of traffic interaction such as intersections, ramps, etc. become more difficult to implement.

For these reasons, TRANSIMS currently uses the "intermediate" approach, intermediate between site-based and vehicle-based. The project is very suited for the use of parallel non-vectorizing supercomputers. Yet, mostly for reasons of client concerns, it currently runs on SparcStations which are coupled by a local area network. Supercomputer implementations are in preparation and will be needed for systematic testing of the dynamics of the simulation.

The reasons why parallel supercomputers are more advantageous compared to coupled workstations lie in the much faster communications network. As a rough number, we can simulate about 50 000 vehicles in real time on a typical desktop CPU. Connecting several of these CPUs via LAN allows simulating larger systems with the same computing speed. However, it is difficult to get faster than 10 times real time with this computational set-up: The communication between workstations is too slow on a LAN.

Dedicated supercomputers are here, in practical terms, about a factor of 10 faster. That means that one can use additional CPNs for enhancing computing speed, not for just making the system larger. A simulation 100 times faster than reality is easily still efficient (see the Intel iPSC example in Fig. 3).

10   Other approaches

The cellular automaton approach is obviously not the only approach to traffic simulation. From a systematic point of view, one has to recognize that one has trade-offs between resolution, fidelity, and scale [25]. Resolution refers to the smallest entities which are resolved in the simulation. For example, a microscopic traffic model resolves individual vehicles. Fidelity refers to the detail with which each entity is modeled. For example, the cellular automaton model has a low fidelity driving dynamics. Scale refers to the system size or temporal scale one can simulate. For example, a model with high resolution and high fidelity will usually run slow, thus only allowing the simulation of a small area and/or short time period.

In consequence, one could try to make the driving dynamics more realistic while preserving the resolution. An obvious option is to use continuous instead of discrete space. Yet, as long as one does not employ an event-driven update (e.g. [26]), one is still stuck with a coarse-grained time step. And all the well-understood techniques from discretized differential equations, i.e. to look for the limit of an infinitely short time step, do not work for traffic, at least not in a straightforward way: In driving, there is a delay between changes in the surroundings and actual changes in the vehicle behavior (given by reaction times, time to move the foot to the break pedal, vehicle inertia, etc.), which is of the order of one second. In consequence, as long as one does not model this delay explicitly but just uses information from the last time-step for the update, time-steps shorter than one second actually lead to *less realistic* traffic dynamics [27, 28, 29]. Moreover, just replacing discrete by continuous space while retaining the one second time-step does not automatically lead to more realistic traffic dynamics. For example, the published fundamental diagram of the PARAMICS project [30] lacks the typical variance structure real world traffic shows under the same circumstances.

PARAMICS is an implementation specially targeted for parallel supercomputers, and has been used on large scale traffic networks [31, 32]. It reaches 250 000 veh sec/sec on a 16k Connection Machine 200 with 512 Floating point units, and 360 000 veh sec/sec on 32 nodes of a Cray T3D [30].

The traditional TRAF/NETSIM [33] uses continuous space and a time-stepped update. A supercomputer implementation on a Cray XMP reached a real time limit of 1000 km [34], which should translate into approximately 13 300 veh sec/sec. Yet, the implementation did not vectorize, so that the simulation does not run much slower on a desktop computer.

Other microscopic models using continuous space and a time-stepped update are, e.g., the Wiedemann model [35], AS [36], MISSION [37, 38], VISSIM [39], THOREAU [40], SISTM [41, 42], and Release 2 INTEGRATION [43, 44]. They have different degrees of fidelity. Most of them have more or less well documented dynamics, but computational speeds are hard to find. Quite in general, it seems that 10 000 veh sec/sec is an upper bound on speeds for these models on a current desktop computer.

For certain questions one may get around a detailed microsimulation. For example, in situations with only very few junctions or intersections, fluid-dynamical approaches may be useful [45, 46, 47, 48, 49], especially for "corridor problems", where all traffic heads for a common destination such as a Central Business District [50, 51]. Sometimes, the traffic dynamics between intersections can be neglected [52].

Schwerdtfeger proposed a model where individual vehicles are moved according to averaged fluid-dynamical rules [53]. DYNASMART [54] and Release 1 of INTEGRATION [55] use similar methods. An early version of DYNASMART reports a computational speed of approx. 600 000 vehicle seconds per second on a CRAY X-MP/24 [56]; since the degree or vecorization is not reported, it is impossible to compare this to other results. Yet, the microsimulation approach, which resolves each individual traveler, is by far the most general and most robust approach. Also, note that INTEGRATION moved to a microscopic approach in Release 2 [44].

11 Some preliminary TRANSIMS results

The TRANSIMS team is currently working on a so-called case study to evaluate and enhance functionality on a practical example and in close collaboration with

Fig. 9. Simulation of the plans which were underlying Fig. 8. The figure shows only a small part of the simulated area, near the intersection of the two freeways. Note that the freeway coming from the south ("Dallas North Tollway") only extends as a much smaller street to the north. In consequence, congestion builds up on all roads which have traffic heading for that road. Note (see Fig. 8) that demand for this northern part of the Dallas North Tollway was higher than capacity.

makes sense.) In consequence, an interim method to generate plans has been designed. It works as follows:

- NCTCOG (North Central Texas Council Of Governments) provides the TRANSIMS team with time-dependent origin-destination data on a zone basis. This data comes out of more conventional models and is used unchecked.
- TRANSIMS generates individual trips from these aggregate numbers and assigns more specific starting and ending points inside the zones for these trips ("population generation").
- The planner generates route plans for these trips, i.e. for each trip a "reasonable" sequence of links (streets) is generated which connects the starting and the ending points. Only car traffic is assumed (Dallas does not have public transit.)
- Finally, these trips are executed in the microsimulation.

Currently (August 1996), the main problem is to find a reasonable set of plans. In other words: Unrealistic features in the microsimulation are currently dominated by unrealistic features in the plans. Figs. 8 and 9 depict the problem and at the same time visualize some parts of the methodology. Fig. 8 shows a superposition of all plans for the period between 17h and 17:30h. Dark links mean that demand is higher than capacity, i.e. more vehicles want to go through these links in that half hour period than what is physically possible.

Fig. 9 shows a simulation of that set of plans. The figure shows only a small part of the simulated area, near the intersection of the two freeways. Note that the freeway coming from the south ("Dallas North Tollway") only extends as a much smaller street to the north. In consequence, congestion builds up on all roads which have traffic heading for that road. Note (see Fig. 8) that demand for this northern part of the Dallas North Tollway was higher than capacity.

That is, demand which is much higher than capacity for the northern extension of the Dallas North Tollway leads to significant congestion starting on the entrances to that road, and that congestion dominates the dynamics. The reason why this is unreasonable is simple: If this would happen in reality, people would notice and find other ways north. We are currently in the process to investigate relaxation procedures for plansets which emulate this people's behavior and thus should avoid such dominating congestion structures.

Acknowledgements

Much of this paper would not have been possible without the continuous work of all members of the TRANSIMS team. In particular, Dick Beckman has developed the population generation methods. Doug Anson, Myron Stein, and Madhav Marathe have developed the route planning methods. Paula Stretz and the TRANSIMS software team undertook most of the actual implementation. James Dearing and Deborah Kubiceck provided the graphics facilities for the plans and for the microsimulation.

References

1. J. Cohen and F. Kelly. A paradox of congestion in a queueing network. *J. Appl. Probability*, 27:730–734, 1990.
2. M. Williams. Personal communication.
3. L. Smith, R. Beckman, D. Anson, K. Nagel, and M. Williams. Transims: Transportation analysis and simulation system. In *Proc. 5th Nat. Transportation Planning Methods Applications Conference*, Seattle, 1995.
4. J. Morrison and V. Loose. Transims model design criteria as derived from federal legislation. Technical report, Los Alamos National Laboratory, 1995.
5. C. Barrett, K. Berkbigler, L. Smith, V. Loose, R. Beckman, J. Davis, D. Roberts, and M. Williams. An operational description of transims. Technical report, Los Alamos National Laboratory, 1995.
6. L.L. Smith. TRANSIMS Travelogue. *Travel model improvement program newsletter*, 2–5, 1995–1996.
7. TRANSIMS web page. http://www-transims.tsasa.lanl.gov/.
8. R. Schwarzmann. Das EUROTOPP Modell, SCOPE/VIKTORIA-Bericht. Technical report, Institut für Verkehrswesen, TH Karlsruhe, July 1992.
9. K. Nagel. Freeway traffic, cellular automata, and some (self-organizing) criticality. In R.A. de Groot and J. Nadrchal, editors, *Physics Computing '92*, page 419. World Scientific, 1993.
10. K. Nagel and M. Schreckenberg. A cellular automaton model for freeway traffic. *J. Phys. I France*, 2:2221, 1992.
11. K. Nagel. Particle hopping models and traffic flow theory. *Phys. Rev. E*, 53(5):4655, 1996.
12. M. Rickert, K. Nagel, M. Schreckenberg, and A. Latour. Two lane traffic simulations using cellular automata. *Physica A*, 1996. In press.
13. P. Wagner. Traffic simulations using cellular automata: Comparison with reality. In D.E.Wolf, M.Schreckenberg, and A.Bachem, editors, *Traffic and Granular Flow*. World Scientific, Singapore, 1996.
14. P. Wagner, S. Krauss, and C. Gawron. A continuous limit of the Nagel-Schreckenberg model. *Phys. Rev. E*, 1996. Submitted.
15. D.W. Heermann. *Computer simulation methods in theoretical physics*. Springer, Heidelberg, 1986.
16. D. Stauffer. Computer simulations of cellular automata. *J. Phys. A*, 24:909–927, 1991.
17. K. Nagel and A. Schleicher. Microscopic traffic modeling on parallel high performance computers. *Parallel Computing*, 20:125–146, 1994.
18. N. Ito. Non-equilibrium critical relaxation and interface energy of the Ising model. *Physica A*, 1:196, 93.
19. M. Rickert, P. Wagner, and Ch. Gawron. Real-time traffic simulation of the German Autobahn Network. In *Proceedings of the 4th PASA Workshop*, 1996. In press.
20. M. Rickert. Estimating parallel efficiency of large-scale traffic simulations. In preparation, 1996.
21. M. Rickert and P. Wagner. Parallel real-time implementation of large-scale, route-plan-driven traffic simulation. *Int.J.Mod.Phys.C*, 1996. In press.
22. R. Lüling, B. Monien, and F. Ramme. Load balancing in large networks: A comparative study. In *3rd IEEE Symposium On Parallel And Distributed Processing*, pages 686–689, 1991.

23. M.W. Olesen. Personal communication.
24. D. Roberts. Personal communication.
25. C.L. Barrett. Personal communication.
26. T. Blanchard and T. Lake. Conservative spatial simulation. In A. Pave, editor, *European Simulation Multiconference*, page 515. The Society for Computer Simulation, Istanbul, 1993. See also other papers in the same proceedings.
27. C.L. Barrett, S. Eubank, K. Nagel, J. Riordan, and M. Wolinsky. Issues in the representation of traffic using multi-resolution cellular automata. LA-UR 95-2658, Los Alamos, 1995.
28. R. Wiedemann. Personal communication.
29. P. Wagner. Personal communication.
30. G.D.B. Cameron and C.I.D. Duncan. PARAMICS–Parallel microscopic simulation of road traffic. *J.Supercomputing*, 1996. In press.
31. C.I.D. Duncan. PARAMICS wide area microscopic simulation of ATT and traffic management. In J.I. Soliman and D. Roller, editors, *Proceedings of the 28th International Symposium on Automotive Technology and Automation (ISATA)*, page 475. Automotive Automation Ltd, Croydon, England, 1995. Paper No. 95ATS044.
32. PARAMICS Web site. http://www.epcc.ed.ac.uk/epcc-projects/paramics/.
33. U.S.Department of Transportation, Federal Highway Administration. *TRAF user reference guide*, 1992. Publication No. FHWA-RD-92-060.
34. H.S. Mahmassani, R. Jayakrishnan, and R. Herman. Network traffic flow theory: Microscopic simulation experiments on supercomputers. *Transpn. Res. A*, 24A (2):149, 1990.
35. R. Wiedemann. Simulation des Straßenverkehrsflusses. Heft 8, Institut für Verkehrswesen der Universität Karlsruhe, 1974.
36. T. Benz. The microscopic traffic simulator AS (Autobahn Simulator). In A. Pave, editor, *European Simulation Multiconference*, page 486. The Society for Computer Simulation, Istanbul, 1993.
37. R. Wiedemann. Modelling of RTI-elements on multi-lane roads. In *Advanced telematics in road transport, Proceedings of the DRIVE conference*, volume 2. Elsevier, 1991.
38. R. Wiedemann. Simulation des Verkehrsablaufs – Beschreibung des Staus. In *Beiträge zur Theorie des Straßenverkehrs*. Forschungsgesellschaft für Straßen- und Verkehrswesen, Köln, 1995.
39. M. Fellendorf. VISSIM. PTV system GmbH, Pforzheimer Str. 15, 76277 Karlsruhe, Germany.
40. W.P. Niedringhaus and P. Wang. IVHS traffic modeling using parallel computing. In *IEEE - IEE Vehicle Navigation and Information Systems Conference VNIS '93, Ottawa*, 1993.
41. SISTM. Transportation Research Laboratory, Old Wokingham Rd, Crowthorne, Berkshire RG11 6AU.
42. M. McDonald and M. A. Brackstone. Simulation of lane usage characteristics on 3 lane motorways. In *Proceedings of the 27th International Symposium on Automotive Technology and Automation (ISATA)*, 1994.
43. M. Van Aerde, B. Hellinga, M. Baker, and H. Rakha. INTEGRATION: An overview of traffic simulation features. *Transportation Research Records*, in press.
44. M. Van Aerde et al. *INTEGRATION (Release 2) User's Guide*, 1995.
45. H.J. Payne. FREEFLO: A macroscopic simulation model of freeway traffic. *Transportation Research Record 722*, 1979.

46. R.D. Kuehne and R. Beckschulte. Non-linearity and stochastics of unstable traffic flow. In C.F. Daganzo, editor, *Proceedings of 12th Int. Symposium on the Theory of Traffic Flow and Transportation*, page 367. Elsevier, Amsterdam, The Netherlands, 1993.

47. B.S. Kerner and P. Konhäuser. Cluster effect in initially homogenous traffic flow. *Phys. Rev. E*, 48(4):R2335–2338, 1993.

48. D. Helbing. Improved fluid-dynamic model for vehicular traffic. *Phys. Rev. E*, 51(4):3164, 1995.

49. C.F. Daganzo. Requiem for second-order fluid approximations fo traffic flow. *Transpn. Res. B*, 29B(4):277, 1995.

50. M. Kuwahara M and G.F. Newell. Queue evolution on freeways leading to a single core city during the morning peak. In Gartner N H and Wilson N H, editors, *Transportation and traffic theory*. Elsevier, 1987.

51. R.H.M. Emmerink, K.W. Axhausen, P. Nijkamp, and P. Rietveld. Effects of information in road transport networks with recurrent congestion. *Transportation*, 22:21, 1995.

52. H.P. Simão and W.B. Powell. Numerical methods for simulating transient, stochastic queueing networks. *Transportation Science*, 26:296, 1992.

53. T. Schwerdtfeger. *Makroskopisches Simulationmodell für Schnellstraßennetze mit Berücksichtigung von Einzelfahrzeugen (DYNEMO)*. PhD thesis, TH Karlsruhe, 1987.

54. H.S. Mahmassani, T. Hu, and R. Jayakrishnan. Dynamic traffic assignment and simulation for advanced network informatics (DYNASMART). In N.H. Gartner and G. Improta, editors, *Urban traffic networks: Dynamic flow modeling and control*. Springer, Berlin/New York, 1995.

55. Transportation Systems Research Group, Queens' University and M. Van Aerde and Associates, Ltd. *INTEGRATION: A model for simulating IVHS in integrated traffic networks, User's guide for model version 1.5e*, 1994.

56. R. Jayakrishnan and H.S. Mahmassani. Dynamic simulation-assignment methodology to evaluate in-vehicle information strategies in urban traffic networks. In O. Balci, R.P. Sadowski, and R.E. Nance, editors, *Proceedings of the 1990 Winter Simulation Conference*, 1990.

Parallel Method for Automatic Shape Determination via the Evolution of Morphological Sequences

Akihiko NAKAGAWA   Andrea KUTICS

Japan Systems Co., Ltd.
2-31-24, Ikejiri, Setagaya-ku, Tokyo 154, JAPAN
ranaka@ppp.bekkoame.or.jp, andi@ppp.bekkoame.or.jp

Abstract : A parallel evolutionary method for object shape determination is proposed by automatically generating morphological operator and operation sequences. Artificial individuals built up from binary morphological operators and operations undergo recombination and mutation processes for producing new generations. The normalized correlation between the generated shape and the corresponding input image region is calculated for fitness. This method requires no preliminary knowledge of the object shape and also no constraints are used for image background and smoothness. The parallel evolutionary approach provides a fast and directed search on large number of possible morphological sequences and the method can be applied on a wide range of images. The morphological operations are implemented by low level image processing steps and executed as parallel tasks by applying both data and algorithmic parallelization. As a concrete application, this method is utilized for the shape determination of skin objects in a system consisting of a camera device connected to a grid architecture of transputer nodes.

1 Introduction

Automatic determination of the shape of objects on a digitized image is a fairly complicated and time-consuming task. Generally, object shapes can be described in two main ways: determining the object boundaries or separating the regions associated with them. Various techniques developed for both approaches can be found in the literature. In the case of boundary detection, edge point candidates are first detected by applying Sobel, Canny or Laplacian filters or statistical edge determination approaches[1, 2]. In most cases, a thresholding post-processing is required on the obtained gradient image to provide thin boundary lines. The problem with these methods is that it is very difficult to find the appropriate threshold values automatically. For region-based segmentation, histogram-based approaches, region growing, split and merge, surface fittings, and texture based methods[3] are most commonly applied. Histogram-based methods suffer from the mentioned threshold-finding difficulties. In most of these approaches, severe constraints for smoothness and object size are often used to achieve robustness and to avoid the difficulties originating from the changing image background structure, low contrast and unknown object shape. Also, these methods use a large number of neighborhood operations for filtering and pattern matching on the entire image space that can result in a significant time delay. In another approach, the objects of interest in the image can be analyzed directly through their geometrical features utilized by morphological methods. Morphological operations[4] apply structuring elements and image translation, union and intersection processes to produce information about the shape of image objects. One of the most important characteristic of morphological methods is that they require the exact specification of the structuring elements. In order to use morphological operators effectively and in order to preserve the original objects' shape in the presence of noise, multiple structuring elements and repeated operations have to be applied in most cases. However, extracting shapes by repeated morphological operations is also a difficult task. Examples of using iterative morphological operators by multiple structuring elements for eliminating noise can be found in the literature[5]. The selection of the structuring elements and the way they have to be combined to perform an effective morphological filtering is usually achieved by a trial approach. Also, the obtained method is usually specific to a certain application. A general method for morphological filter design is proposed by Song and Delp[6] where they apply union of openings and intersection of closings by four lines and L shaped structuring elements. Considerable effort has been made to decompose complex structuring elements. Jones and Salbe[7] analyzed the decomposition of opening and closing operations for single and multiple structuring elements and demonstrated that these operations can be broken down into equivalent morphological filter sets. Theoretical work applying predicate logic for describing morphological operators was proposed by Joo, Haralick and Shapiro[8], but it has not been implemented in an algorithm. The purpose of these decomposition algorithms[9, 10] is to simplify the calculation and consequently the cost of computation and also to help the implementation of morphological operations by hardware. All of these techniques are very helpful in understanding the behavior of various kinds of morphological operators. However it is rather difficult to automatically design effective morphological operators suitable for direct, automatic shape extraction.

In this paper, we introduce an evolutionary approach for the generation of morphological sequences utilized for automatic shape determination. The object shape is produced by the best fit sequence of morphological operators and operation structures applied on roughly segmented object regions. As a concrete application, the method is applied for the determination of the shape of skin deformities. Extracting the shape of skin objects, like speckles and melanomas etc. has a great medical and also cosmetic importance, however, it is rather difficult to find an appropriate imaging technique considering even high cost magnetic resonance imaging and ultrasonic systems. The reason of this is that it is difficult and time consuming to obtain cross sections in few mm wide intervals on broad skin regions. Ultraviolet and visible images are taken by a CCD camera and used as input for shape determination of objects on arbitrarily selected skin areas. In order to achieve a fast system, the method is implemented parallel on a grid architecture of T800 transputers[11].

2 Parallel evolution of morphological operator and operation sequences

Evolutionary methods[12] were originally proposed to provide solutions for complex scheduling and optimization problems[13]. In these methods, the Darwinian theory of evolution is used as the model for problem solving. The evolutionary computation model includes a set of solutions acting as artificial individuals in a population. These individuals are mapped to computer data structures and operators that mimic the natural reproduction are defined to produce offsprings for further generations. By applying selection and genetic manipulation processes on generations of individuals, this approach carries out a directed search on the space of possible solutions to find the best fit individual hopefully representing a near optimal solution.

2.1 Preprocessing and Creating Starting Population

In our system, a digitized gray-level image is taken as input. First a rough preliminary segmentation process is carried out by means of separated two-dimensional Laplacian of Gaussian operations[14]. The obtained rough object areas are then used as the basis of morphological operations. These areas then undergo morphological procedures that are determined by using the evolutionary approach. In order to carry out the preprocessing, the image data is first split into small two-dimensional regions and loaded to the subtransputers in a treelike fashion[15]. Two horizontally adjacent nodes save the same data segment. During the preprocessing, the separated LoG operation steps are carried out parallel on these neighborhood nodes. A segment size of about 64x64 is applied to 512x512 size images, to avoid the problem of communication overhead caused by data reloading. Then, after data summing, a rearrangement of data segments is performed by the main transputer node to provide segments containing only one

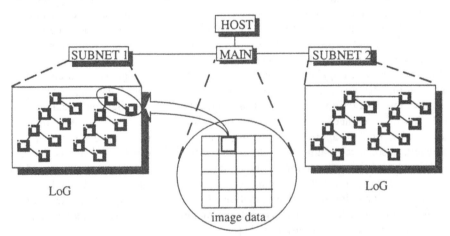

Fig. 1. Calculation outline of preprocessing (LoG)

or a few objects for the further processing steps. The outline of the preprocessing is shown in Fig. 1.

A starting population of solutions for the shape determination task has to be created next. Small size morphological structuring elements are utilized as features. A solution is built up from a series of small structuring elements with a maximum size of 3x3 and the corresponding series of low level morphological operations such as dilations and erosions. These solutions can act well as chromosomes, because the most important shape characteristics of the objects are represented in these series. This is a very rough model, as each individual is represented with only one chromosome to describe its characteristics. The starting operator and operation series are randomly generated. These operators then act on the rough segmented areas that are obtained during the preprocessing.

2.2 Mapping to data structures

Each solution that represents an artificial individual is mapped to a data structure in order to accomplish the evolutionary method computationally. This mapping is also necessary for converting new individuals from the evolutionary method into a form that the genetic model can evaluate. This data structure has to be designed very carefully as we must provide a data representation that is capable to transfer important characteristics from the current set of individuals to their offsprings by genetic type operators. Data objects consisting of two list structures are generated. The first list that maps the morphological operator series is built up from small two-dimensional arrays in order to provide an appropriate representation of the connectivity constraints and preserve the two-dimensional nature of the structuring elements. The operator series that consists of dilation and erosion operators, each assigned to a given member of the structuring element series is mapped to a list of binary numbers, by encoding the erosion operation with binary zero and the dilation operation with binary one. The data representation structure of the individuals is illustrated in Fig. 2.

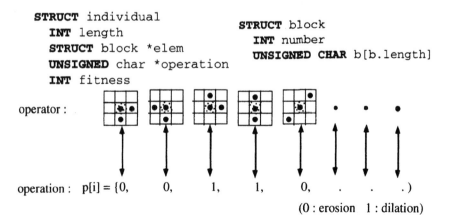

```
STRUCT individual
    INT length
    STRUCT block *elem
    UNSIGNED char *operation
    INT fitness

STRUCT block
    INT number
    UNSIGNED CHAR b[b.length]
```

operator :

operation : p[i] = {0,   0,   1,   1,   0,   .   .   .)

(0 : erosion   1 : dilation)

Fig. 2. Representation of individuals (the concrete data structure is written in occam)

2.3 Evaluation and selection of individuals

Each individual acting as a solution for the determination of the object shape has to be evaluated to give a measure of its fitness. Beside the representation of individuals, selection of an appropriate fitness function is another key issue in the method because the convergence highly depends on this function. Here we define the fitness of individuals as the normalized correlation between a region of the image obtained by applying the sequence of morphological operations coded in a given individual and the corresponding region of the original image. The normalized correlation can be expressed as the follows:

$$
nc = \frac{\left[N \sum_i I_i F_i - \left(\sum_i I_i \right) \left(\sum_i F_i \right) \right]}{\sqrt{\left[N \sum_i I_i^2 - \left(\sum_i I_i \right)^2 \right] \left[N \sum_i F_i^2 - \left(\sum_i F_i \right)^2 \right]}} \tag{1}
$$

where

N is the total number of pixels
I_i is an image pixel
F_i is the corresponding pixel in the calculated image

Using the information of fitness, a selection is carried out next, to provide a set of individuals that will be manipulated for producing offsprings and implemented by calculating the fitness distribution of the individuals. This selection is carried out in a roulette wheel like fashion[16]. First the global fitness value is calculated by summing the fitness values for each individual. Then we can calculate the selection probability for the individuals on the basis of fitness that will determine the slot sizes of the roulette wheel. The selection can be carried out by generating as many random numbers in the [0, 1] interval as the population size. Any time when a random number falls into a given probability interval, the individual associated with that particular probability interval is selected.

The parallel implementation of this calculation is based on the special characteristic of the morphological operations. The morphological dilation and erosion operations can be calculated by low level image operations of image shifting, union, and intersection[17] and can be accomplished independently. In our case, a morphological operation can be characterized by a 3x3 array. The shifting directions are determined by eight elements of a given operator. Carrying out these independent steps in a parallel fashion would result in a nearly eight-fold speedup comparing to the time requirements of the sequential processing with a given operator. However, we have a series of operators to be applied in a predetermined sequence. In order to accomplish this whole process effectively on the transputer network, a pipeline-based algorithm is carried out. This pipeline approach is applied in a dynamic way, which means that the scheduling of the processors carrying out a given calculation step on a particular data segment is arranged according to the morphological operation to be processed next. As we use only binary morphological operators in this work, the operator arrays contain only binary 0 or 1 data values in their locations. According to the eight significant locations of an operator, eight transputer nodes can process all of the shifting steps. However, this would cause a number of idling nodes, as no shifting process is necessary in the presence of a zero value at a given location in the operator. This problem can be avoided by allocating processors only to the nonzero elements of the operator. The number of the necessary image shifting steps, i.e., the number of necessary nodes varies from 1 to 8 according to the number of 1 values at the positions of the array. Thus, the separated shifting steps are carried out parallel on a given data segment by a block of adjacent nodes allocated in the necessary number. Finally union or intersection steps according to the given operation (dilation or erosion) are carried out. The input data loading and output data collection are supervised by the main transputer, and data communication tasks are executed on each subnode for data getting and passing. In Fig. 6., the processing ratio versus pipeline stages is drawn. Pipeline stage 1/8 means that the processing is carried out sequentially on one processor. Pipeline stage 1 depicts the processing time ratio in the case when the eight shiftings (max) and the union or intersection steps are carried out parallel on eight transputer nodes for only one operator at a time and the other twenty four nodes are idling. The further pipeline stages (from 4 to 32) show the processing time ratio in the cases when the transputer nodes are allocated dynamically to carry out the morphological operation steps without leaving nodes idling. In the case of 4 pipeline stages each allocated processor block has to execute eight shifting steps. While in the case of 32 pipeline stages each processor is working on a different operator in the operator series that contains only one significant position.

2.4 Generating Offspring by Crossover and Mutation

Two kinds of manipulation processes are carried out on the selected pool of individuals to generate new members for further generations: crossover and mutation. These operators are designed to transmit the most significant features from parents to offspring. In order to use morphological structures as features, the method is required to evolve the features that utilize the two-dimensional characteristic of the structuring elements, as it is mentioned previously. Both manipulation operations are accomplished with a given probability only, i.e., each selected chromosome is given an equal chance for manipulation similarly to natural reproduction. In both cases, first a randomization is carried out to give a probability to each selected individual to undergo manipulation and then the selected individuals are fed to a buffer.

In the case of a crossover, new individuals are created by recombining randomly selected parts of two parent individuals. Initially the pairs of individuals are also randomly selected. Then one of the two parts of the individuals, the morphological operator or the operation series is randomly

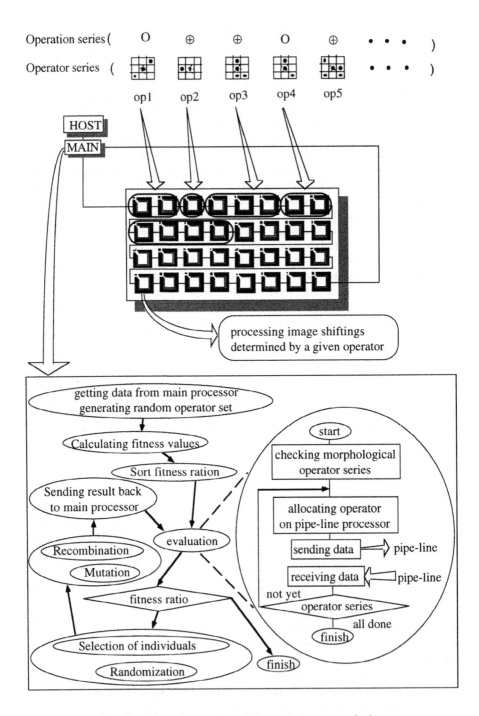

Fig. 3. Calculation outline of the evolutionary method

chosen. When the structuring element series is chosen, the crossover is carried out in two steps. First a structuring element in the operator series is selected and then a crossing point as a location in the two-dimensional structure of the selected structuring element is chosen. The crossover is accomplished by changing the counterparts of the two series. In the case of the operation series, the crossover is carried out by changing the two parts of the operation series by randomly determining a crossing point, i.e., a location in the operation series.

Applying only recombination operations might not lead us to a satisfactorily good solution as we cannot guarantee that all of the structuring elements included in a nearly optimal solution are already represented in different individuals of the starting population. In order to avoid this problem, a mutation is carried out with a relatively low probability to randomly generate new structuring elements or create new operation order. Such a mutation is carried out in a way similar to the crossover case. First an individual, and then a given part of that particular individual is chosen. If the operator part of the individual is selected, the location to change a bit inside a structuring element is determined in two steps, too. In the case of the operation series, an operation location inside the operation series is randomly determined and the appropriate bit is changed in its data structure.

A new generation is created by replacing the less fit part of the population with the newly created individuals. The replacing ratio is determined on the basis of the fitness distribution and it is usually between 0.3 and 0.5. The rest of the algorithm is merely repeating the evaluation, selection and manipulation processes in a cycle until either the required fitness is reached, or no more progress can be observed.

The selection and manipulation processes are accomplished in sequence on the main node. The manipulation steps, such as the crossovers and mutations on the individuals and individual pairs are carried out as concurrent processes. The outline of the whole calculation is illustrated in Fig. 3.

3 Application system

The method is implemented for the detection of skin spots such as speckles and melanomas and other lesions of different pigmentation levels. Extracting object shapes originating from different skin deformities has a great importance in the investigation and treatment of skin diseases. It may also be helpful for the early recognition of skin cancer by extracting and determining the shape of objects that might be cancerous and are hardly visible under normal lighting condition. At a further stage of development, it is possible to classify these objects on the basis of the shape characteristics extracted by means of the morphological sequences. Another important field of the skin spot extraction is the cosmetic industry where it can be very useful in the development and testing of various skin-care products and also to check the results of cosmetic treatments applied to different skin types. In our application system, input images are taken by a CCD camera using ultraviolet or visible-blue filtering in the (300-400) nm and (400-500) nm wavelength bands. The reason of using ultraviolet or blue filtered pictures is that

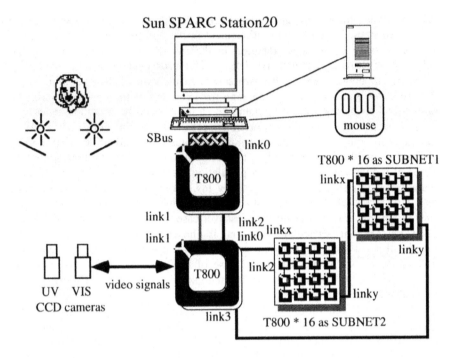

Fig. 4. System architecture

detailed information can be obtained about the objects located on the surface as the ultraviolet light is directly reflected from the surface of the skin without penetration. The skin spots with various pigmentation levels appear as dark regions on a lighter gray-level background depending on the models' skin type. The image taken by the camera system first undergo an 8 bit A/D conversion and is fed to the transputer network containing a main node and a grid network of 32 T800 transputers. The main transputer is connected to a root transputer board that is located in the host Sun SPARC Station 20 via serial interface. The user interface for controlling data acquisition, network booting, program loading and data processing is executed on the host machine. The system architecture is shown in Fig. 4. Actual implementation for detecting the shape of speckles in an ultraviolet filtered picture of a male face is shown in Fig. 5. Picture (a) represents the original image. The roughly segmented starting areas are shown as dark regions on picture (b). An intermediate state of the calculation and the determined shapes are presented on picture (c) and (d) respectively. The final shapes are then also depicted by applying a shading and a two-fold magnification on picture (e) and (f). A population size of 100 and individuals with a length between 5 and 40 structuring elements and operations were used in this case. The crossover probability was set to 0.3, while mutations were accomplished with a probability of 0.05 only. The fitness ratio varied between 0.6 and 0.94 after producing generations in a maximum number of 200. In some cases, solutions

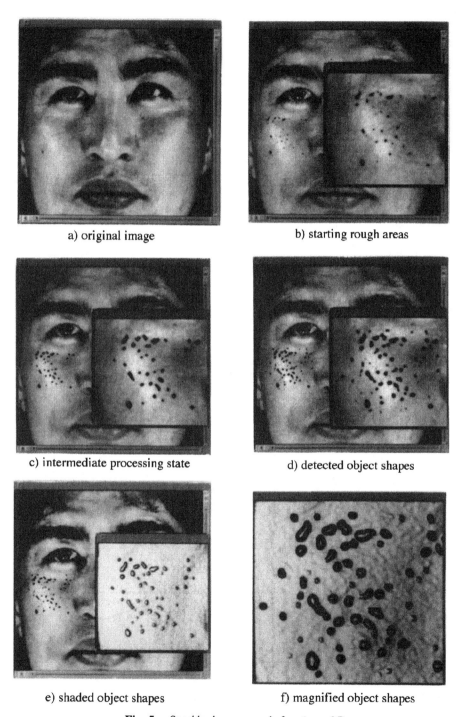

a) original image

b) starting rough areas

c) intermediate processing state

d) detected object shapes

e) shaded object shapes

f) magnified object shapes

Fig. 5. Speckle shapes on male face (age : 35)

only approximating the optimum could be obtained. However, these solutions could sometimes be reached after producing less then 50 generations. In order to produce higher fitness values and reach them through fewer generations, the use of a wider scale of morphological operations should be considered. The average processing time required to carry out the method for a 512x512 size image containing 10 to 100 objects is less than 750 ms on 32 transputers and less than 1260 ms on 16 nodes. The processing time necessary to carry out the processes on the network can vary upon the size, number and the geometrical distribution of the objects. In the case of images with large number of skin objects of different sizes and unbalanced distribution, the processing time can be relatively high due to frequent data loading resulting in a communication overhead. The program modules running on the transputer system are written in occam 2 language, while the user and controlling interface are written in C on the host Sun workstation and implemented by threads.

4 Conclusion

In this paper, a new parallel method for object shape determination is proposed by automatically generating morphological operator and operation sequences. In this evolutionary approach no constraints are used for image background and smoothness. Neither preliminary knowledge of the object shape is required. The algorithm is built up of decomposed, low level image processing operations such as image shifting, union and intersection that provide a good opportunity of parallelization. Applying a pipeline-based processing where a block of processors

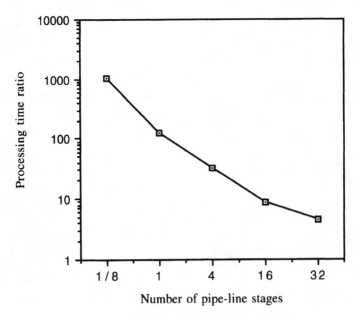

Fig. 6. Processing time ratio versus pipeline stages

executes the decomposed morphological operations in a parallel fashion ensures an efficient calculation by eliminating the problem of idling nodes. Some 15 times speedup can be achieved by the method compared to the performance of a single host machine. There is no guarantee, that the optimal solution is found in all cases, but the evolutionary approach provides a fast, global and directed search on the space of possible solutions to reach a nearly optimal one. The entire space of shapes that can be obtained by applying a series of morphological operations with 3x3 size structuring elements can be attained by carrying out crossover and mutation processes. Since this space can be very large, this feature adds a real advantage to this method. It can also be applied for shape determination on a wide range of images. Using gray scale morphology and implementing the approach for parallel morphological decomposition is considered as future work.

References

1. R. C. Gonzalez, R. E. Woods: Digital Image Processing. Reading MA: Addison Wesley 1992.
2. G. J . Awcock, R. Thomas: Applied Image Processing: McGraw-Hill 1996.
3. R. M. Haralick, L.G. Shapiro: Computer and Robot Vision, New York: Addison-Wesley 1992.
4. C. Giardina, E. Dougherty: Morphological Methods in Image and Signal Processing. Englewood Cliffs: Prentice-Hall 1988
5. J. Serra: Image Analysis and Mathematical Morphology. New York: Academic Press 1988
6. J. Song, E. J. Delp: The Analysis of Morphological Filters with Multiple Structuring Elements. Computer Vision, Graphics and Image Processing: Vol. 50, pp. 308-328 (1990)
7. R. Salbe, I. Jones: The Design of Morphological Filters Using Multiple Structuring Elements, Part I.: Openings and Closings. Pattern Recognition Letters: Vol. 13, pp. 123-129 (1992)
8. H. Joo, R. M. Haralick, L.G. Shapiro: Toward the Automatic Generation of Mathematical Morphology Procedures Using Predicate Logic. Proceedings on the Third International Conference on Computer Vision: Osaka, pp. 156-165 (1990)
9. I. Pitas, A. Venetsanopoulos: Morphological shape decomposition. IEEE Transactions on Pattern Analysis and Machine Intelligence 12, pp.38-45 (1990)
10. X. Zhuang, R. M. Haralick: Morphological structuring element decomposition. Computer Vision Graphics and Image Processing, Vol. 35, pp. 370-382 (1986)
11. INMOS Limited: The transputer applications notebook, architecture and software. London: Prentice Hall 1989
12. J. Holland: Adaptation in natural and artificial systems Ann Arbor: University of Michigan Press 1975
13. D. E. Goldberg: Genetic Algorithms in Search, Optimization, and Machine Learning: Addison-Wesley, Reading, 1989

14. A. Huertas, G. Medioni: Detection of intensity changes with subpixel accuracy using Laplacian of Gaussian masks. IEEE Transactions on Pattern Analysis and Machine Intelligence 7, pp. 651-664 (1986)
15. A. Kutics, A. Nakagawa, M. Date: Use of Transputers for the Fast Detection Of Medical Objects Transputer/Occam 6, Proceedings of the 16th Transputer/Occam International Conference, Japan, pp. 300-313 (1994)
16. L. Chambers: Practical Handbook of Genetic algorithms. Boca Raton: CRC Press 1995
17. R. M. Haralick, S. Sternberg, X. Zhuang: Image analysis using mathematical morphology. IEEE Transactions on Pattern Analysis and Machine Intelligence 7, pp. (1986)

Principal Component Analysis on Vector Computers

Dora B. Heras, José Carlos Cabaleiro, Vicente Blanco,
Pablo Costas and Francisco F. Rivera

Dept. Electrónica y Computación
Univ. Santiago de Compostela
España

Abstract. Principal component analysis is a classical multivariate technique used as a basic tool in the field of image processing. Due to the iterative character and the high computational cost of these algorithms over conventional computers, they are good candidates for pipelined processing. In this work we analyse these algorithms from the viewpoint of vectorization and present an efficient implementation on the Fujitsu VP–2400/10. We systematically applied different code transformations to the algorithm making use of the vectorial capabilities of the system. In particular we have tested a number of vectorization techniques that optimize the reuse of the vector registers, exploit all levels of the memory hierarchy, and utilize the pipelined units in parallel (concurrency between them). We have considered images of 32 × 32 pixels and have divided the algorithm into three different stages. The speedups obtained for the native vectorizing compiler were 1.3, 1.3 and 7.9 for the different stages. These speedups were multiplied by factors of 5, 50 and 55 respectively, after applying our code transformations. The best improvement was achieved in the third stage of the algorithm, which is the most time consumming.

1 Introduction

Currently, the most detailed knowledge on microscopical particle morphology is obtained by the combination of electron microscopy, digital image processing and three–dimensional reconstruction methods. The individual image of a particle only provides a measurement of such a particle. Averaging is required because the individual images of a particle obtained by electron microscope contain a large amount of noise, allowing the significant part of the image to be extracted only from a set of repeated measurements. The resulting *average image* will be a better approximation than any image used to obtain it. In such cases, you have to separate the different particle populations that are in the biochemical preparation before obtaining a significant average image or using the images for three-dimensional reconstruction. If the set of images represents

---

* This work was supported in part by the CICYT under grant TIC92-0942-C03-03 and Xunta de Galicia under grant XUGA20606B93.

the same type of particle, the *average image* is obtained and three–dimensional reconstruction is performed. However, usually several populations are mixed. In order to distinguish them, principal component analysis is used among other methods. A subsequent automatic clustering process would enable the separation of the different populations.

The notion of principal components of a sample was introduced as a statistical tool for reducing multivariate data encountered in applied statistical research to a smaller dimensionality [9]. He defined a "plane of closest fit" as a subspace which minimizes the sum of squares of the distances between pairs of points containing data. The term "principal components" was later applied in order to analyse covariance and correlation structures. Since then, it has become increasingly popular in multivariate statistical theory and applications. The mathematical treatment of principal component analysis is based on characteristic roots and vectors of positive definite symmetric matrices.

In fact, for a principal component analysis, it is not necessary to calculate all the eigenvalues and eigenvectors of the variance-covariance (or correlation) matrix. Only a few (those of largest modulus) containing a large percentage of the total variance have to be calculated.

We present an implementation of a principal component analysis algorithm on a Fujitsu VP–2400/10 vector computer. A parallel implementation (exploiting spatial parallelism) can be found in [3, 4]. In the following sections we describe the algorithm and the quality of the automatically vectorized code, as well as the proposed improvement on this vectorial code.

2 Principal component analysis

The principal component analysis method consists of six stages [10], which are shown schematically in Figure 1:

principal component analysis
{

    1. represent the M images as P point vectors

    2. normalize the M image vectors

    3. generate the covariance matrix C

    4. obtain the k eigenvalues of largest modulus of C and their corresponding eigenvectors

    5. obtain the M vectors as a linear combination of the k eigenvectors

    6. represent the k eigenvectors as images

}

We have considered a population of M images with $N \times N$ pixels. An alignment process of the image samples is initially carried out in order to present them in a common reference system.

Among the different methods for performing the alignment, we used the *reference-free* alignment method [17]. This is an algorithm in which the accuracy

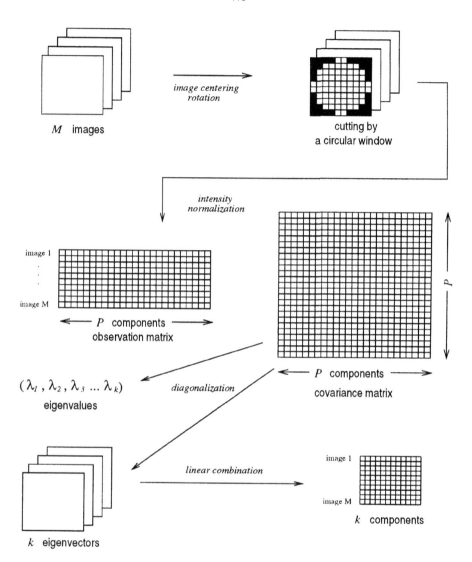

Fig. 1. *Principal component analysis steps*

is iteratively refined until convergence is achieved. During the alignment process the shift and rotational alignments, necessary in each iteration, are accumulated in transformation matrices. The application of these shift and rotational alignments to the images is performed after the reference–free alignment process is concluded. The two-dimensional transformation matrices obtained are used in a bilinear interpolation procedure. The alignment process was not included in the code that we vectorized (we considered it as a preprocessing step). In this work, the centering step consists in applying the previously obtained transformation matrices.

After the alignment phase has been carried out, a circular mask eliminates non-significant parts of the image (step 1 in Figure 1) and performs the intensity normalization process (step 2 in Figure 1). At this point, each image is represented by P *pixels* $(P < N^2)$ as a vector $\mathbf{x}_i$:

$$\mathbf{x}_i = \begin{pmatrix} x_{i1} \\ x_{i2} \\ \vdots \\ x_{ij} \\ \vdots \\ x_{iP} \end{pmatrix} \tag{1}$$

where element x_{ij} corresponds to the j-th *pixel* in image i. The mean vector is defined as

$$\mathbf{m} = \mathcal{E}\{\mathbf{x}\} \tag{2}$$

where $\mathcal{E}\{\mathbf{x}\}$ is the expected value of $\mathbf{x}$ [12]. This expected value is calculated as the expected value corresponding to each vector component.

The normalized vectors, i.e. vectors with zero average and unitary standard deviation after the intensity normalization process, constitute the so-called "observation matrix" in Figure 1.

The objective is to find a set of features with discriminating capability, i.e. they must separate the existing classes as clearly as possible. Also, they must be reliable i.e. the particles belonging to the same class must present the smallest dispersion possible. Finally, they must be uncorrelated. Characteristics that are strongly interdependent must never be used, because they do not add discriminating information and generate redundancy.

The decorrelation is measured between pairs of characteristics of the corresponding covariance matrix. The next step is to compute the covariance matrix in order to obtain the significant properties of the sample matrix.

$$C = \mathcal{E}\{(\mathbf{x} - \mathbf{m})(\mathbf{x} - \mathbf{m})^T\} \tag{3}$$

This covariance matrix is a $P \times P$ real symmetric matrix. c_{ii} is the *variance* of x_i, the i-th component in each vector in the population. c_{ij} is the covariance between i-th and j-th components in each vector. If elements x_i and x_j are not correlated, their covariance is zero and consequently $c_{ij} = c_{ji} = 0$

Using N vectors of characteristics (N images) of the population, equations (2) and (3) can be expressed as:

$$\mathbf{m} = \frac{1}{M} \sum_{k=1}^{M} \mathbf{x}_k \tag{4}$$

and

$$C = \frac{1}{M} \sum_{k=1}^{M} \mathbf{x}_k \mathbf{x}_k^T - \mathbf{mm}^T \qquad (5)$$

Regarding reliability, it can be quantified by analysing the principal diagonal values of the covariance matrix, since they are the variances of the corresponding characteristics. The larger the difference between these elements the greater the statistical dispersion.

At this point, a linear transformation is performed in order to increase the discriminating capabilities of the features. This can be achieved by diagonalizing the covariance matrix and selecting the features with the highest variance values, thus reducing the dimensionality of the problem [10, 19, 16, 2].

It is always possible to find a set of P (the size of the vectors representing the images after eliminating non-significant images) eigenvalues and orthonormal eigenvectors because matrix C is real and symmetric. The eigenvectors define a new base in the P dimensional space and are not correlated [11].

Sorting the eigenvectors obtained by decreasing values of eigenvalues, the k features with higher discriminant capabilities will be stored in the first positions of the covariance matrix. Generally, a high percentage of the variance is accumulated in the first 8 or 10 elements of this matrix, and consequentelly each initial vector can be represented as a linear combination of these eigenvectors.

The most time-consuming step of principal component analysis is eigenvalue and eigenvector computation. We used the Hotelling procedure [14] (based on the power iteration method) to solve this problem. This procedure permits the extraction of the eigenvalues sorted in decreasing order, which means we can stop the process when the accumulated variance exceeds some preset value. In our case, we stop the algorithm when 98% of the variance is obtained or when the number of eigenvalues is 20 (whichever is first). Neverthless, in most practical cases of macromolecule analysis, it is sufficient 80% or 85% of the acumulated variance.

To show this procedure of separation between classes of images representing different particles, we applied the proccess described above to sets of images. We used 50 artificial images of 32×32 pixels, with added noise, such as the images that are displayed in Figure 2. The principal component analysis shows that 90% of the variance is accumulated by the first 12 eigenvalues. More especifically the first eigenvalue accumulates 62% of such variance. In Figure 3 the eigenvector corresponding to the first eigenvalue is displayed as an image. This eigenvector (autoimage) shows the characteristic which distinguishes between the two groups of images. According to this, it is easy to identify two groups of images in the set: one group has images rotated at a fixed angle with respect to the images belonging to the other group.

3 Vectorization

To seek concurrency in computations, spatial and temporal parallelism are applied. Vector computers basically improve temporal parallelism. Each operation

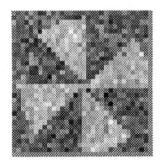 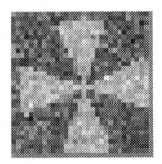

Fig. 2. *Artificial images*

Fig. 3. *Image corresponding to the first eigenvector*

is split into different functional units placed on separate hardware structures called *stages*. All together they constitute a *pipeline unit*. The concept of pipelining can be applied to both the Control Unit, splitting the different stages in the execution of an instruction, and the Processing Units [13]. *Vectorization* is the process by means of which the compiler of a vectorial computer encodes a sequence of operations and data in such a way that they can be executed as vectorial operations in the pipelining units. A variety of transformations called vectorization techniques can be applied to the code. Their efficient application requires a previous examination of the architectural characteristics of the computer.

3.1 Fujitsu VP–2400/10

We carried out the implementation of the method in a Fujitsu VP–2400/10 vector supercomputer. This is a register-to-register computer where all the vector operations, except for loading and storage, operate with vectors that are stored in registers. The structure of this computer is shown in Figure 4. We find one scalar unit and one vector unit, which access the main memory directly through a 2*8 byte data bus. The main memory is structured in interleaved modules.

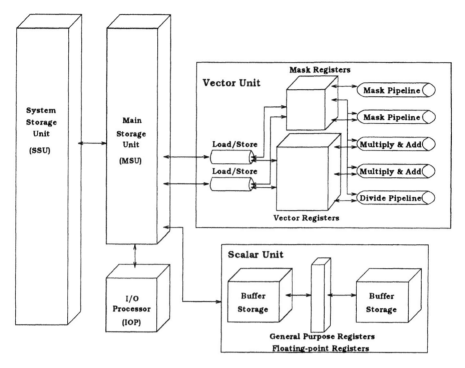

Fig. 4. *Structure of the Fujitsu VP–2400/10*

To optimize the vectorial execution of the Principal Component algorithm, we focused on the following elements [5]:

1. **Vectorial Registers:** These are the highest level in the memory hierarchy and a high velocity access area [15]. There are 256 vectorial registers with 32 elements in the VP–2400. Each element has a length of 8 bytes. Their main characteristic is that their configuration can be modified by hardware [1]. This means that the registers are chained in groups. Therefore, the number of elements per register is increased at the expense of decreasing their total number. To build this configuration, the compiler automatically analyses the program to be executed. We can profit from these registers trying to promote data reuse. In this way, the amount of loads/stores from/to main memory is reduced.

2. **Multiple pipelines:** The VP–2400/10 has seven pipelines: two mask pipelines, two for multiplication/addition, two for loading/storage and one division unit. Six pipelines can work concurrently. There is also the possibility of having longer pipelines as the result of chaining some units [5]. Thus, a number of vectorial instructions can be joined and the execution time can be reduced.

3. **Main memory:** In the VP–2400 series the main memory is interleaved. It is composed of 4 elements, although there is only one working element in

the computer we used. There are 4 modules per element and each module is split into 4 different 16 Mb segments. Each segment is formed by 16 ways of 1 Mb. The word length is 8 bytes. Therefore, there are 512 Mb of memory per module. Loads and stores of up to 8 bytes per module can be perfomed in each clock cycle (concurrent accesses to the 4 modules).

3.2   Vectorial optimization techniques

Several vectorization techniques are well known [7, 8, 5], and in the particular case of principal component analysis, great benefits can be obtained by applying the following strategies:

1. **Vectorize the longest loop**, in order to minimize the effect of the start up time. To do so, it may be neccessary to exchange loops in the code. And a restructuring of the code might also be required.
2. **Execute in the scalar unit the loops that are too short**. In this kind of loops there is no trade–off between the start–up cost of the pipelining units and the savings obtained by performing vectorial instructions. A great number of iterations would be neccessary to make this trade–off effective.
3. **Loop fusion**. Some loops are fully vectorized and even so do not make an eficcient use of the hardware. The reason is that they do not have enough density of operations, which means they do not profit from the simultaneous operation of some pipeline units.
4. **Optimize memory access**, exploiting data locality. Higher performance in the use of the interleaved memory is obtained when data is stored in order as much as possible. The higher the separation between consecutive accesses, the lower this performance is. A restructuring of the code may be neccessary to optimize memory access because it allows, for example, accessing by columns, and this is the way in which matrices are stored in FORTRAN.
5. **Exploit spatial parallelism**, applying any modifications to the code that increase the density of operations in a nesting. In this way, an adavantage is obtained from the parallelism inherent to the architectural structure of the VP–2400/10. One of these modifications is called **unroll**. It consists in replicating the body of a loop repeatedly (the number of times is called the depth of the unroll) by correctly modifying the control index of the loop. The optimal depth of the unroll depends on the kind and number of operations that are performed in the loop body. It is specially useful in the case of reduction operations [8]. In this kind of instructions the improvement comes not only from the increase of concurrency but from the decrease in the number of accesses to memory[7].

3.3   Vectorization of the algorithm

We focus our improvement on the three most time–consuming steps of the principal component analysis algorithm [6]. Splitting into three independent steps we can independently analyse the computational load asociated with each step. These are:

1. *Image centering* which includes image translation and rotation by a given position and angle which is previously calculated by means of the alignment proccess, image vector conversion and intensity normalization. The computational time necessary for preforming the alingment proccess is not included in this step of the algorithm. The *image centering* step only includes the aplication of the transformation matrices calculated in the previous process of alignment.

   With the previous steps we are preparing the images before the **principal component analysis**.

2. *Covariance matrix calculation* from the sample matrix. This is the third stage of the algorithm in Sect. 2.

3. *Eigenvalue and eigenvector calculation* and more specifically the matrix-vector product, as we will show by means of an example in section 3.4. This computation is performed in the fourth stage of the algorithm that is described in Sect. 2.

3.4 Results

In Table 1 we display the execution time for the sequential and automatically vectorized code versions of all the stages of the algorithm. All the measured times are obtained using 10 synthetic images of 32×32 pixels. The situation of considering synthetic images is not a restriction because we only measure computational costs.The centering stage shows the execution time for just one image.

ALGORITHM STAGE	SEQUENTIAL	AUTOMATICALLY V.	SPEEDUP
centering (per image)	1130 μs	905 μs	1.3
covariance matrix	364480 μs	286050 μs	1.3
matrix-vector product	184760 μs	23500 μs	7.9

Table 1. Runtime for the sequential and automatically vectorized codes

Applying the vectorizing techniques mentioned above, the improvement achieved is significant, as shown in Table 5. Each stage of the algorithm was optimized as follows:

1. *Image centering stage.* In order to improve this computation, the techniques that offer increases in performance are optimization of memory accesses and loop fusion.

The first change applied was to remove intermediate variables, in such a way that the swapping between loops can be performed in an immediate manner. This swapping is used to access the matrices column–wise, obtaining a lower access latency.

Finally, a loop fusion was performed converting the two–dimensional space of iterations into a one–dimensional space.

In Table 2, the runtime results obtained for both techniques are shown.

VECTORIZING TECHNIQUE	VECTORIZED	SPEEDUP
automatic vectorization	905 μs	1.3
memory access optimization	180 μs	5.9
loop fusion	175 μs	6.5

Table 2. Runtime for the automatically vectorized and optimized vectorial codes for the *image centering stage*

2. *Covariance matrix computation.* The first modification we considered was to vectorize the longest loop to exploit spatial parallelism. It is necessary to replace the original three–dimensional nested loop structure into a two–dimensional one and to vectorize one of them.

Automatic vectorization performs a row–wise access to the covariance matrix. This fact originates a poorly efficient memory access that can be overcome if we force a column–wise access. This optimization can be performed in a straight way because the covariance matrix is symmetric. In addition we could compute just the triangular half of this matrix, but the inner loop would have a non–constant length that would originate a decrease in the vector code.

Finally, we have analyzed the influence of the unroll on the loops that immediately cover the vectorized loop. The unroll was optimal in the case of two iterations. The improvement obtained using this strategy is mainly due to the fact that the operation is a reduction, minimizing the associated dependency.

The results of applying all these techniques are summarized in Table 3.

3. *Eigenvalue calculation* The first optimization that we applied in this stage was a swapping between nested loops. This can be performed after removing intermediate variables used in the sequential code, optimizing the memory access. The best result applying unroll was obtained with three iterations, exploiting spatial parallelism.

The results obtained in applying both strategies are shown in Table 4.

In order to justify the fact of considering the matrix-vector product time instead of the total eigenvector and eigenvalue calculation we can examine the

VECTORIZING TECHNIQUE	VECTORIZED	SPEEDUP
automatic vectorization	286050 μs	1.3
exploiting spatial parallelism	74000 μs	4.9
memory access optimization	22030 μs	17
vectorize the longest loop	12000 μs	30
exploiting spatial parallelism (unroll)	5750 μs	63

Table 3. Runtime for the automatically vectorized and optimized vectorial codes for the *covariance matrix computation stage*

VECTORIZING TECHNIQUE	VECTORIZED	SPEEDUP
automatic vectorization	23500 μs	7.9
memory access optimization	540 μs	342
exploiting spatial parallelism (unroll)	430 μs	430

Table 4. Runtime for the automatically vectorized and optimized vectorial codes for the *eigenvalue calculation stage*

following example. We solved a real problem, by considering the principal component analysis for 50 images of 32×32 pixels in an homogeneous sample from electronic microscope observations [10]. The total execution time was 5.43 ms. The diagonalization procedure of the covariance matrix required 8085 iterations to compute 20 eigenvalues and eigenvectors and this took 4.14 ms, i.e. 510 μs per iteration. Comparing with the matrix-vector product execution time (430 μs), we can see that the time required by the rest of the operations in each iteration is short. That is why we only considered the time of the matrix-vector product instead of the total time for the third stage. These results are displayed in Table 6.

ALGORITHM STAGE	AUTOMATICALLY V.	VECTORIZED	SPEEDUP
centering (per image)	905 μs	175 μs	6.5
covariance matrix	286050 μs	5750 μs	63
matrix-vector product	23500 μs	430 μs	430

Table 5. Runtime for the automatically vectorized and optimized vectorial codes

total execution time	5.43 ms
total diagonalization procedure time	4.14 ms
time per iteration	510 μs
matrix–vector product per iteration	430 μs

Table 6. Principal component analysis using 50 32×32 pixel images from electronic microscopy

4 Conclusions

The speedup achieved in the image centering stage is the lowest of the whole algorithm. The reason is that the code of this stage consists of conditional instructions inside small loops. The speedup obtained for the second stage is multiplied by a factor of 50 with respect to automatic vectorization. This factor is 55 for the diagonalization of the covariance matrix. It is the most expensive stage and in adition, it is the best vectorized stage.

In view of these results, and comparing the speedup obtained by means of automatic vectorization with the speedup obtained applying different vectorizing techniques, we can conclude that even though the vectorizing compiler is able to recognize the vectorizable code, the participation of a programmer is required. This programmer uses a number of code restructurations in order to take maximum advantage of the architectural features of vector supercomputers. In our case, it is of special significance to observe how the speedup is modified when *memory access optimization* and *unroll* are applied.

It is difficult to evaluate the whole principal component analysis to obtain general conclusions about its performance. The main reason is that the total runtime depends on the number and size of the images to be processed, and also on the number of iterations needed to satisfy the convergence criterion of the Hotelling algorithm, which depends on the closeness of the eigenvalues. These experiments show that the principal components analysis technique is a good candidate for exploiting temporal parallelism. The reasons are the nesting structure of the code, especially suitable for vectorizing, and the high computational time required for the code. Significant improvements can be achieved on this kind of vectorial supercomputer systems. This same conclusion applies to other image processing algorithms which are structured as nested loops of high dimensionality.

References

1. R. Allen and K. Keneddy. Vector register allocation. *IEEE Transactions on computers*, 41(10):1290–1317, 1992.
2. J.-P. Bretaudiere and J. Frank. Reconstitution of molecule images analysed by correspondence analysis: a tool for structural interpretation. *Journal of Microscopy*, 144:1–14, 1986.
3. J. Cabaleiro. *Análisis, Caracterización y recontrucción 3D de macromoléculas en multiprocesadores*. PhD thesis, Dept. Electrónica y Computación, Univ. Santiago de Compostela, may 1994. In Spanish.
4. J. Cabaleiro, J. Carazo, and E. Zapata. Parallel algorithm for principal component analysis based on hotelling procedure. In P. Milligan and A. Nuñez, editors, *EUROMICRO Workshop On Parallel and Distributed Processing*, pages 144–149, Gran Canaria, 1993. IEEE Computer Society Press.
5. H. Cheng. Vector pipelining, chaining, and speed on the IBM 3090 and cray X-MP. *IEEE Computer*, 22(9):31–46, sep 1989.

6. P. Costas. Análisis en componentes principales sobre computadores vectoriales. Master's thesis, Dept. Electrónica y Computación, Univ. Santiago de Compostela, jul 1994. In English.

7. W. Cowell and C. Thompson. Transforming FORTRAN DO loops to improve performance on vector architectures. *ACM Transactions on Mathematical Software*, 12(4):324–353, 1986.

8. J. Dongarra and S. Eisenstat. Squeezing the most out of an algorithm in Cray FORTRAN. *ACM Transactions on Mathematical Software*, 10(3):219–230, 1984.

9. B. Flury. *Common Principal Components and Related Multivariate Models*. John Wiley & Sons, 1988.

10. J. Frank, M. Radermacher, T. Wagenknecht, and A. Verschoor. Studying ribosome structure by electron microscopy and computer-image processing. In *Methods in Enzymology*, pages 3–35, San Diego, 1988. Academic Press. vol. 164.

11. G. H. Golub and C. F. V. Loan. *Matrix Computations*. The Johns Hopkins University Press, Baltimore and London, 1989.

12. R. C. González and R. E. Woods. *Digital Image Processing*. Adison-Wesley Publishing Company, Massachusetts, 1992.

13. D. Hennesy, J.L. & Patterson. *Arquitectura de computadores*. McGraw–Hill, 1993.

14. H. Hotelling. Analysis of a complex of statistical variables into principal components. *Journal of Educational Psychology*, 24:417–441, 1933.

15. K. Hwang. *Advanced computer architecture*. McGraw–Hill, 1993.

16. D. Maravall and Gómez-Allende. *Reconocimiento de formas y visi'on artificial*. RA-MA Editorial, Madrid, 1993.

17. P. Penczek, M. Radermarcher, and J. Frank. Three–dimensional reconstruction of single particles embedded in ice. *Ultramicroscopy*, 40:33–53, 1992.

18. C. Polycronopoulos. *Parallel Programming and Compilers*. Kluwer Academic Publishers, 1988.

19. M. Tatsuoka. *Multivariate Analysis: Techniques for Educational and Psychological Research*. John Wiley & Sons, University of Illinois, 1971.

Functional Programming and Parallel Processing

(Invited paper)

Rafael Dueire Lins*

Departamento de Informática
Universidade Federal de Pernambuco
Recife - PE - BRAZIL

Abstract

Functional languages belong to a neat and very high-level programming paradigm. A functional program is a set of function definitions. The λ-Calculus, a theory of functions under recursion, offers a solid theoretic background to functional programming. In 1978, John Backus pointed at the functional programming as a natural candidate to solve *"the software crisis"*.

One of the many promises of functional programming was the possibility of extracting parallelism. This paper analyses the evolution of the functional programming paradigm under this outlook.

1 Introduction

In his 1978 Turing Award Lecture [6], John Backus draw the attention of the computer science community to functional programming as offering an alternative to the problems of programming known as "the software crisis" and as a way to break the von-Neumann bottleneck between processors and memory, which he perceived to be a limiting factor in the design of new computer architectures. Functional languages are, in essence, a nicer syntax to the λ-Calculus, a theory of functions [24, 7]. This mathematical background makes proving programs correct a much easier task than in imperative programming, a programming paradigm that closely reflects the features of the machine architecture.

There are several possible ways to exploit the parallelism present in a functional program. Each system select a number of them, and it is therefore difficult to isolate the effect of a single technique on overall performance, even when concrete performance results are available. The partitioning of a program can be done either *implicitly* or *explicitly*. In the first case the system decides which tasks should be created, while with explicit partitioning the programmer is left with the problem of determining which expressions should be created as tasks.

---

*also: Honorary Lecturer, The University of Kent at Canterbury, Kent - U.K.

In either case, the partitioning could be *static*, in which case the number of tasks which will be created at runtime is predetermined, or *dynamic*, in which case tasks are created depending on factors such as the overall runtime load, or load control annotations. Tasks may be placed on the processor creating the task, on the processor owning the data which the task requires, or on some other processor. Task placement may also be explicit or implicit, static or dynamic.

The best partition of a program to a machine is the one that maximises the available parallelism, up to the number of processors available, while minimising the parallel overhead. It is important to try to predict the duration or *granularity* of each task to allow an equal distribution of work (load-balancing). Communication and task creation overheads can be minimised by adopting coarser-grained tasks, taking care not to increase idle time in processors.

One can divide the history of functional programming into two phases. The first period, which comprises the 1980s and before, corresponds to the time in which parallelism was sought as a way to make functional languages run as fast as imperative ones. The second period is the time in which "real" parallel processing can find an alternative in functional programming.

This paper overviews the relationship between parallelism and functional programming. Emphasis is given to the aspects of the functional programming paradigm that either were, have been, or are thought of as yielding efficient parallel implementations. It starts with a brief introduction to functional programming and its theoretical background. As the history of parallel functional programming implementation is closely intertwined with developments in sequential compiler technology, we overview its most important milestones. Garbage collection has always been linked to functional programming, thus parallel dynamic memory management algorithms are covered. We analyse the promises made by the functional programming community and give an account of what has been delivered so far.

2 Functional Programming: Taxonomy

The concept of what is a functional language has evolved with time, but its main feature is that a program is a set of definitions of higer-order functions – functions are not only passed as parameters to other functions, but can also be the result of the evaluation of a given function. According to this definition, LISP [82] was the first functional programming language.

Two other features are also used to classify functional languages: the existence (or absence) of destructive assignment and their evaluation mechanism. If a language has no destructive assignment, the value of any sub-expression is static, it is said to be "pure" or to enjoy the property of *referential transparency*. If the arguments to a function are passed by value, i.e. all the arguments are evaluated before the function itself, it is called "strict". Non-strict languages may be implemented using either a *data-driven* (also known as dataflow) or a *demand-driven* approach. Demand driven languages evaluate arguments as required by the function. If the result of the argument is recomputed each time it is needed, this evaluation mechanism is called *call-by-name*. Conversely, if the result is shared the evaluation mechanism is called *call-by-need*, and the language is said to be *lazy* or *procrastinating*. Non-strict languages have the advantage that only expressions which must be evaluated to give the program result actually are evaluated. They allow the use of infinite data structures, such as infinite sets and lists, which are described by a formation law. These structures can be seen as intensionally defined.

LISP and SML [84] are examples of strict impure functional languages. HOPE [20] and OPAL [35] are strict pure functional languages. While Miranda[1] [113], Haskell [56], pH [88], and Id [87] are instances of pure functional languages.

3   An Overview of Miranda

Miranda [113] is a polymorphically typed pure lazy functional language developed by David Turner. Its syntactic elegance made it a standard in functional programming. In this section, we use Miranda to present the most important aspects of the functional programming style.

Functional programs consist of definitions of functions and other objects. The execution of a program in a functional language consists of the evaluation of an expression, and we can see this as proceeding by successive rewriting of an expression until it takes printable (or normal) form. (A normal form is one that cannot be rewritten further.) For example, if we say

```
fac n = n * fac(n-1)   , n>0      (1)
      = 1               , otherwise
```

then

```
fac 2
```

is rewritten thus

$$
\begin{aligned}
\texttt{fac 2} \;\Rightarrow\;& 2 * fac\ (2-1) \quad (2)\\
\Rightarrow\;& 2 * fac\ 1\\
\Rightarrow\;& 2 * (1 * fac\ (1-1))\\
\Rightarrow\;& 2 * (1 * fac\ 0)\\
\Rightarrow\;& 2 * (1 * 1)\\
\Rightarrow\;& 2 * 1\\
\Rightarrow\;& 2
\end{aligned}
$$

The definition (1) is used in the first line of (2) where we have to substitute the actual value 2 for the variable n - this process of parameter passing forms the major overhead in rewriting implementations of functional languages.

---

[1] Miranda is a trademark of Research Software Ltd.

3.1 Pattern-matching

Pattern-matching is another interesting simplification to the syntax of expressions in functional languages. Using pattern-matching one can place guards on the left hand side of a function definition. For instance, using pattern-matching the function definition above can be expressed more neatly as

```
fac 0 = 1
fac n = n * fac(n-1)
```

In the general case pattern-matching can become quite complex. Miranda allows pattern-matching on algebraic data-types in general. In these cases pattern-matching not only gives guards as presented in the example above but also selectors which will decompose members of algebraic data-types.

3.2 Introducing new data types

The primitive or basic types in Miranda are: *num, char,* and *bool.* The type *num* comprises both integers and floating point numbers. As elements of the type *bool* we have the truth values *True* and *False.* The type *char* comprises the characters of the ASCII character set. In-built in Miranda there are also more complex data type constructors such as 'list' and 'tuple'.

A new algebraic type in Miranda is introduced by an equation using the symbol '::='. For instance, the type 'tree' can be introduced by the declaration of the type constructors as

```
tree ::= Leaf num | Node num tree tree
```

Leaf and *Node* have type *num* $\to$ *tree* and $((num \to tree) \to tree) \to tree$, respectively.

(The $\to$ operator denotes curryfication[2] and is left-associative). In Miranda the definition mechanism uses pattern matching on constructors. For instance, we can define a function which gives the 'weight' of a tree as

```
wt (Leaf n) = n
wt (Node n t1 t2) = (wt t1)+(wt t2)+n
```

The type system of Miranda subsumes Pascal enumerated types, records, variant records, and some uses of pointers which are involved in the construction of recursive dynamic data structures.

3.3 ZF-expressions

The general form of a ZF-expression in Miranda is

$$\{exp \; ; \; qualifiers\}$$

---

[2] After the logician H.B.Curry. Curryfication means the transformation in which a n-ary function becomes n functions to one argument

where there can be any positive number of qualifiers. Each qualifier is either a generator or a filter. A generator is a list of values which will be attributed to a specific variable in the expression *exp*. A filter is an arbitrary boolean expression used to restrict further the range of a generator.

ZF-expressions are a convenient and elegant syntactic 'sugaring' of a class of expressions. For example, the list of all the squares of the integer numbers between 1 and 10 can be expressed in Miranda by the following ZF-expression:

```
{ x*x; x <- (1..10)}
```

3.4 Local definitions

The general form of a definition will be:

$$
f\ x_0 \ldots x_n =\quad e_0 \quad where \\
a^1 v_1^1 \ldots v_p^1 = d^1 \\
\vdots \\
a^m v_1^m \ldots v_q^m = d^m
$$

The definitions of the functions $a^1, \cdots, a^m$ within the *where* block will be mutually recursive, in general. Their scope is restricted to the expression e_0 (and to the right-hand sides of their definitions, as they are recursive). Local definitions are used both to define auxiliary functions and to hold values of intermediate computations which may be referenced a number of times in the expression e_0.

The definitions of $a^1, \cdots, a^m$ may themselves contain local definitions: as is customary, a local definition will obscure a more global one.

4 Theoretical Background

In this section we present a brief introduction to the λ-Calculus, Combinatory Logic and Categorical Combinators. This is fundamental in understanding the complexity of the computational models of functional languages, their advantages and drawbacks.

4.1 The λ-Calculus

The λ-Calculus was developed by Alonzo Church [24] as part of a system for higher-order logic, with the aim of providing foundations for mathematics following the Peano-Russell school. It was shown inconsistent as a foundational theory, but the subset of it that dealt only with functions was proved to be sound. In 1936, Turing [111] tied together the concepts of λ-definability and computability. Due to its simple syntax and semantics the λ-Calculus has been widely used to provide semantics to programming languages of all paradigms.

Functional languages are nothing but a nicer syntax to the λ-Calculus. The direct implementation of the λ-Calculus has also been used as platform for functional languages.

4.1.1 Syntax

The λ-Calculus can be seen as a very simple programming language, in which there are only four constructors: variables, constants, application, and abstraction.

The following BNF expresses the syntax of the λ-Calculus:

$$
\begin{array}{lll}
< expr > & ::= < variable > & \textit{Variable names} \\
& | < constant > & \textit{Built-in constants} \\
& | < expr >< expr > & \textit{Applications} \\
& | \lambda < variable >.< expr > & \textit{Lambda abstractions}
\end{array}
$$

The identity function, a function that returns whatever gets as input,

$$id\ x = x$$

is expressed in the λ-Calculus as:

$$id \longrightarrow \lambda\,x.x$$

Similarly, the projection function,

$$Proj\ x\ y\ z = y$$

is translated into the λ-Calculus as:

$$Proj \longrightarrow \lambda\,x\lambda\,y\lambda\,z.y$$

A lambda abstraction always consists of the λ, the formal paramenter, the . and the body of the function. In general lambda abstractions are *anonymous* functions, but in practice one makes a *label* association as denoted by the $\longrightarrow$ in the examples above. Application is represented by juxtaposition and is left associative.

4.1.2 Operational Semantics

The rules of "playing" with λ-expressions are extremely simple. There are three operations known as α, β, and η conversion. The latter brings *intensionality* to the λ-Calculus and is of no interest to the implementation of functional languages. We present below the other two conversions.

Bound and Free Variables: Let x be a variable and M, N and O be λ-expressions. A variable x is said to be *free* in a λ-expression M, recursively, if:

1. $M = x$,

2. $M = \lambda y.N$ and if (x occurs free in N) and ($x \neq y$),

3. $M = N\ O$ and if x occurs free either in N or in O.

A non-free variable is called *bound*. In the expression $(\lambda x.\lambda y.(xx(yx)(\lambda y.z)y))$, variables x and y are bound to the closest $\lambda\,x$ and $\lambda\,y$, respectively, while variable z occurs free.

α-Conversion: The exchange of the *name* of a bound variable is called α-conversion. For instance, λ-expressions $\lambda x.\lambda y.xx$ and $\lambda z.\lambda y.zz$ present the same behaviour, despite the fact their variables have different names.

β-Conversion: The substitution of real parameters for formal ones is performed in the λ-Calculus by β-conversion, which can be expressed as,

$$(\lambda x.t)s \xrightarrow{\beta} [s/x]t$$

where [s/x]t means t with every free occurrence of x replaced by s.
In a simple example:

$$(\lambda x.ax)5 \xrightarrow{\beta} [5/x](ax) = a5$$

Now, in a more complex one:

$$
\begin{aligned}
(\lambda x \lambda y \lambda z.xyz)(\lambda a \lambda b \lambda c.a(bc))(\lambda f.f) \quad &\xrightarrow{\beta} \quad (\lambda y \lambda z.(\lambda a \lambda b \lambda c.a(bc))yz)(\lambda f.f) \\
&\xrightarrow{\beta} \quad (\lambda z.(\lambda a \lambda b \lambda c.a(bc))(\lambda f.f)z) \\
&\xrightarrow{\beta} \quad (\lambda z.(\lambda b \lambda c(\lambda f.f)(bc))z)
\end{aligned}
$$

The Church-Rosser theorems give consistency to the λ-Calculus as a theory of functions by stating that:

1. Normal forms are *unique* (modulo α-conversion).

2. β-Reducing the leftmost-outermost expression at each point of a reduction sequence leads to normal form, if it exists.

Capture: The use of variable names may bring problems with β-reduction. For instance, in the λ-expression,

$$(\lambda x.(\lambda y.yx))y$$

the rightmost y occurs free. If a straightforward β-reduction takes place one gets,

$$[y/x](\lambda y.yx) \xrightarrow{\beta} \lambda y.yy$$

Observe that the free variable was captured, i.e. it became bound, implying that the new expression has lost its original meaning. Performing a suitable α-conversion before β-reduction solves the problem of capture, but brings in a high operational cost.

4.2   Combinatory Logic

Around 1920 Moses Schönfinkel draw the attention of the logic community to the convenience of eliminating variables from first-order logic, by transforming it into an applicative combination of constant functions or **combinators**. A few years later, Haskell B. Curry rediscovered Combinatory Logic and built a deductive theory. Very early the similarities between the λ-Calculus and Combinatory Logic were spotted.

Combinatory logic naturally avoids the problem of captures and it is one of the key ideas in the implementation of functional programming languages.

4.2.1 Bracket Abstraction

The translation of a λ-expression into SK combinators is performed by an algorithm called "bracket abstraction". Let x be a variable. The bracket abstraction of x, denoted by $[x]$ is:

$$[x]\theta = \text{if (x does not occur freen in } \theta \text{)}$$
$$\text{then } K\theta$$
$$\text{else if } (\theta = x)$$
$$\text{then } SKK$$
$$\text{else if } (\theta = \gamma x) \text{ and}$$
$$\text{(x not free in } \gamma \text{)}$$
$$\text{then } \gamma$$
$$\text{else } (\theta = \delta\gamma)$$
$$\text{else } S([x]\delta)([x]\gamma)$$

Below we see the example of the translation of a λ-expression into SK-combinators:

$$
\begin{aligned}
Y &= (\lambda f.(\lambda y.f(yy))(\lambda y.f(yy))) \\
&= [f]([y]f(yy))([y]f(yy)) \\
&= [f](S([y]f)([y]yy))(S([y]f)([y]yy)) \\
&= [f](S(Kf)(S([y]y)([y]y)))(S(Kf)(S([y]y)([y]y))) \\
&= [f](S(Kf)(S(SKK)(SKK)))(S(Kf)(S(SKK)(SKK))) \\
&= S([f]S(Kf)(S(SKK)(SKK)))([f]S(Kf)(S(SKK)(SKK))) \\
&= S(S([f]S(Kf))([f]S(SKK)(SKK)))\;(\ldots) \\
&= S(S(S([f]S)([f]Kf))(K(S(SKK)(SKK))))\;(\ldots) \\
&= S(S(S(KS)(S([f]K)([f]f)))(K(S(SKK)(SKK))))\;(\ldots) \\
&= S(S(S(KS)(S(KK)(SKK)))(K(S(SKK)(SKK)))) \\
&\qquad (S(S(KS)(S(KK)(SKK)))(K(S(SKK)(SKK))))
\end{aligned}
$$

One can observe an (exponential) explosion in code size in the SK-bracket abstraction of a λ-expression.

4.2.2 Reduction Rules

The only operation in combinatory logic is application, which is denoted by juxtaposition and is left-associative. The evaluation of combinators follow simple rules, such as the ones for the S and K combinators below:

$$K\ a\ b\ =\ a$$
$$S\ a\ b\ c\ =\ a\ c\ (b\ c)$$

The translation of a combinatory expression into its λ-equivalent is trivial.

4.3   Categorical Combinators

Categorical combinators form a formal system similar to combinatory Logic. The original system was developed by Curien [31] inspired by the equivalence of the theories of typed λ-Calculus and cartesian closed categories as shown by Lambek [67] and Scott [103].

Curien chooses particular orientations of the axioms and deduces equations of a cartesian closed category different selections of them will generate several different rewriting systems for reducing the code generated by the compilation algorithm above. The system which he calls CCL_β simulates λ-Calculus β-reduction by a sequence of elementary reduction steps of rewritings on the categorical code. A number of optimisations were introduced to Categorical Combinators [70, 79], which made the system more efficient for functional language implementation by reducing the complexity of pattern-matching. Further details can be found in [70, 79] and are out of the scope of this paper.

5   Interpreting Functional Languages

To understand the way parallel implementations of functional languages evolved one needs to have an idea on how sequential implementations were made, as the former were based on the latter.

Code interpretation is a much simpler implementation technique than compilation. Until 1989 the implementations of functional languages were made by building interpreters to some kind of variant of the theories of functions presented above.

In this section we present the two most important of them: Landin's SECD machine and the Combinator Graph Reduction Machine by David Turner.

5.1   The SECD Machine

The very first implementation technique for functional languages was the SECD machine, developed in 1964 by Peter Landin [69].

Landin's SECD machine is a generalisation of the abstract machine underlying implementations of imperative programming languages like Pascal. It was used to give the operational semantics of Algol. It is named after its four components: **S**tack, **E**nvironment, **C**ode, and **D**ump.

As all functions were curryied the granularity of computation was very fine. Variable capture was a problem and its avoidance implied in scanning the body of every λ-expression before replacing each formal parameter by the corresponding real parameter.

5.2 Turner's Graph Reducer

The method of translation of functional languages into combinators, was first explored by Turner in [112]. Turner used a larger set of combinators than the ones in Curry's Combinatory Logic. Turner avoided the generation of the K combinator with a new bracket abstraction algorithm, which yielded quadratic code expansion. Turner rewrites the leftmost-outermost combinator at each point in the code. When no further rewriting can take place the expression is said to be in *normal form*.

Turner was also the first to propose a graph reduction machine to interpret combinatory expressions. A stack was used to find the leftmost-outermost combinator, the next to be reduced. A local *unwind* was performed after each combinator reduction, avoiding pointer following from the root of the graph. An explicit fixpoint combinator was introduced to implement recursion. Instead of following its rewriting law:

$$Y \ x \Rightarrow x \ (Y \ x)$$

Turner used a strategy of performing the graph rewriting to an application of x to itself.

6 From Interpretation to Compilation

Language compilation is a much faster implementation technique than interpretation. An important step towards the the efficient implementation of strict functional languages was Cardelli's FAM [23].

Johnsson [59] developed a strategy for compiling lazy functional languages, described as an abstract stack machine, called the G-Machine. The G-machine merged the best features of the SECD and Turner machines.

6.1 The G-Machine

The G-Machine [59, 77] was developed by Johnsson and Augustsson, in the *Chalmers Institute of Technology*, Götemburg, Sweden. With the aim of providing efficient implementation of lazy functional languages in von-Neumann machines.

The original G-Machine is very simple. We can say that this machine works as an interpreter with *lazy graph generation*, i.e. the original graph is replaced by code which when executed generates a graph to be interpreted.

Suppose, for example, that we want to evaluate the following expression, which returns the list of the squares of each Natural number.

$$list \ 0, \text{ where}$$
$$list \ n = square \ n : list \ (suc \ n)$$

In the definition above, the symbol : denotes the infix list construction operator and *square* and *suc* are pre-defined functions and are defined as follows,

$$square\ x\ =\ x \times x$$
$$suc\ x\ =\ x + 1$$

The expression, *list* 0, will be represented in the G-Machine as in Figure 1(a).

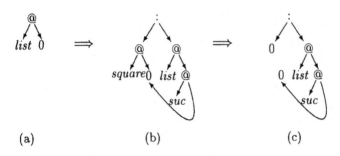

(a)   (b)   (c)

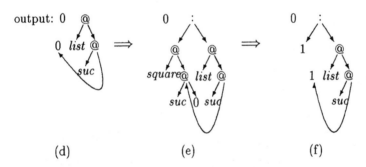

(d)   (e)   (f)

Figure 1: Example of the Evaluation Mechanism

Using the definition of the function *list* as a rewriting law, this expression is reduced to Figure 1(b). The integer 0 is the shared argument to *square* and *suc*. The resulting list expression is in canonical form, but neither its head nor its tail are. After reducing a graph to its canonical form, the next step is printing the result. Printing a list means to print its head and then to print his tail. As only ground type expressions can be printed, the machine will, first, reduce the head of the list, print it and then reduce its tail and print it. Using the definition of *square* as a rewriting law, the head of the list is then rewritten (for shorter) to the square of the integer 0, that is also 0. This rewriting step can be seen in Figure 1(c).

The sequence of reductions presented schematically above present "what" the G-Machine does. Now, as the head of list is on canonical form, it can be printed and removed from the graph (Figure 1(d)).

The evaluation of the tail continues in a similar fashion(Figure 1(e)). Please observe that computation does not terminate. Now the argument to *square* and *suc* is the expression *suc* 0. Again, the machine will try to print it. The definition of *square* reduces the expression to the square of the integer denoted by the expression *suc* 0. First, the graph representing *suc* 0 is reduced to its canonical form, the integer 1. As *suc* 0 is shared, all expressions that reference it will benefit from the reduction accomplished. Then, *square* is applied on the resulting expression, yielding the integer 1. The graph after these rewritings is as in Figure 1(f).

Again, as the head is now in canonical form, it can be printed and dropped from the graph. Reduction continues in the same manner.

The G-machine [59] provided the first fast implementation for lazy functional languages. The code generated by the G-Machine when executed produces time and space performance at least an order of magnitude faster than interpreted functional languages. Lazy graph generation makes the graph to be generated only if needed (in the case of higher-order functions and lazy evaluation). It also opens the possibility of a myriad of code optimisation strategies, avoiding graph generation at all. The original G-Machine implemented at Chalmers, by Johnsson and his colleagues [59], generated code in VAX-780 Assembly language, which made implementation extremely hard and machine dependent. It was common sense in the community of implementation of functional languages that assembly language implementation was the price to pay if one wanted efficiency. Chalmers LML compiler is still a reference in terms of performance of lazy functional languages. The G-machine way of controlling the execution flow and evaluation was followed by most of other implementations, even the ones based on different abstract machines as the Spineless G-Machine [19], the Spineless Tagless G-machine [95], TIM [40], and GM-C [86].

6.2 The Categorical Machines

As we mentioned before category theory offers another theory of functions which is used to implement functional languages. An approach to the execution of categorical combinators which uses a stack machine is described in Reference [27]. Categorical multi-combinators [70] are a generalisation of the concepts of categorical combinators, in which the equivalent of several β-reductions is performed in a single rewriting step.

The author of this paper has made several implementations of compiled functional languages [86, 78, 109, 79], based on categorical multi-combinators. All these implementations were made in C, thus were portable and simpler than the assembly ones. C was used as a macro-assembler and all "execution flow control" was made on a higher-level abstract machine.

ΓCMC [75, 73] was the first abstract machine for the implementation of lazy functional languages to transfer the execution flow control to C, as much as possible. ΓCMC translates strict functions on all its arguments that produce results of ground type directly into procedures in C to take advantage of the efficient context switching in modern architectures based on RISC, which is able to implement function calls at a very low cost. All arithmetic expressions are also translated directed into C code. ΓCMC glues together procedure calls, unevaluated expressions and functions, data-structures, etc. Categorical Multi-Combinators [70, 79, 76] served as a basis for the evaluation model of the ΓCMC abstract machine.

7 FP and Parallelism in the 1980s and before

The idea of parallel functional programming dates back to 1975 when Burge [16] suggested the technique of evaluating function arguments in parallel, with the possibility of functions absorbing unevaluated arguments and perhaps also exploiting speculative evaluation. In that same year, Berkling[11] suggested the application of functional languages to parallel processing.

The first attempt to exploit the parallelism of functional programs targeted either the evaluation of actual parameters before replacing them by the formal parameters or was done at combinator argument level. These strategies aimed at very fine grained parallelism and worked at a very high level abstract machines, such as graph reducers.

During this period novel architectures were seen as the way out of the inefficiency of functional languages, if compared to imperative ones such as C or FORTRAN. This led to a spate of designs for special-purpose machines, many of which were parallel. ALICE (Applicative Language Idealised Computing Engine) [33], designed at Imperial College in 1981, by Darlington and Reeve was the first and most famous reduction machine. The prototype ALICE comprised 40 Transputer-based processing agents and packet pools, connected by a multistage switching network. The eventual aim was to build ALICE in VLSI, but the only ALICEs built used stock hardware. The performance of this machine, as all the others from this period, was ultimately disappointing [51]. The granularity of computations was too fine and the use of many small packets also degraded performance to some extent. ALICE served as the basis for other more recent architectures such as the ICL Flagship [118, 116, 39] and EDS/Goldrush designs [115].

Parallel combinator reducers such as COBWEB [50], Burroughs' NORMA [99], and SKIM [108] are examples of other early machines which were built, but were overtaken by the widespread adoption of compilation techniques inaugurated by the G-machine. Machines such as Magó's FFP [80], Rediflow [62] and COBWEB [2] have failed to come to fruition.

One interesting proposal of parallel architecture was the graph reducer GRIP (Graph Reduction in Parallel) [94, 93], which is built from a network of distributed conventional processors. It has a two-level bus structure and incorporates fast packet-switching hardware for message routing. Intelligent memory units would operate on globally shared graph and spark task pools. It uses microcoded CPUs for the intelligent memory units and PALs for the packet-switching hardware. This gave a degree of design flexibility which allowed to undergo a transformation from a parallel abstract machine interpreter to a machine running compiled Haskell directly. Even with this new configuration the performance of the parallel GRIP machine is slower than the single node (sequential) version of the Haskell compiler.

On the language side a number of ideas were tried. Serial Combinators [54] were intended to be inserted by an automatically partitioning compiler. The Alfafa project implemented serial combinators for the Intel iPSC [43], but the performance figures were disappointing. Buckwheat re-implemented this model for the Encore Multimax, as shared-memory multiprocessor but got only a relative speedup [44].

Annotations for parallelism have been proposed by many authors. Burton's @p to indicate parallel function application [22] was one of the first and simplest ideas. Burn's work [18, 17] on evaluation transformers and Hope+ [64] on Flagship used complex annotations to control the precise degree of evaluation through strictness. Hudak's work [53, 57] on *para-functional programming* for functional languages with annotations that preserve the functional semantics, uses the idea of *schedules* of events, which includes explicit demands and process creation/termination, plus sequential and parallel compositions as well as task mappings to particular processors. Caliban [63] provided a separate functional language for task placement, in which a program comprises a process part and a wiring part. An implementation of Caliban for the Meiko transputer surface [29] was made but no performance figures were published [48].

The idea of trying to capture patterns of parallel computation, such as divide-and-conquer or pipelining, was named as *algorithmic skeletons* by Cole [25]. Ideally, the same skeleton can be used for different architectures: it is necessary only to change the implementation of the skeleton in order to get good performance portability. The user's program itself is unchanged, both textually and semantically. ZAPP [21] can be seen as an early example of skeletons. The use of skeletons seems to be promising and is being continued at several sites.

8   Benchmarking

Due to the inefficiency of the implementations of functional languages and also the limitations of architectures available, the benchmarks used either in sequential or parallel machines were too small. Implementation performance were measured in *fibs per second*, the number of Fibonacci calculations performed in one second. Most implementations were able to calculate **fib 20** as their largest Fibonacci number.

A classic divide-and-conquer program, a variant on the naive Fibonacci program, called **nfib** was used to benchmark parallel implementations:

```
nfib n = if n <= 1 then 1
         else 1 + nfib(n-1) + nfib(n-2)
```

Since the two recursive calls to **nfib** are independent, they can each be executed in parallel. If it is done naively an exponentially large number ($O(2^n)$) of tasks is created, in equal number to the result of the program.

9 Dataflow

Single-assignment languages or *dataflow* may have either strict or non-strict semantics. The evaluation of all the arguments to functions is performed before evaluating the application.

The TTDA (Tagged Token Dataflow Architecture) designed at MIT [4] was never realised in hardware. The dataflow machine built at Manchester [46, 47] demonstrated good relative speedup for several small benchmark programs, and provided much basic data on task scheduling. The prototype machine with 12 function units ran dataflow programs at about 25% of the speed of a VAX 11/780 running C [46]. The Monsoon [91], the SIGMA-1 [119], the P-RISC [5] and *T [89] are examples of dataflow architectures. The last two of them are derived from conventional RISC machines.

Dataflow programs typically produce much fine-grained parallelism, which must be managed carefully to avoid memory exhaustion [30]. Despite the attempts to increase granularity by combining instructions into larger blocks, the results are still relatively fine-grained [101].

10 Parallel Garbage Collection

Dynamic memory management has been linked to functional programming since the very beginning. Mark-scan, the first garbage collection algorithm appeared with LISP in 1960. As functional languages have always been eager consumers of space resources garbage collection is mandatory. The idea of using a parallel garbage collector to gain efficiency has always been present in the minds of implementors of parallel functional machines.

In uniprocessors, the Mark-Scan garbage collection algorithm works in two phases: if the user process requests cells when the free-list is empty, the garbage collection routine is invoked. All the cells in the transitive closure of root are marked before scanning the entire heap and returning unmarked cells to the free-list. Mark-Scan is a stop/start algorithm: its disadvantage is that the user process is suspended while the garbage collector runs. Its major disadvantage is the unpredictability of the garbage collection interludes, which makes it hard to design systems to meet real-time requirements.

10.1 Shared Memory Architectures

Donald Knuth credits Marvin Minsky for first suggesting parallelism as a way to avoid suspension of operations (Exercise 2.3.5-12, pp.422 in [65]). Parallelism need not imply concurrency. Garbage collection could occur, for example, during keyboard input, as long as it could be suspended on short notice to continue processing on the input and later be resumed without losing all the previously expended effort.

The model followed by architectures that tried to exploit parallelism of garbage collection consists of two processors, called the *mutator* and the *collec-*

tor, working in parallel and sharing the same the workspace (which is organised as a heap of cells). The mutator does all the 'useful' work, modifying the connectivity of the data structure and the values of data fields within that structure. The collector is solely responsible for identifying and re-cycling garbage cells. Its algorithm is based on Mark-Scan but runs continuously in two phases. In the first phase active cells are identified and marked; in the second phase the heap is scanned and cells that are known to be certain garbage are returned to the free-list. The collector makes no change to shape of the heap other than to link garbage cells into the free-list. It does, however, use and make changes to a mark field within each cell. These mark fields are also used by the mutator; indeed, they provide all the necessary communication between the two processes. Note that this architectural model is can easily adapt to work with Turner's graph reducer.

Guy Steele's Multiprocessing Compactifying Garbage Collection algorithm was the first published parallel and concurrent architecture for garbage collection [106, 107]. In addition to freeing unused storage, Steele's algorithm compacted remaining list structures to give better performance in a virtual memory environment. The mass of detail presented by Steele contributed to make understanding his ideas difficult.

Independently, Dijkstra proposed a similar scheme in some unpublished notes [36], later published in [37]. Dijkstra and his colleagues tackled this problem 'as one of the more challenging — and hopefully instructive — problems' in parallel programming. Their architecture attracted considerable interest in the computer science community. Woodger's scenario showed that if the granularity of the coarse grain algorithm was made finer, a bug would appear; in describing his proof of the algorithm Gries reported that he had 'seen five purported solutions to this problem, either in print or ready to be submitted for publication' each of which contained errors [42]. A correct version of the algorithm appeared in [38]. Wadler showed that, for time-sharing rather than multiprocessor systems, such algorithms require a greater percentage of processor time than classical sequential collection does [114].

Kung and Song developed a version of Dijkstra's algorithm which used four colours but did not need to trace the free-list [66], and Ben-Ari, who considered it 'one of the most difficult concurrent programs ever studied', presented several parallel mark-scan algorithms based on it but with much simpler proofs of correctness than those presented by Kung and Song, Gries and Dijkstra *et al.* [8, 9]. Ben-Ari's algorithms used only two colours. Gries accredits Stenning for an unpublished version of the on-the-fly algorithm which also used only two colours. Lamport generalised the architecture for using multiple processes [68]. The on-the-fly algorithm was also implemented in hardware/software in the Intel iAPX-432 microprocessor and iMAX operating system [98].

An efficient algorithm for concurrent collection was proposed by Appel-Ellis-Li [3] for machines that support virtual memory. This algorithm uses paging information provided by the operating system to synchronise the operations between processors.

The previously mentioned algorithms are based on marking and scanning the workspace. A different strategy is offered by reference counting [26]. The first reference counting algorithm for shared memory architectures is presented by Kakuta-Nakamura-Iida in reference [61]. A much simpler architecture was proposed by Lins [71] and later generalised into a multiprocessor architecture described in [72].

11   Promises, Failures and New Hopes

As we mentioned in the previous section, during the past decade the possibility of having hardware implementations (using microcoding, for instance) or parallelism were seen as the solution to the inefficiency of the implementations of functional languages. Purity of programming style was the name of the game. No one was prepared to compromise in any point neither to make programs more efficient nor to get closer to the needs of applications. The functional programming community used to claim that functional programs:

- had a higher level of abstraction and semantic elegance due to the use of higher-order functions and lazy evaluation.

- easier to write than their imperative counterparts, because the programs were the algorithms themselves.

- easier to read due to its very compact notation, reaching in some cases less than 10% of the size of their imperative equivalent.

- easier to prove correct as functional languages were based on the λ-Calculus.

- easier to go parallel due to referential transparency.

- executable specifications, providing simulation of programs for free.

The cold war during the 1980s made the governments of countries such as the U.S.A., U.K., France, Germany, and Italy make massive investments in functional programming and formal methods. Software reliability was fundamental to enterprises such as the Star War programme.

The end of the cold war meant that reliability was not the most important feature one could expect from software anymore. Research teams had to be redirected towards products which ought to be ready for sale within a 6 to 12 month horizon. The promises of the past were not delivered on time and general pessimism was cast on previous hopes:

- higher-order functions and lazy evaluation perform symbolic computation imposing that large data-structures are kept unevaluated, resulting in inefficient sequential code.

- functional programs were unable to cope with Input/Output easily.

- function composition can make programs extremely intricate to read in true spaghetti style.

- the size of large functional programs was comparable to the imperative equivalent.

- proving the correctness of functional programs, although much easier than imperative ones was not easy at all and would only be possible with proof assistant tools.

- to exploit referential transparency as a way to obtain parallelism one gets too fine granularity of code yielding parallel inefficiency.

In February 1993 the Department of Trade and Industry (DTI) of the British Government organised a workshop in London to know if functional programming was dead and buried or if there were ways of resurrecting it. The FP community came to the conclusion that a number of important steps had been made:

- we had learned *how to compile* functional languages in sequential machines.

- the need for large benchmarks was clear. Benchmark suits such as *NoFib* and *Pseudoknot* [41] were developed and allow comparing different implementations.

- the concept of *monads* [85] made possible to express I/O and the concept of state in true functional style.

- a number of "real-world" applications gave very good results:

  - The experiment of David Turner at AMACO (U.S.A.) in translating old oil reservoir programs written in FORTRAN into Miranda, an interpreted functional language, yielded a performance gain of around 1.000%.
  - At ECRC (Munchen) the team led by Mike Reeve obtained a performance gain of about 300% by translating FORTRAN code into ML of the program for simulating chemical pollution in Venice bay.
  - Programs for numeric computation written in the single-assignment functional language Sisal [83], developed at Lawrence Livermore National Labs., rival those for Parallel FORTRAN.

Faster and faster the FP community saw that parallelism was not the way to make functional programs run efficiently, but the other way round functional programming had something good to offer to the parallel processing community.

12 FP and Parallelism in the 1990s

As we mentioned before, only recently one can consider that we have learned how to compile functional languages for uniprocessor machines. The experience gained with the machines built in the 1980s proved that building special purpose hardware is costly and, most of times, is too slow a process to meet the advances of commercial general purpose machines. This is the main reason why parallel functional programming has concentrated in language implementation on commercially available hardware. Parallel garbage collection algorithms are also sought as ways to improve the performance of parallel programs.

Recent implementations of parallel functional languages are being built on top of fast sequential functional compilers. A number of ideas are being experimented. Implementations of skeletons are starting to appear [32, 15], which seem to be promising. ZAPP has achieved good results for divide-and-conquer parallelism on networks of transputers [45, 81]. Some recent approaches to exploit data parallelism in functional languages are POD comprehensions [52], which aim to combine data parallelism with lazy evaluation, bidirectional fold and s-can [90], and the data parallel language NESL, which provides a mechanism for nesting parallelism [13, 14].

Two new versions of Haskell seem to be very promising: Concurrent Haskell [96] and GUM [110]. Concurrent Haskell is a concurrent extension to Haskell, which aims to provide a more expressive substrate upon which to build sophisticated I/O-performing programs, notably ones that support graphical user interfaces for which the usefulness of concurrency is well established. The goal of the designers of Concurrent Haskell is to attain implicit, semantically transparent parallelism, but the version available now uses explicit parallelism.

GUM is a portable, parallel implementation of Haskell, which uses the PVM communications harness. As a result, GUM is available both on shared-memory (Sun SPARCserver multiprocessors) and distributed-memory (networks of workstations) architectures. The high message-latency of distributed machines is ameliorated by sending messages asynchronously, and by sending large packets of related data in each message. Initial performance figures demonstrate absolute speedups relative to the best sequential compiler technology.

12.1 Parallel Garbage Collection

Garbage collection [60], which was originally linked to LISP and functional languages, has gained importance with its use by object-oriented languages and a wide variety of software platforms and tools.

12.1.1 Shared Memory Machines

The parallel garbage collection algorithms developed in the 1970s and 1980s for shared memory architectures are still being implemented in the machines built now.

A number of optimisations were made to Appel-Ellis-Li's algorithm [3] and these seem to be one of the most promising directions in the area [60].

12.1.2 Distributed Systems

A number of algorithms use mark-scan in distributed architectures. Hudak and Keller's algorithm [55] works in two steps. First, all processors cooperate to mark all accessible cells, and then all processors collect unmarked cells. Ali [1]

presents algorithms which allow each processor to mark-scan its own heap independently. At the end of such a local garbage collection the processor informs
all other processors which remote pointers it retains, and the other processors
then treat these as roots that must be marked during their own garbage collection. Hughes [58] gives an algorithm which has lower storage overheads than
Ali's, but is also likely to take longer to recover remotely-referenced garbage.
Neither Ali's nor Hughes' algorithms are truly real-time since any particular
computation may be delayed for a long time while its processor does a garbage
collection. The algorithm described in reference [105] uses mark-scan locally in
each processor. External references are avoided by transferring an object (data) between processors whenever a local mark-scan discovers that there are no
locally held references to the object, thereby increasing the locality of objects.
This algorithm can present a very high communication overhead, depending on
the size of the objects.

Two new algorithms make reference counting suitable for use in looselycoupled multiprocessor architectures: Weighted Reference Counting [12, 117].
A cyclic version of it is presented in [74].

Garbage collection in distributed systems is still in the very beginning [60].
Hierarchical schemes seem to be more promising in delivering good performance.

13 Further Reading

Parallel functional programming is a broad area. Roe's thesis [100] presents a
useful introduction up to 1991 with examples of different styles of parallelis-
m. Schreiner's annotated bibliography [102], with over 400 entries, is relatively
complete and highly useful. The works of Ben-Dyke [10] and Hammond [48],
despite their overoptimism sometimes, provide useful guidelines the area and
were heavily consulted when writing this paper.

14 Conclusions

A number of important conclusions can be drawn from what has been presented
in the relationship between functional programming and parallel processing:

- The "natural" parallelism from referential transparency, parallel argumen-
  t evaluation, parallel combinator evaluation, and even parallel function
  evaluation, are too fine grained to compensate any degree of interprocessor communication. These are still to be analysed if they bring any special
  speed-up in superscalar machines.

- Special purpose architectures are hard to meet the development of general
  purpose commercially available hardware, in general.

- Do not go parallel prematurely. Get the maximum performance of sequential implementations first.

- Try to use benchmarks as close to prospective applications as possible.

The functional programming paradigm has been highly important and influential to many branches of computer science. No doubt it still has a lot to offer to software engineering:

- It is still easier to prove functional programs correct than in any other paradigm. We envisage that in the near future proof assistant tools will make this task much easier.

- Functional programs are executable specifications that provide prototypes for free.

- Program transformation is much easier due to the "algebraic" nature of functions. This may allow new code optimisation mechanisms to bring better performance in sequential and parallel architectures.

15 Acknowledgements

The author is grateful to Ricardo Lima and Genésio Cruz Neto for their comments and help.

This work was sponsored by CNPq (Brazil) research grants 52.2248/95-0, 52.2429/94-7 and 680075/94-1 (PROTEM).

References

[1] K.A.M.Ali. *Object-Oriented Storage Management and Garbage Collection in Distributed Processing Systems*. PhD thesis, Royal Institute of Technology, Stockholm, December 1984.

[2] P. Anderson, C. L. Hankin, P. R. J. Kelly, P. E. Osmon, and M. J. Shute. COBWEB-2: Structured Specification of a Wafer Scale Supercomputer. In *PARLE '87*, pages 51-67. Springer-Verlag LNCS 258, 1987.

[3] A.W. Appel, J.R. Ellis, and K.Li. Real-time concurrent collection on stock multiprocessors. *ACM SIGPLAN Notices*, 23(7):11–20, 1988.

[4] Arvind, V. Kathail, and K. K. Pingali. A Dataflow Architecture with Tagged Tokens. Technical Report LCS Memo TM-174, MIT, 1980.

[5] Arvind and R. S. Nikhil. Can Dataflow Subsume von Neumann Computing? Technical Report CSG Memo 292, MIT, November 1988.

[6] J.Backus. Can Programming be Liberated from the von Neumann Style? A Functional Style and Its Algebra of Programs. *Comm. ACM* 21 (8): 613-641, August 1978.

[7] H.P. Barendregt *The Lambda Calculus its Syntax and Semantics Studies in Logic and the Foundation of Mathematics* North-Holland (1984 - 2nd. ed.)

[8] M.Ben-Ari. On-the-fly garbage collection: new algorithms inspired by program proofs. In M. Nielsen and E. M. Schmidt, editors, *Automata, languages and programming. Ninth colloquium (Aarhus, Denmark)*, pages 14–22, New Yokr , July 12–16 1982. Springer-Verlag.

[9] M.Ben-Ari. Algorithms for on-the-fly garbage collection. *ACM Transactions on Programming Languages and Systems*, 6(3):333–344, July 1984.

[10] A. D. Ben-Dyke. The History of Parallel Functional Programming. FTPable from ftp.cs.bham.ac.uk, August 1994.

[11] K. J. Berkling. Reduction Languages for Reduction Machines. In 2nd. Annual *ACM Symp. on Comp. Arch.*, pages 133-140. ACM/IEEE 75CH0916-7C, 1975.

[12] D. I. Bevan. Distributed Garbage Collection using Reference Counting. In *PARLE '87*, pages 176-187. Springer Verlag LNCS 259, 1987.

[13] G. E. Blelloch. NESL: A Nested Data-Parallel Language (Version 2.6). Technical Report CMUCS-93-129, School of Computer Science, Carnegie Mellon University, April 1993.

[14] G. E. Blelloch, S. Chatterjee, J. C. Hardwick, J.Sipelstein, and M. Zagha. Implementation of a Portable Nested Data-Parallel Language. In *Principles and Practices of Parallel Programming*, pages 102-111, 1993.

[15] T. A. Bratvold. A Skeleton-Based Parallelising Compiler for ML. In (IO4), pages 23-34.

[16] W. H. Burge. *Recursive Programming Techniques*. Addison-Wesley, 1975.

[17] G. L. Burn. *Lazy Functional Languages. Abstract Interpretation and Compilation*. Research Monographs in Parallel and Distributed Computing. Pitman, 1991.

[18] G. L. Burn. Evaluation Transformers - A Model for the Parallel Evaluation of Functional Languages (Extended Abstract). In *FPCA '87*, pages 446-470. Springer-Verlag LNCS 274, 1987.

[19] G.L.Burns, S.L.Peyton Jones and J.D.Robson. The spineless g-machine. In *Proc.ACM Conference on Lisp and Functional Programming*, pages 244–258, Snowbird, USA, 1988.

[20] R. M. Burstall, D. B. MacQueen, and D. T, Sannella. Hope. Technical Report CSR-62-80, Edinburgh University, 1980.

[21] F. W. Burton and M. R. Sleep. Executing functional programs on a virtual tree of processors. *In FPCA '81*, pages 187-194, 1981.

[22] Warren Burton. Annotations to Control Parallelism and Reduction Order in the Distributed Evaluation of Functional Programs, ACM *TOPLAS*, 6(2),1984.20

[23] L.Cardelli. The functional abstract machine. *Polymorphism*, 1, 1983.

[24] A.Church, A set of postulates for the foundation of logic, *Annals of Math.* (2)33, pp.346–366.

[25] M. I. Cole. *Algorithmic Skeletons: Structured Management of Parallel Computation*. Research Monographs in Parallel and Distributed Computing. Pitman, 1989.

[26] G.E. Collins. A method for overlapping and erasure of lists. *Communications of the ACM*, 3(12):655–657, Dec. 1960.

[27] G.Cousineau P-L.Curien and M.Mauny. The categorical abstract machine. In J-P.Jouannaud, editor, *Functional Programming Languages and Computer Architecture*. SLNCS 201, 1985.

[28] S. Cox, H. Glaser, and M. J. Reeve. Implementing Functional Languages ou the Transputer. In [34], pages 287-295, 1989

[29] S. Cox, S.-Y. Huang, P. H. J. Kelly, J. Liu, and F. Taylor. Program Transformation for Static Process Network. In *PARLE '92*, pages 497-512. Springer-Verlag LNCS 605, 1992.

[30] D. Culler and Arvind. Resource Requirements of Dataflow Programs. In *15th. Annual ACM Symp.* on Comp. Arch., 1988.

[31] P-L.Curien. *Categorical Combinators, Sequential Algorithms and Functional Programming*. Research Notes in Theoretical Computer Science. Pitman Publishing Ltd., 1986.

[32] J. Darlington, A, J. Field, P. G. Harrison, P. H. J. Kelly, D. W. N. Sharp, Q. Wu, and R. L. While. Parallel Programming using Skeleton Functions. In *PARLE '93*, pages 146-160. Springer-Verlag LNCS 694, 1993.

[33] J. Darlington and M. J. Reeve. ALICE: A Multiple-Processor Reduction Machine for the Parallel Evaluation of Applicative Languages. In *FPCA '81*, pages 65-76, 1981.

[34] M. K. Davis and R, J. M. Hughes, editors. *Glasgow Workshop on Functional Programming*. Springer-Verlag WICS, 1989.

[35] K. Didrich, A. Fett, C. Gerke, W. Grieskamp, and P. Pepper. OPAL: Design and Implementation of an Algebraic Programming Language, In J. Gutknecht, editor, Programming *Languages and System Architectures, Zurich, Switzerland*, pages 228-244. Springer-Verlag LNCS 782, March 1994.

[36] E.W. Dijkstra. Notes on a real time garbage collection system. From a conversation with D.E.Knuth (private collection of D.E.Knuth), 1975.

[37] E.W. Dijkstra, L.Lamport, A.J. Martin, C.S. Scholten, and E.F.M.Steffens. On-the-fly garbage collection: An exercise in cooperation. In *Lecture Notes in Computer Science, No. 46*. Springer-Verlag, New York, 1976.

[38] E.W. Dijkstra, L.Lamport, A.J. Martin, C.S. Scholten, and E.F.M.Steffens. On-the-fly garbage collection: An exercise in cooperation. *Communications of the ACM*, 21(11):965–975, November 1978.

[39] J. Darlington et al. An Introduction to the FLAGSHIP Programing Environment. In *CONPAR '88, Manchester*. Cambridge University Press, 1988.

[40] J.Fairbairn and S.Wray. TIM: A simple, lazy abstract machine to execute supercombinators. In *Proceedings of Third International Conference on Fuctinal Programming and Computer Architecture*, pages 34–45. LNCS 274, Springer Verlag, 1987.

[41] P. Hartel et al. Pseudoknot: a Float-Intensive Benchmark for Functional Compilers. to appear in *J. of Functional Programming*, 1996.

[42] D.Gries. An exercise in proving parallel programs correct. *Communications of the ACM*, 20(12):921–930, Dec. 1977.

[43] B. Goldberg and P. Hudak. Alfalfa: Distributed Graph Reduction on a Hypercube Multiprocessor. In *Workshop* on Graph *Reduction, Santa FE, New Mexico*, pages 94-113. Springer-Verlag LNCS 279, September 1986.

[44] B. F. Goldberg. Multiprocessor Execution of Functional Programs. Intl. *Journal of Parallel* Programming, 17(5):425- 473,1988.

[45] R. G. Goldsmith, D. L. McBurney, and M. R. Sleep. Parallel Execution of Concurrent Clean on ZAPP. In [104], chapter 21.

[46] J. R. Gurd, C. C, Kirkham, and I. Watson. The Manchester Prototype Dataflow Computer. *Comm. ACM*, 28(1):34-52, January 1985.

[47] J. R. Gurd, C.C. Kirkham, and J. R. W. Glauert. A Multilayered Data Flow Computer Architecture. Technical report, Manchester University, 1978.

[48] K.Hammond. Parallel Functional Programming: An Introduction. FTPable from ftp.dcs.glasgow.ac.uk, August 1994.

[49] K. Hammond and J. T. O'Donnell, editors. *Glasgow Workshop on Functional Programming*. Springer-Verlag WICS, 1993.

[50] Chris L. Hankin, P. E. Osmon, and M. J. Shute. COBWEB - a combinator reduction architecture. In *FPCA '85*, pages 99-112, September 1985.

[51] P. G. Harrison and M. J. Reeve. The Parallel Graph Reduction Machine, Alice. In *Work-shop on Graph Reduction, Santa FE, New Mexico*, pages 181-202. Springer-Verlag LNCS 279, September 1986.

[52] J. M. D. Hill. The AIM is Laziness in a Data Parallel Language, In [49], pages 83-99.

[53] P. Hudak. Para-Functional Programming in Haskell. In Boleslaw K. B. K. Szymanski, editor,*Parallel Functional Languages and Compilers*, Fontier Series, chapter 5, pages 159-196. ACM Press, 1991.

[54] P. Hudak and B. Goldberg. Serial Combinators: "Optimal" Grains of Parallelism. In *FPCA '85*, pages 382-399, September 1985.

[55] P. Hudak and R. M. Keller. Garbage Collection and Task Deletion in Distributed Applicative Systems. In *ACM Symp. on Lisp and Functional Programming*, pages 168-178, 1982.

[56] P. Hudak, S. L. Peyton Jones, and P. L. Wadler. Report on the Programming Language Haskell: a Non-Strict, Purely Functional Language. *Special Issue of SIGPLAN Notices*, 16(5), May 1992.

[57] P. Hudak and L. Smith. Para-functional Programming: A Paradigm for Programming Multiprocessor Systems. In *ACM POPL*, pages 243-254, January 1986.

[58] R. J. M. Hughes. A Distributed Garbage Collection Algorithm. In *FPCA '85*, pages 256-272, September 1985.

[59] T.Johnsson. *Compiling Lazy Functional Languages.* PhD thesis, Chalmers Tekniska Högskola, Göteborg, Sweend, January 1987.

[60] R.E.Jones and R.D.Lins. Garbage Collection: Algorithms for Automatic Dynamic Memory Management, John Wiley and Sons, 1996, ISBN 0 471 94148 4.

[61] K.Kakuta, H.Nakamura, and S.Iida. Parallel reference counting algorithm. *Information Processing Letters*, 23(1):33–37, 1986.

[62] R. M, Keller, F.C.H. Lin, and J. Tanaka, Rediflow multiprocessing. In *IEEE Compcon*, pages 410-417, February 1984.

[63] P. Kelly. *Functional Programming for Looselycoupled Multiprocessors.* Research Monographs in Parallel and Distributed Computing. Pitman, 1989.

[64] J. M. Kewley and K. Glynn. Evaluation Annotations for Hope+. In [34], pages 329-337.

[65] D.E. Knuth. *The art of computer programming*, volume I: Fundamental algorithms, chapter 2. Addison-Wesley, Reading, Ma., 2nd edition, 1973.

[66] H.T. Kung and S.W. Song. An efficient parallel garbage collection system and its correctness proof. In *IEEE Symposium on Foundations of Computer Science*, pages 120–131. IEEE, 1977.

[67] J.Lambek. From lambda-calculus to cartesian closed categories. In J.P.Seldin and J.R.Hindley, editors, *in To H.B.Curry: Essays on Combinatory Logic, Lambda-Calculus and Formalism*. Academic Press, 1980.

[68] L.Lamport. Garbage collection with multiple processes: An exercise in parallelism. In *Proceedings of the 1976 International Conference on Parallel Processing*, pages 50–54, 1976.

[69] P.J.Landin. The mechanical evaluation of expressions, Computer Journal, 6(4):308-320, 1964.

[70] R.D.Lins. Categorical multi-combinators. In Gilles Kahn, editor, *Functional Programming Languages and Computer Architecture*, pages 60–79. Springer-Verlag, September 1987. LNCS 274.

[71] R.D. Lins. A shared memory architecture for parallel cyclic reference counting. *Microprocessing and Microprogramming*, 34:31–35, September 1991.

[72] R.D. Lins. A multi-processor shared memory architecture for parallel cyclic reference counting. *Microprocessing and Microprogramming*, 35:563–568, September 1992.

[73] R.D.Lins, G.G.Cruz Neto & R.F.Lima. Implementing and Optimising ΓCMC, Proceedings of Euromicro'94, pp.353-361, IEEE Computer Society Press, Sep. 1994.

[74] R.D.Lins and R.E.Jones. Cyclic Weighted Reference Counting, Proceedings of *WP&DP'93*, K.Boyanov (editor), North-Holland, 1993.

[75] R.D.Lins & B.O.Lira. ΓCMC: A Novel Way of Implementing Functional Languages, *Journal of Programming Languages*, 1:19-39, Chapmann & Hall, January 1993.

[76] R.D.Lins, S.J.Thompson and S.Peyton Jones, On the Equivalence between CM-C and TIM, *Journal of Functional Programming*, 4(1):47-63, Cambridge University Press, January/1994.

[77] R.D.Lins & P.G.Soares. Some Performance Figures for the G-Machine and its Optimisations. Microprocessing and Microprogramming 37(1993) 163-166, North-Holland.

[78] R.D.Lins & S.J.Thompson. CM-CM: A categorical multi-combinator machine. In *Proceedings of XVI LatinoAmerican Conference on Informatics*, Assuncion, Paraguay, September 1990.

[79] R.D.Lins & S.J.Thompson. Implementing SASL using categorical multi-combinators. *Software — Practice and Experience*, 20(8):1137–1165, November 1990.

[80] G. A. Magoo and D. F. Stanat. The FFP Machine. In *High-Level Language Computer Architectures*, pages 430-468, 1989.

[81] D. L. McBurney and M. R. Sleep. Transputer-Based Experiments with the ZAPP Architecture. In *PARLE '87*, pages 242 259. Springer-Verlag LNCS 258, 1987.

[82] J.McCarthy. Recursive functions of symbolic expressions and their computation by machine. *Communications of the ACM*, 3:184–195, 1960.

[83] J, McGraw. *SISAL: Streams and Iterations in a Single-Assignment Language: Reference Manual version 1.2.* Lawrence Livermore Natl. Lab., 1985. Manual M-146, Revision 1.

[84] R.Milner. Standard ML proposal. *The ML/LCF/Hope Newsletter*, 1(3), January 1984.

[85] E.Moggi. Computational lambda calculus and monads, in Logic in Computer Science, California, IEEE Press, June 1989.

[86] M.A.Musicante & R.D.Lins. GMC: A Graph Multi-Combinator Machine. *Microprocessing and Microprogramming*, 31:31–35, North-Holland, April 1991.

[87] R. S. Nikhil. Id (version 90.1) reference manual. Technical Report CSG Memo 284-2, Lab. for Computer Science, MIT, July 1991.

[88] R. S. Nikhil, Arvind, and J. Hicks. pH Language Proposal (Preliminary), 1st. September 1993. Electronic communication ou comp.lang.functional.

[89] R. S. Nikhil G. M. Papadopoulos, and Arvind. *T: A Multithreaded Massively Parallel Architecture. In *19th. ACM Annual Symp. on Comp. Arch.*, pages 156-167, 1992.

[90] J. T. O'Donnell. Bidirectional Fold and Scan. In [49], pages 193-200.

[91] G. M. Papadopoulos. *Implementation of a General Purpose Dataflow Multiprocessor*. PhD thesis, Laboratory for Computer Science, MIT, August 1988.

[92] N. Perry. Hope+. Technical Report IC/FPR/LANG/2.5.1/7 Issue 5, Imperial College, London, February 1988.

[93] S. L. Peyton Jones, C. Clack, and J. Salkid. High-Performance Parallel Graph Reduction. In *PARLE '89*, pages 193-206, Eindhoven, The Netherlands, June 12-16, 1989. Springer-Verlag LNCS 365,

[94] S. L. Peyton Jones, C. Clack, J. Salkild, and M. Hardie. GRIP - a High-Performance Architecture for Parallel Graph Reduction. In *FPCA '87*, pages 98-112. Springer-Verlag LNCS 274, 1987.

[95] S.L.Peyton Jones and J.Salkild. The spineless tagless g-machine. In *Proc.ACM Conference on Functional Programming Languages and Computer Architecture*, pages 184–201, Snowbird, USA, 1989.

[96] S.L.Peyton Jones, A.Gordon and S.Finne. Concurrent Haskell. In Proc. of 23rd ACM Symposium on Principles of Programming Languages (POPL'96), Florida, 1996.

[97] M. J. Plasmeijer and M. C. J. D. van Eekelen, editors. *Proc. 5th. Intl. Workshop on Parallel Impl. of Funct. Langs.* Nijmegen, 1993.

[98] F.J. Pollack, G.W. Cox, D.W. Hammerstein, K.C. Kahn, K.K. Lai, and J.R. Rattner. Supporting Ada memory management in the iAPX–432. In *Proceedings of the Symposium on Architectural Support for Programming Languages and Operating Systems*, pages 117–131. SIGPLAN Notices (ACM) 17,4, 1982.

[99] H. Richards. An Overview of Burroughs NORMA. Technical report, Austin Research Centre, Burroughs Corp., January 1985.

[100] P. Roe. Parallel Programming *using Functional Languages*. PhD thesis, Glasgow University, April 1991.

[101] V. Sarkar and J. Hennessy. Partitioning Parallel Programs for Macro-Dataflow. In *ACM Symp. on Lisp and Functional Programming*, pages 202–211, 1986.

[102] W. Schreiner. Parallel Functional Programming - an Annotated Bibliography, Technical Report 93-24, RISC-Linz, Johannes Kepler University, Linz, Austria, May 1993.

[103] D.Scott. Relating theories of the lambda-calculus. In J.P.Seldin and J.R.Hindley, editors, *in To H.B.Curry: Essays on Combinatory Logic, Lambda-Calculus and Formalism*. Academic Press, 1980.

[104] M. R. Sleep, M. J. Plasmeijer, and M. C. J. D. van Eekelen, editors. Tem *Graph Rewriting: Theory and Practice*. Wiley, 1993.

[105] M.Shapiro, O.Gruber and D.Plainfossé. A garbage detection protocol for a realistic distributed object-support system. Technical Report 1320, Rapports de Recherche, INRIA-Rocqencourt, Novembre 1990.

[106] G.L. Steele. Multiprocessing compactifying garbage collection. *Communications of the ACM*, 18(9):495–508, September 1975.

[107] G.L. Steele. Corrigendum: Multiprocessing compactifying garbage collection. *Communications of the ACM*, 19(6):354, June 1976.

[108] W. R. Stoye. The *Implementation of Functional Languages using Custom Hardware*. PhD thesis, University of Cambridge, 1985.

[109] S.J.Thompson & R.D.Lins. The Categorical Multi-Combinator Machine: CMCM, *The Computer Journal*, vol 35(2): 170-176, Cambridge University Press, April 1992.

[110] P.W.Trinder, K.Hammond, J.S.Mattson Jr., A.S.Partridge. GUM: a portable parallel implementation of Haskell, Proc. of Programming Language Design and Implementation, Philadelfia, USA, May, 1996.

[111] A.M.Turing, On computable numbers with an application to the Entschei-dungsproblem, *Proc. London Math. Soc.* 42, pp 230–265.

[112] D.A. Turner. A new implementation technique for applicative languages. *Software — Practice and Experience*, 9, 1979.

[113] D.A.Turner. Functional Programming as executable Specifications, *Phil. Transactions of the Royal Society of London 312*, pp.363-388, 1984.

[114] P.L. Wadler. Analysis of an algorithm for real-time garbage collection. *Communications of the ACM*, 19(9):491–500, September 1976.

[115] I. Watson. Simulation of a Physical EDS Machine Architecture. Technical report, Department of Computer Science, University of Manchester, UK, September 1989.

[116] I. Watson, V. Woods, P. Watson, R. Banach, M. Greenberg, and J, Sargeant. Flagship: a Parallel Architecture for Declarative Programming. In *15th. Annual ACM Symp, on Comp. Arch.*, page 124, 1988.

[117] P. Watson and I. Watson. An Efficient Garbage Collection Scheme for Parallel Computer Architectures. In *PARLE '87*, pages 432-443. Springer Verlag LNCS 259, 1987.

[118] P. Watson and I. Watson. Evaluating Functional Programs on the FLAG-SHIP Machine. In *FPCA '87*, pages 80-97. Springer-Verlag LNCS 274, September 1987.

[119] T. Yuba, T. Shimada, K. Hiraki, and H. Kashiwagi. SIGMA-1: A Dataflow Computer for Scientific Computations. *Computer Physics Communications*, pages 141-148, 1985.

A Scalable Implementation of an Interactive Increasing Realism Ray-Tracing Algorithm

A. Augusto de Sousa[1,2], and F. Nunes Ferreira[1,2]

[1] INESC, Pr. Mompilher 22, Apt. 4433, 4007 Porto CODEX, Portugal
[2] FEUP, R. dos Bragas, 4099 Porto CODEX, Portugal

Abstract. Ray-Tracing is a well known algorithm that simulates the reflection and refraction of light rays in the objects surfaces, beginning with rays sent from the view point. Each ray has to be tested against objects in the 3D scene and this is very time consuming. This paper describes one parallel solution. It is oriented to a general network, where each processor manages data and tasks and shares them with the others to optimise common resources utilisation. In this context, a Virtual Sharing Memory (VSM) is used and an efficient load balancing strategy can be implemented. With this approach, the amount of messages in the network tends to increase and strategies to reduce them are needed. Schemes to optimise remote accesses to data have been developed, namely a special type of multicast message addressing named *InPathTo*.

1. Introduction

Ray-Tracing is a well known algorithm used in image synthesis of 3D scenes [1]. It simulates the reflection and refraction of light rays in the objects surfaces. Instead of sending rays directly from the light sources, it uses a different approach, based on the rays that are more important to the observer. One ray is sent from the view point through each image sample (at least one per pixel) and is tested against objects in the scene to detect the first intersected. From the point of intersection, new rays (usually known by shadow feelers) are sent to light sources and reflected and transmitted rays are generated and recursively processed as the initial ray. A shading tree is defined with the reflected and transmitted rays and one shading tree characterises the colour of each image sample.

One serious problem of ray-tracing is the large processing time necessary to create one single image. This is due to the intersection tests that are performed, for each ray, with the objects in the scene. In this context, several techniques are known to reduce the amount and the complexity of the intersection tests [2], [3], [4], [5], [6], [7].

The parallelization of the ray-tracing algorithm is also an obvious solution for the problem. In [8] a taxonomy of ray-tracing parallelization methods is presented and several are described (a similar description for recently published works can be found in [9]). The most efficient and simple methods seem to fall in the category Image Space Subdivision (an image is subdivided in distinct regions or tasks and a set of regions is allocated to a particular processor) with Distribution of Database (objects are, in some way, scattered through the processors distributed memory). Examples of such implementations are presented in [10] and in [11].

In [12], a new approach to ray-tracing is presented. Named *Interactive Increasing Realism Ray-tracing Algorithm* (IIRRA), it intends to give the user the ability to follow the image generation, in a progressive manner, instead of seeing the complete image at the end of the processing. The main goal is to create initial images with fewer samples and shading trees with limited depth, under user interactive control [13].

A parallel implementation of IIRRA, based in a Transputer architecture is described in [14]. It is basically a *Farm of Processors*, defining a light ray as an individual task to be carried out by a worker. The system evaluation, in terms of Speed-Up and Efficiency, denotes some problems of bottleneck produced by the processors related to common services, namely coordination (farmer), stock of rays (rays database) and visualisation. In short, the efficiency decreases for simple scenes or for architectures with many processors (workers). Also the 3D scene database is replicated through all processors, so the utilisation of the system memory is not optimised and creates some problems with large scenes.

This paper presents an evolution of a such system that intends to solve the mentioned problems. The general system architecture, including its main processes is presented in section 2. Sections 3 to 5 are dedicated to the description of the main processes, respectively, Task Managers, Routers and Data Managers. They include an analysis of a special addressing type, named *InPathTo*, that was defined to limit the amount of messages necessary to perform remote accesses to data and tasks. Results related to *InPathTo* addressing and data management are presented in section 6 and the final conclusions in section 7.

2. System Architecture

Regarding a better scalability, a new approach to the Parallel Implementation of IIRRA is necessary. The main goals are [9]:

- to increase tasks size to reduce the communications overhead;
- to distribute tasks management among processors to reduce the communications and to reduce the bottleneck created by common services processors;
- to make a better use of the system memory.

Under the graphics point of view, the way the image is created in a progressive manner must also be improved by taking more information from neighbour samples[1]. This leads to a solution based on image cells, each one characterised by several samples and, if necessary, recursively subdivided in smaller cells. Each cell is coloured through the interpolation of its samples.

This solution in conjunction with the tasks size problem suggests that a task should be defined as an image cell, with its own samples. Since all shading trees related to the samples grow in depth, it can be concluded that one image cell or task is rarely given as terminated.

---

[1] The independence of samples used in the above version did not permit the concentration of computational effort in the image regions with more detail and is responsible for a strong mosaic effect.

In this context, IIRRA is completely different from other image space subdivision ray-tracing parallel systems, where one task starts, processes and terminates after the correspondent screen update.

In [15], one parallel system (AMP) for image synthesis through the radiosity algorithm [16], [17] is presented. Like in IIRRA, one task is kept in one processor that, periodically, executes micro-tasks on it and activates a sequence of accesses to several data items. Based on the AMP system, the general architecture of IIRRA is presented in Fig. 1. Each processor runs a set of processes: Application (AP, ray-tracing), Data Manager (DM), Task Manager (TM) and Router (R). Each processor communicates with the others through a network of channels.

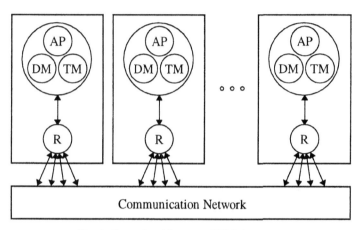

Fig. 1. General architecture of IIRRA system

Processes DM in different processors can communicate with one another to share data as processes TM do to share tasks. Both types can use remote access strategies through the network to obtain items that are needed by local applications:

- if the item location is fixed in one processor, this one is addressed;
- if the item has no fixed location, then a message is sent to several processors, starting with the neighbours of the asking processor.

Each message is routed through the network by routers, so none of the other processes needs to know the network configuration.

3. Task Managers

As mentioned above, one task corresponds to an image cell (initial cells are defined by the coordinator processor) and, according to the notion of increasing realism, each cell can be subdivided in smaller ones, creating new tasks. Hence, the number of tasks tends to increase until cells are classified as minimal or as coherent.

Since image cells contribute differently to image quality, they have to be classified and sorted by importance, in the databases related to the Task Managers. To minimise problems like amount of memory and efficiency of sorting algorithms, each TM manipulates two databases of tasks, IDB (Interactive Data Base) and NIDB (Non Interactive Data Base). The system runs in two steps, Interactive and Non Interactive Mode, exploring both databases [9].

The management of tasks is then completely distributed among all the processors, *i.e.*, each worker classifies and sorts its own tasks. The main problem that subsists is Load Balancing.

Dynamic load balancing methods tend to present better results than *a priori* methods, with the cost of an increase in the inter-processors communications [18].

A priori methods can efficiently be adopted when a prevision of the computational effort, necessary for each task, can be pre-determined. Thus, tasks can be distributed through the processors in such a way that the sum of the computational efforts is the same for all processors.

In [11], a method to perform an *a priori* load balancing in parallel ray-tracing systems with image space subdivision is presented. The method is based in a sub-sampling of the entire image to obtain information about the complexity distribution, using a fixed density of exhaustively processed samples. This is not adequate for IIRRA because, in this system, the concentration of samples and the level of detail per sample is variable in space and time.

The *a prirori* method adopted in IIRRA is known by Scattered Decomposition and is based in the complexity coherence that can be found in neighbour samples [8]. Consider a system made of P processors:

- In a macro-cell composed by P image cells of the same size, the maximum coherence for the set of cells tends to be obtained when the macro-cell is a square with $\sqrt{P} \cdot \sqrt{P}$ image cells;
- Assigning each image cell to a different processor, a good estimation for the image complexity distribution is obtained inside the macro-cell;
- Replication of the above for all macro-cells tends to produce a good *a priori* load balancing.

An example is presented in Fig. 2 for a system composed by sixteen processors.

1	2	3	4	1	2	3	4	1	2	3	4	1	2	3	4
5	6	7	8	5	6	7	8	5	6	7	8	5	6	7	8
9	10	11	12	9	10	11	12	9	10	11	12	9	10	11	12
13	14	15	16	13	14	15	16	13	14	15	16	13	14	15	16
1	2	3	4	1	2	3	4	1	2	3	4	1	2	3	4
5	6	7	8	5	6	7	8	5	6	7	8	5	6	7	8
9	10	11	12	9	10	11	12	9	10	11	12	9	10	11	12
13	14	15	16	13	14	15	16	13	14	15	16	13	14	15	16
1	2	3	4	1	2	3	4	1	2	3	4	1	2	3	4
5	6	7	8	5	6	7	8	5	6	7	8	5	6	7	8
9	10	11	12	9	10	11	12	9	10	11	12	9	10	11	12
13	14	15	16	13	14	15	16	13	14	15	16	13	14	15	16

Fig. 2. Example of Scattered Decomposition for *a priori* load balancing in Ray-Tracing

The efficiency of scattered decomposition is limited by the image cells dimensions. For load balancing purposes, image cells should be the smallest possible but, in terms of progressive image quality, large dimensions are more adequate. It is clear that a compromise between both is necessary and, in this situation, an unbalanced load of processors results. A more accurate solution for the problem can only be obtained by dynamic load balancing and is still under study.

Nevertheless, the above *a priori* method is still useful for the system efficiency. With a negligible cost in terms of pre-processing time, it performs a good initial task distribution that permits a decrease in the amount of messages exchanged between processors to perform dynamic load balancing.

4. Routers

Sharing information among processors requires a more complex solution in what concerns the routers. At any time, one message can be sent by any processor to any other or to a group of processors. Deadlock problems can arise and a study of such situation is presented in [19].

Since each processor contains several independent processes, with different responsibilities, it is desirable that one message address is composed by two levels, the first designating the processor and the second one process inside the processor. Unicast, Multicast or Broadcast addressing can be used at both levels [20].

Broadcast messages are automatically distributed through the network by means of replication performed by each router. Nevertheless, if the network is represented by a cyclic graph, a direct implementation of this strategy could originate an undetermined number of messages.

The adopted solution for the problem is based in the definition of a Graph Spanning Tree. Since a spanning tree contains all the nodes of a graph but does not define any cycle, the replication of a message through its branches guarantees the visit of every node with a finite number of messages. The additional information needed by a single router is the set of its channels that belong to the network spanning tree.

In terms of multicasting, a special type of addressing, named *InPathTo*, is being used to optimise remote accesses to data or tasks items in the context of distributed memory. The idea behind *InPathTo* addressing is simple: being n the distance (measured in number of channels) between two processors P_0 and P_n, if P_0 asks P_n for an item, the total amount of messages generated in the path between P_0 and P_n, is n; nevertheless, if the same item could be obtained from an intermediate processor in the path, the amount of messages generated would be reduced.

In Fig. 3, $S_{P,i}$ represents the probability of processor P_i satisfying the request made by processor P_0. Since P_0 sends a request message to processor P_n, then $S_{P,0}=0$ and $S_{P,n}=1$.

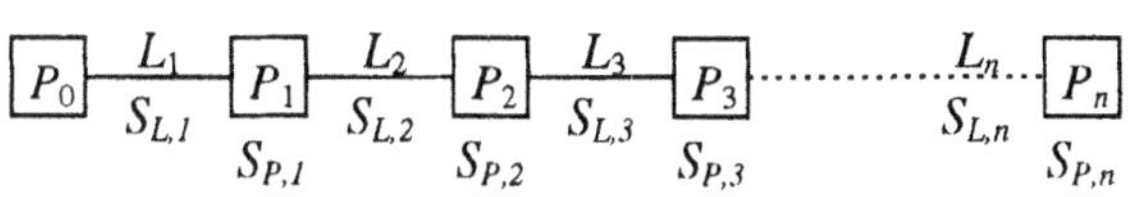

Fig. 3. Remote access from the processor P_0 to the processor P_n

If P_i cannot satisfy the request, then the message has to be re-sent. The probability $S_{L,i}$ of a message being re-sent by processor P_{i-1}, through channel L_i is

$$S_{L,i} = S_{L,i-1} \cdot (1 - S_{P,i-1}) = \prod_{k=0}^{i-1} (1 - S_{P,k}) \tag{1}$$

The value $S_{L,i}$ has another interpretation. It corresponds to the number of messages that cross channel L_i, per message sent by processor P_0.

On the other hand, the total distance that one remote access request will finally cover is equivalent to the total number of messages generated through the path, so it can be calculated by the sum of $S_{L,i}$,

$$D = \sum_{i=1}^{n} \prod_{k=0}^{i-1} (1 - S_{P,k}) \tag{2}$$

Therefore and assuming an equal probability for each processor $S_{P,k}=S_P$ (except for P_0 and P_n), the distance covered by the request message is, on average,

$$\overline{D} = \frac{1 - (1 - S_P)^n}{S_P} \tag{3}$$

In (3), a null S_P corresponds to the impossibility of any intermediate processor to acknowledge the requesting message. In other words, the requested item will be available only in the final processor P_n and, as a consequence, the distance covered by the message will be n. In the opposite situation, $S_P=1$, the requested item is available in the first processor next to P_0 and the distance is reduced to 1. This behaviour is clearly demonstrated in the curves of Fig. 4, for several values of n.

It can be seen that a reduction in the distance $\overline{D}$ is less noticeable in the cases where the total distance n is smaller, unless S_P can take large values (close to unit). But in large networks, with large values of n, a strong reduction in $\overline{D}$ can be obtained, even with small probabilities S_P. From this observation, an *InPathTo* addressing scheme reduces the amount of messages travelling in the network and, consequently, the latency in accessing remote data.

When remotely accessing an item, the requesting process marks the correspondent message as *InPathTo* and delivers it to the router. Then,

1. The router in the next processor (found in the path to the addressed processor) verifies the existence of a local process with the identifier of the addressed process; if not, then goes to step 3.
2. The local process replies to the message if it has the requested data and, in this case, no more messages are generated; if the data is not present, then it is responsible for delivering the message to the router.
3. In the impossibility of getting a local answer, the router re-sends the message to the next processor, in the direction of the addressed processor.

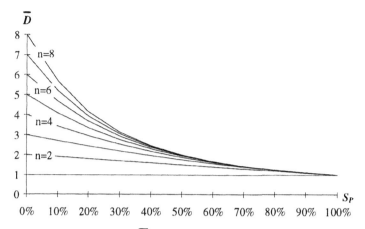

Fig. 4. Variation of average distance $\overline{D}$, with the probability S_P of any intermediate processor to acknowledge the initial request

These three points are repeated until an intermediate processor can reply or, in the worst situation, the addressed processor is hit.

The probability of one requesting message being acknowledged by an intermediate processor is significant only if objects are maintained in local memories, for some time, after their use. As a conclusion, the efficiency of the *InPathTo* scheme to access remote data is very influenced by the Data Manager of each processor.

5. Data Managers

High dimension problems, in multicomputer parallel systems, require data distribution among the local memories of the processors. Accesses to remote data are then necessary but, due to slow communications through the network, they must be avoided. On the other hand, accesses to local data are much faster. Strategies based on a hierarchy of memory are then necessary to balance local and remote accesses.

Ray-tracing presents some characteristics that facilitate a parallel implementation based on a Virtual Shared Memory (VSM). A caching mechanism can be efficient because only a small part of the entire objects database is necessary at a certain moment and because the database is read-only, which eliminates the problem of data coherence. A data fault in any cache generates one access to remote data in the other processors or, as a last resort, to the system auxiliary memory, in disk.

In [11], one ray-tracing system based on a VSM philosophy, in a distributed memory parallel computer, is presented. Every local data memory is divided in two blocks, one resident and the other used as a cache. All the resident blocks are seen as a contiguous space, organised in pages and the entire database of objects is distributed among the resident blocks. Each processor is responsible for the objects stored in its resident block.

One memory management module located in each processor tries to obtain an object, by order of preference: in the resident block, in the local cache and, if missed, in the processor that is responsible for the object. The method does not have any strategy to make the initial distribution of the objects and this is, perhaps, its main problem.

In [8], a different scheme, based only in cache memory, is used. It concentrates in questions like data locality [21] and coherence in ray-tracing [22], to attain an efficient management of an objects cache, located in each processor. No initial distribution of objects is necessary and the absence of a resident block permits the utilisation of larger caches.

This approach is adequate for the IIRRA system but, as coherence is more difficult to explore, one can preview some loss of efficiency. Actualy, it is possible to explore some forms of coherence, related to objects in the neighbourhood of light sources [5], but other forms of coherence [22], [10], are more dificult to explore.

In IIRRA, accesses to the database of objects are made accordingly to the VSM model, following a memory hierarchy in four levels: local memory, local cache, remote cache and auxiliary memory in disk:

The local memory. Embedded in the code, it is used to store small size data that are very often accessed. These data are, in principle, replicated in all the processors.

Local cache. Data that were remotely accessed can be used again in the future, so they must be stored in memory and managed by an adequate replacement method.

Remote Cache. When an item is not found in any of the two levels above, one remote access message is sent in the direction of the coordinator processor, but trying to obtain a reply in an intermediate processor (using the *InPathTo* facilities). The coordinator's cache is larger than the others and works as a system cache between the other processors and the auxiliary memory.

Auxiliary memory. Implemented in disk, it gives the possibility to process 3D scenes with a large number of objects. It is accessed only when an object is not found in the coordinator cache.

Following this VSM model, each processor has its own cache. Every worker's cache is managed to optimise the accesses made by local applications, *i.e.*, at any moment, the objects that are maintained in memory are those that, statistically, have more chances of being accessed by one of the local applications (the replacement strategy used is the Least Recently Used - LRU that is well swited to software implementations [21], [23]). The contents of one cache can also be used to reply to a remote access, but this can not modify its management.

By the contrary, the cache located in the coordinator processor is managed to optimise the accesses by the other processors. This creates some complementarity between the contents of the coordinator's cache and the other caches, avoiding the superposition and minimising the number of accesses to the auxiliary memory.

Sometimes it happens that one remote access message arrives (in the context of *InPathTo* addressing) to one processor that already asked the network for the same item. Re-sending this message is redundant and is avoided by means of a data structure that associates a list of queries (WRQu - Waiting for Reply Queue) to each data item.

466

Whenever one data manager performs a remote cache access to an item, a list WRQu is created for that item. While the corresponding reply is not received, any possible message received from other processor, asking for the same item, will add one element to the list. The contents of this element is the address of the asking process, so it is possible, after the reception of the data item, to send a reply message to every process that is waiting for it.

6. Results

The VSM described was developed for the IIRRA system. However, it has some characteristics that permits its use in several other applications.

To evaluate the behaviour of the VSM for a generic application, it is necessary to determine the access time to a set of common data items. However, the locality of data is strongly dependent of the application. One reasonable approximation can be obtained with random accesses to data, using different cache sizes.

The curve in Fig. 5 represents the variation of access times to about one thousand data items (randomly selected from the objects data base) versus the size of caches (expressed in percentage of the total data base size). The system was composed by nine processors T800 connected as a tree and all remote accesses were directed to the root (coordinator).

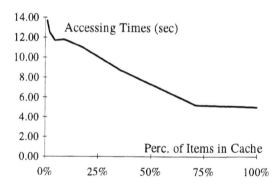

Fig. 5. Random access times to about one thousand data items, in seconds, as a function of the percentage of items that are maintained in cache

It must be noted that data items are accessed in a random sequence and thus, several accesses are related to the same object. In these cases, augmenting caches increases the probability of the item being found in the local cache or in the caches of the intermediate processors during a remote access.

The presented results were obtained starting with initial situations, *i.e.*, with empty caches at the beginning of each test. It is also important to evaluate the system behaviour in normal working, *i.e.*, starting each test with some set of objects in cache.

Figure 6 represents the evolution of the time necessary to randomly access about one thousand objects, repeated six times, for three cases of cache sizes (10%, 70% and 100% of the total objects data base).

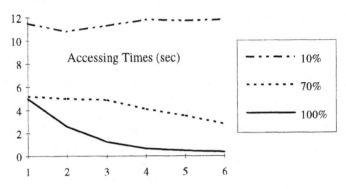

Fig. 6. Accessing times in simulation of normal working

As expected, small caches associated to a small locality of data result in sensibly constant accessing times. With larger caches (70% of the entire data base), access times decrease to about a half in 6000 accesses.

Obviously, a full cache (100% of the entire data base) is not a normal situation, but permits the observation of the system tendency in large locality of data conditions. The curve tends to a value less than 0.4 sec, that is obtained when all objects are already in the caches.

As a conclusion, even in the better conditions of a large locality of data, the average accessing time per item will be larger than 0.4 msec. This time is actually very large, mainly because the majority of the objects are still present in the local memory.

This problem is justified by the algorithms used to manage the cache, that are completely implemented in software. Actually, the maintenance of all the data structure related to the LRU replacement strategy and the WRQu lists is very time consuming and can only be efficient if implemented in hardware.

7. Conclusions

In this work, a description of a new approach to ray-tracing in a transputer based architecture is made. Special care is taken to the scalability problem and to the way the system distributed memory is used. The general architecture is presented, including the main processes of each processor, as well as the responsibilities of each process.

In terms of load balancing, dynamic strategies are under study, so a static method, named Scattered Decomposition, currently implemented, is described. The workload is distributed through processors according to the concept of image complexity coherence.

For 3D scenes composed of a large number of objects, it is not feasible to replicate the entire scene in all processors. This problem is solved by a Virtual Shared Memory system following a memory hierarchy in four levels. Each processor maintains its own cache and performs remote accesses to other processors, in the worst case to disk, whenever a data fault is detected. Redundancy of remote access messages is also reduced.

Even with the reduction of remote accesses redundancy that was presented, the above strategy tends to produce a large amount of messages. Special care is taken in the implementation of routers, to reduce this problem and a special type of multicast addressing, named *InPathTo*, is described. The main advantage in its use is shown to be a reduction on the amount of remote access messages, mainly for systems composed by a large number of processors which encourages the scalability of the system.

The principle in the basis of the *InPathTo* addressing is to interrogate intermediate processors for a certain item that is being remotely addressed in a far processor. There is some probability of the item to be found in one of the intermediate processors and, specially in large diameter networks, the distance covered by the message can be strongly reduced.

Results have been particulary encouraging. As an example, if the probability per processor to reply to a remote access message is only 20%, then an initial distance of 8 channels can be reduced to approximately 4 channels which represents a reduction of about 50% in the number of messages. Nevertheless, the global access time to data is not good, mainly because all the management of caches is made by software.

An obvious topic to future work is the implementation, in hardware, of that management. Other future work has to be related with dinamic load balancing methods that permits the share of tasks using the same or an extended version of the *InPathTo* scheme.

References

1. Whitted, T.: An Improved Illumination Model for Shaded Display. Communications of ACM 23,6 (1980) 343-349
2. Kay, T., Kajiya, J.: Ray Tracing Complex Scenes. ACM Computer Graphics (Proceedings of SIGGRAPH'86) 20,4 (1986) 269-278
3. Glassner, A.: Space Subdivision for Fast Ray Tracing. IEEE Computer Graphics & Applications 4,10 (1984)
4. Fujimoto, A., Tanaka, T., Iwata, K.: ARTS: Accelerated Ray-Tracing System. IEEE Computer Graphics & Applications 6,4 (1986) 16-26
5. Haines, E., Greenberg, D.: The Light Buffer: A Shadow-Testing Accelerator. IEEE Computer Graphics & Applications 6,9 (1986) 6-16
6. Arvo, J., Kirk, D.: Fast Ray Tracing by Ray Classification. ACM Computer Graphics (Proceedings of SIGGRAPH'87) 21,4 (1987) 55-64
7. Hall, R., Greenberg, D.: A Testbed for Realistic Image Synthesis. IEEE Computer Graphics & Applications 3,8 (1983) 10-20
8. Green, S.: Parallel Processing for Computer Graphics. PITMAN (1991)
9. Sousa, A.: Image Synthesis, Progressive Control of the Level of Realism and Parallel Architectures. PhD Thesis, FEUP, Porto (1995) (in Portuguese)

10. Green, S., Paddon, D.: Exploiting Coherence for Multiprocessor Ray-Tracing. IEEE Computer Graphics & Applications 9,6 (1989) 12-26

11. Badouel, D., Bouatouch, K., Priol, T.: Distributing Data and Control for Ray-Tracing in Parallel, IEEE Computer Graphics & Applications 14,4 (1994) 69-77

12. Sousa, A., Costa, A., Ferreira, F.: Interactive Ray-Tracing for Image Production with Increasing Realism. Editors Vandoni, C., Duce, D. (Proceedings of EUROGRAPHICS'90, Montreux, 1990) North-Holland (1990) 449-457

13. Leitão, J., Sousa, A., Costa, A., Ferreira, F.: Strategies and Implementation of Ray-Tracing with Increasing Realism. Graphics Modelling and Visualisation in Science & Technology (Proceedings of Workshop on Graphics Modelling and Visualisation in Science an Technology, Darmstadt, April 92) Springer-Verlag (1993)

14. Sousa, A., Ferreira, F.: A Parallel Implementation of an Interactive Ray-Tracing Algorithm. Computing Systems in Engineering 6,4/5 (Proceedings of 1st International Meeting on Vector and Parallel Processing, Porto, September 1993) Pergamon Press, Elsevier Science Ltd (1995) 409-414

15. Chalmers, A.: A Minimum Path System for Parallel Processing. PhD Thesis, University of Bristol (1991)

16. Goral, C., Torrance, K., Greenberg, D., Battaile, B.: Modelling the Interaction of Light Between Diffuse Surfaces. ACM Computer Graphics (Proceedings of SIGGRAPH'84) 18,3 (1984) 213-222

17. Cohen, M., Chen, S., Wallace, J., Greenberg, D.: A Progressive Refinement Approach to Fast Radiosity Image Generation. ACM Computer Graphics (Proceedings of SIGGRAPH'88) 22,4 (1988)

18. Kuchen, H., Wagener, A.: Comparison of Dynamyc Load Balancing Strategies. Elsevier Sciences Publisher, North-Holland (1991) 303-314

19. Sousa, A., Ferreira, F.: On Writing a Router for Message Passing in a Transputer Network. Computing Systems in Engineering 6,4/5 (Proceedings of 1st International Meeting on Vector and Parallel Processing, Porto, September 1993), Pergamon Press, Elsevier Science Ltd (1995) 471-476

20. Hwang, K.: Advanced Computer Architecture: Parallelism, Scalability, Programmability. McGraw-Hill (1993)

21. Silberschatz, A., Peterson, J.: Operating Systems Concepts. 2nd Edition, Addison-Wesley (1988)

22. Arvo, J., Kirk, D.: A Survey of Ray Tracing Acceleration Techniques. SIGGRAPH'88 Course Notes: Introduction to Ray Tracing (1988)

23. Deitel, H.: An Introduction to Operating Systems. Addison-Wesley (1984)

List of Authors

Springer
and the
environment

At Springer we firmly believe that an international science publisher has a special obligation to the environment, and our corporate policies consistently reflect this conviction.

We also expect our business partners – paper mills, printers, packaging manufacturers, etc. – to commit themselves to using materials and production processes that do not harm the environment. The paper in this book is made from low- or no-chlorine pulp and is acid free, in conformance with international standards for paper permanency.

Lecture Notes in Computer Science

For information about Vols. 1–1141

please contact your bookseller or Springer-Verlag

Vol. 1180: V. Chandru, V. Vinay (Eds.), Foundations of Software Technology and Theoretical Computer Science. Proceedings, 1996. XI, 387 pages. 1996.

Vol. 1181: D. Bjørner, M. Broy, I.V. Pottosin (Eds.), Perspectives of System Informatics. Proceedings, 1996. XVII, 447 pages. 1996.

Vol. 1182: W. Hasan, Optimization of SQL Queries for Parallel Machines. XVIII, 133 pages. 1996.

Vol. 1183: A. Wierse, G.G. Grinstein, U. Lang (Eds.), Database Issues for Data Visualization. Proceedings, 1995. XIV, 219 pages. 1996.

Vol. 1184: J. Waśniewski, J. Dongarra, K. Madsen, D. Olesen (Eds.), Applied Parallel Computing. Proceedings, 1996. XIII, 722 pages. 1996.

Vol. 1185: G. Ventre, J. Domingo-Pascual, A. Danthine (Eds.), Multimedia Telecommunications and Applications. Proceedings, 1996. XII, 267 pages. 1996.

Vol. 1186: F. Afrati, P. Kolaitis (Eds.), Database Theory - ICDT'97. Proceedings, 1997. XIII, 477 pages. 1997.

Vol. 1187: K. Schlechta, Nonmonotonic Logics. IX, 243 pages. 1997. (Subseries LNAI).

Vol. 1188: T. Martin, A.L. Ralescu (Eds.), Fuzzy Logic in Artificial Intelligence. Proceedings, 1995. VIII, 272 pages. 1997. (Subseries LNAI).

Vol. 1189: M. Lomas (Ed.), Security Protocols. Proceedings, 1996. VIII, 203 pages. 1997.

Vol. 1190: S. North (Ed.), Graph Drawing. Proceedings, 1996. XI, 409 pages. 1997.

Vol. 1191: V. Gaede, A. Brodsky, O. Günther, D. Srivastava, V. Vianu, M. Wallace (Eds.), Constraint Databases and Applications. Proceedings, 1996. X, 345 pages. 1996.

Vol. 1192: M. Dam (Ed.), Analysis and Verification of Multiple-Agent Languages. Proceedings, 1996. VIII, 435 pages. 1997.

Vol. 1193: J.P. Müller, M.J. Wooldridge, N.R. Jennings (Eds.), Intelligent Agents III. XV, 401 pages. 1997. (Subseries LNAI).

Vol. 1194: M. Sipper, Evolution of Parallel Cellular Machines. XIII, 199 pages. 1997.

Vol. 1195: R. Trappl, P. Petta (Eds.), Creating Personalities for Synthetic Actors. VII, 251 pages. 1997. (Subseries LNAI).

Vol. 1196: L. Vulkov, J. Waśniewski, P. Yalamov (Eds.), Numerical Analysis and Its Applications. Proceedings, 1996. XIII, 608 pages. 1997.

Vol. 1197: F. d'Amore, P.G. Franciosa, A. Marchetti-Spaccamela (Eds.), Graph-Theoretic Concepts in Computer Science. Proceedings, 1996. XI, 410 pages. 1997.

Vol. 1198: H.S. Nwana, N. Azarmi (Eds.), Software Agents and Soft Computing: Towards Enhancing Machine Intelligence. XIV, 298 pages. 1997. (Subseries LNAI).

Vol. 1199: D.K. Panda, C.B. Stunkel (Eds.), Communication and Architectural Support for Network-Based Parallel Computing. Proceedings, 1997. X, 269 pages. 1997.

Vol. 1200: R. Reischuk, M. Morvan (Eds.), STACS 97. Proceedings, 1997. XIII, 614 pages. 1997.

Vol. 1201: O. Maler (Ed.), Hybrid and Real-Time Systems. Proceedings, 1997. IX, 417 pages. 1997.

Vol. 1202: P. Kandzia, M. Klusch (Eds.), Cooperative Information Agents. Proceedings, 1997. IX, 287 pages. 1997. (Subseries LNAI).

Vol. 1203: G. Bongiovanni, D.P. Bovet, G. Di Battista (Eds.), Algorithms and Complexity. Proceedings, 1997. VIII, 311 pages. 1997.

Vol. 1204: H. Mössenböck (Ed.), Modular Programming Languages. Proceedings, 1997. X, 379 pages. 1997.

Vol. 1205: J. Troccaz, E. Grimson, R. Mösges (Eds.), CVRMed-MRCAS'97. Proceedings, 1997. XIX, 834 pages. 1997.

Vol. 1206: J. Bigün, G. Chollet, G. Borgefors (Eds.), Audio- and Video-based Biometric Person Authentication. Proceedings, 1997. XII, 450 pages. 1997.

Vol. 1207: J. Gallagher (Ed.), Logic Program Synthesis and Transformation. Proceedings, 1996. VII, 325 pages. 1997.

Vol. 1208: S. Ben-David (Ed.), Computational Learning Theory. Proceedings, 1997. VIII, 331 pages. 1997. (Subseries LNAI).

Vol. 1209: L. Cavedon, A. Rao, W. Wobcke (Eds.), Intelligent Agent Systems. Proceedings, 1996. IX, 188 pages. 1997. (Subseries LNAI).

Vol. 1210: P. de Groote, J.R. Hindley (Eds.), Typed Lambda Calculi and Applications. Proceedings, 1997. VIII, 405 pages. 1997.

Vol. 1211: E. Keravnou, C. Garbay, R. Baud, J. Wyatt (Eds.), Artificial Intelligence in Medicine. Proceedings, 1997. XIII, 526 pages. 1997. (Subseries LNAI).

Vol. 1212: J. P. Bowen, M.G. Hinchey, D. Till (Eds.), ZUM '97: The Z Formal Specification Notation. Proceedings, 1997. X, 435 pages. 1997.

Vol. 1213: P. J. Angeline, R. G. Reynolds, J. R. McDonnell, R. Eberhart (Eds.), Evolutionary Programming. Proceedings, 1997. X, 457 pages. 1997.

Vol. 1214: M. Bidoit, M. Dauchet (Eds.), TAPSOFT '97: Theory and Practice of Software Development. Proceedings, 1997. XV, 884 pages. 1997.

Vol. 1215: J. M. L. M. Palma, J. Dongarra (Eds.), Vector and Parallel Processing – VECPAR'96. Proceedings, 1996. XI, 471 pages. 1997.

Vol. 1216: J. Dix, L. Moniz Pereira, T.C. Przymusinski (Eds.), Non-Monotonic Extensions of Logic Programming. Proceedings, 1996. XI, 224 pages. 1997. (Subseries LNAI).

Vol. 1217: E. Brinksma (Ed.), Tools and Algorithms for the Construction and Analysis of Systems. Proceedings, 1997. X, 433 pages. 1997.

Vol. 1218: G. Păum, A. Salomaa (Eds.), New Trends in Formal Languages. IX, 465 pages. 1997.

Vol. 1219: K. Rothermel, R. Popescu-Zeletin (Eds.), Mobile Agents. Proceedings, 1997. VIII, 223 pages. 1997.

Vol. 1220: P. Brezany, Input/Output Intensive Massively Parallel Computing. XIV, 288 pages. 1997.